OXFORD MATHEMATICAL MONOGRAPHS

OXFORD MATHEMATICAL MONOGRAPHS

Hyperbolic Dynamics and Brownian Motion

An Introduction

Jacques Franchi
University of Strasbourg, France

Yves Le Jan
University Paris-Sud (Orsay), France

UNIVERSITY PRESS

UNIVERSITY PRESS

Great Clarendon Street, Oxford, OX2 6DP,
United Kingdom

Oxford University Press is a department of the University of Oxford.
It furthers the University's objective of excellence in research, scholarship,
and education by publishing worldwide. Oxford is a registered trade mark of
Oxford University Press in the UK and in certain other countries

First Edition published in 2012

Impression: 1

British Library Cataloguing in Publication Data

Data available

Library of Congress Cataloging in Publication Data

Library of Congress Control Number: 2012941066

ISBN 978–0–19–965410–9

Printed and bound by
CPI Group (UK) Ltd, Croydon, CR0 4YY

Omnes enim trahimur et ducimur ad cognitionis et scientiae cupiditatem in qua excellere pulchrum putamus (...) In hoc genere et naturali et honesto duo vitia vitanda sunt unum ne incognita pro cognitis habeamus hisque temere assentiamur quod vitium effugere qui volet (...) adhibebit ad considerandas res et tempus et diligentiam. Alterum est vitium quod quidam nimis magnum studium multamque operam in res obscuras atque difficiles conferunt easdemque non necessarias.

(Cicero, *De Officiis*)

Or du Hasard il n'est point de science:
S'il en était, on aurait tort
De l'appeler hasard, ni fortune, ni sort,
Toutes choses très incertaines.

(Jean de La Fontaine)

Contents

Introduction

The idea of this book is to illustrate the interplay between several distinct domains of mathematics, without assuming that the reader is a specialist of any of these domains. We view the book as a minor tribute to the unity of mathematics. Its content can be summarized in three items.

Firstly, this book provides an elementary introduction to hyperbolic geometry, based on the Lorentz group. Recall that in special relativity, space–time is equipped with a pseudo-scalar product, which vanishes on the light cone. The Lorentz group plays a role in relativistic space–time analogous to rotations in Euclidean space; the main difference is the existence of boosts, also called hyperbolic rotations. The light cone can be viewed as the set of straight lines which are asymptotic to the unit pseudo-sphere, namely light rays. Hyperbolic geometry is the geometry of this pseudo-sphere. The boundary of the hyperbolic space is defined as the set of light rays. Special attention is given to the geodesic and horocyclic flows.

Secondly, this book introduces some basic notions of stochastic analysis: the Wiener process, Itô's stochastic integral and some elements of Itô's calculus. This introduction allows us to study linear stochastic differential equations on groups of matrices, and then diffusion processes on homogeneous spaces. In particular, spherical and hyperbolic Brownian motions, diffusions on stable leaves, and relativistic diffusion are constructed in this way.

Thirdly, quotients of hyperbolic space under a discrete group of isometries (i.e., in two dimensions, Riemann surfaces endowed with a negative-constant-curvature metric) are also introduced, and these form the framework in which some elements of hyperbolic dynamics are presented, especially the ergodicity of the geodesic and horocyclic flows. The book culminates in an analysis of the chaotic behaviour of the geodesic flow, which is performed using stochastic analysis methods. The main result is known as Sinai's central limit theorem. Chaotic behaviour arises from the instability of the geodesic flow: small initial perturbations produce a large effect after some time. This follows from the commutation relation between the geodesic flow and either the horocyclic flow or the rotations, which originates from the structure of the Lie algebra of the Lorentz group. The origin of the central limit theorem is De Moivre's theorem on the convergence in distribution of the normalized partial sums of a coin-tossing game to a bell-shaped curve. It is remarkable that such a stochastic behaviour can be observed in the solution of a differential equation.

These methods were presented some years ago in research articles addressed to experienced readers. In this book, the necessary material from group theory, geometry and stochastic analysis is presented in a self-contained and intentionally elementary way.

Only a basic knowledge of linear algebra, calculus and probability theory is required. We have avoided using or defining general notions of geometry such as manifolds and bundles, except for the tangent and frame bundles of hyperbolic space, defined via the embedding in Minkowski space. Of course, readers familiar with hyperbolic geometry will traverse most of the first five chapters rapidly. Those who know stochastic analysis will do the same with the sixth chapter and the beginning of the seventh.

Our approach to hyperbolic geometry is based on special relativity. A key role is played by the Lorentz–Möbius group $PSO(1, d)$, the Iwasawa decomposition, commutation relations, the Haar measure and the hyperbolic Laplacian. We chose to present hyperbolic geometry via relativity to benefit from the physical intuition. Besides, this choice allows us to present, in the stochastic-analysis section, a relativistic process which was introduced by Dudley. We found it rather appropriate, since relativity is generally ignored in treatises on stochastic processes.

There are a lot of good expositions of stochastic analysis. We have tried to make our exposition as short and elementary as possible, with the aim of making it easily available to analysts and geometers who might legitimately be reluctant to have to go through fifty pages of preliminaries before getting to the heart of the subject. Our exposition is closer to Itô and McKean's original work (see [I] and [MK]).

The main results and proofs (at least in the context of this book) are printed in a large font. The reader may initially merely glance at the remaining parts, printed in a smaller font.

Finally, some related results, which are not used here to prove the main results but complete the expositions of hyperbolic geometry and stochastic calculus, are given in the appendices. For the sake of completeness, the appendices also contain a construction of the Wiener measure.

Summary

The first chapter deals with the Lorentz group $PSO(1, d)$, which is the (connected component of the unit in the) linear isometry group of Minkowski space–time. It is isomorphic to the Möbius group of direct hyperbolic isometries of $\mathbb{H}^d$.

The chapter begins with a short, elementary introduction to Lie algebras and associated notions: Lie groups, adjoint actions, the Killing form and Lie derivatives. By 'elementary' we mean in particular that all of these notions are presented in the context of matrices. More precisely, we start with subalgebras of the Lie algebra of square matrices and use the exponential of matrices to define the associated group. Note that this group is not always a Lie group, as it is not necessarily closed. Nevertheless, all the groups which will appear in the following are Lie groups in the usual sense.

Then the Minkowski space $\mathbb{R}^{1,d}$ and its pseudo-metric are introduced, together with the Lorentz–Möbius group $PSO(1, d)$ and the space $\mathbb{F}^d$ of Lorentz frames, on which $PSO(1, d)$ acts both on the right and on the left. We then introduce the affine subgroup $\mathbb{A}^d$, generated by the first boost and the parabolic translations, and we determine the conjugacy classes of $PSO(1, d)$.

The hyperbolic space $\mathbb{H}^d$ is defined as the unit pseudo-sphere of the Minkowski space $\mathbb{R}^{1,d}$. The Iwasawa decomposition of $PSO(1, d)$ is given, and yields Poincaré coordinates in $\mathbb{H}^d$, while the hyperbolic metric is related to the Cartan decomposition of $PSO(1, d)$.

The second chapter presents some basic notions of hyperbolic geometry: geodesics, light rays, tangent bundles etc. These are systematically derived from properties of Minkowski space. For example, the boundary of hyperbolic space is given by the light cone, geodesics by planes intersecting the light cone in two rays, and horospheres by affine hyperplanes parallel to a light ray and intersecting the hyperbolic space. Intrinsic formulae related to the hyperbolic distance are obtained by using only elementary linear algebra within $\mathbb{R}^{1,d}$. Harmonic conjugation is also discussed in this framework. Poincaré coordinates are extended to the boundary $\partial\mathbb{H}^d$.

Then the geodesic and horocyclic flows are defined by the right action of $\mathbb{A}^d$ on frames, and we give physical interpretations. The classical ball and upper-half-space models are presented, and the latter is related to Poincaré coordinates. A crucial commutation relation in the Lorentz–Möbius group is established. Stable leaves and the Busemann function are introduced.

The third chapter deals with operators and measures. The Casimir operator Ξ on $PSO(1, d)$, i.e., the second-order differential operator associated with the Killing form, is the fundamental operator of the theory. It induces the Laplace operator $\mathcal{D}$ on the affine group $\mathbb{A}^d$, and the hyperbolic Laplacian Δ. After proving some fundamental properties of Haar measures on groups, we determine the Haar measure of $PSO(1, d)$. The chapter ends with a presentation of harmonic, Liouville and volume measures. These can all be derived from the Haar measure, and their analytical expressions are derived in this way.

The fourth chapter deals with the geometric theory of Kleinian groups and their fundamental domains. It begins with the example of the parabolic tessellation of the hyperbolic plane by means of $2n$-gons limited by full geodesic lines (i.e., ideal $2n$-gons). Then Dirichlet polyhedra and modular groups are discussed, with $\Gamma(2)$ and $\Gamma(1)$ as the main examples.

In the fifth chapter, we consider measures of Γ-invariant sets, and establish a mixing theorem for the action of the geodesic and horocyclic flows on square-integrable Γ-invariant functions. We derive a Poincaré inequality (i.e., the existence of a spectral gap) for the Laplacian acting on Γ-invariant functions, Γ being a generic cofinite and geometrically finite Kleinian group.

The sixth chapter deals with the basic Itô calculus. Fundamental notions such as predictability, martingales and stopping times are introduced in the discrete case. We then give a short account of the necessary background on martingales and Brownian motion, and deal finally with the basic tools of Itô's calculus: the stochastic integral and the Itô change-of-variable formula.

The seventh chapter is devoted to (left and right) Brownian motions on groups of matrices, which we construct as solutions to linear stochastic differential equations. We establish in particular that the solution of such an equation lives in the subgroup associated with the Lie subalgebra generated by the coefficients of the equation. Reversed processes, Hilbert–Schmidt estimates, approximation by stochastic exponentials, Lyapunov exponents and diffusion processes are also considered. Then we concentrate on important examples: the Heisenberg group, $PSL(2)$, $SO(d)$, $PSO(1, d)$, the affine group $\mathbb{A}^d$ and the Poincaré group $\mathcal{P}^{d+1}$. By means of a projection, we obtain the spherical and hyperbolic Brownian motions, and relativistic diffusion in Minkowski space.

In the eighth chapter, we provide a proof of the Sinai central limit theorem, generalized to the case of a cofinite and geometrically finite Kleinian group. This theorem shows that, asymptotically, geodesics behave chaotically, and yields a quantitative expression for this phenomenon. The method we use is to establish such a result first for Brownian trajectories, which is easier because of their strong independence properties. Then we compare geodesics with Brownian trajectories, by means of a change of contour and time reversal. This requires, in particular, considering diffusion paths on the stable foliation and deriving the existence of a key potential kernel, using the commutation relation of Chapter 2 and the spectral gap presented in Chapter 5.

Acknowledgements

We thank Nicolas Juillet and Athanase Papadopoulos for their careful reading of a preliminary version and for many style corrections, Raymond Séroul for kind technical hints, and the referees for their advice.

List of figures

Chapter One
The Lorentz–Möbius group PSO(1,d)

A large part of this book is centred on a careful analysis of this group of major physical interest, to which this first chapter is mainly devoted.

We begin with a short, elementary introduction to Lie algebras, containing the basic notions which will be useful in the particular case of PSO(1, d): Lie groups, adjoint actions, the Killing form, Lie derivatives and their basic properties. All these notions are presented in the framework of the algebra $\mathcal{M}(d)$ of square matrices, and the exponentials of matrices are used to define the group associated with a Lie subalgebra of $\mathcal{M}(d)$. Note that such a group is not always a Lie group, as it is not necessarily closed. Nevertheless, all the groups which will appear in the following are Lie groups in the usual sense. Lie groups can be thought of as transformations of systems with finitely many degrees of freedom; for example, a curve in SO(3) describes the motion of a solid attached to a fixed point. An infinitesimal motion is described by an element of the corresponding Lie algebra.

Then the Minkowski space $\mathbb{R}^{1,d}$ (which in special relativity, for $d = 3$, describes physical space–time) and its pseudo-metric are introduced, together with the Lorentz–Möbius group PSO(1, d), its Lie algebra so(1, d) and the space $\mathbb{F}^d$ of Lorentz frames, on which PSO(1, d) acts both on the right and on the left.

We then introduce the affine subgroup $\mathbb{A}^d$, generated by the first boost and the parabolic (horizontal) translations, which will play a crucial role in what follows, in several places. We determine the conjugacy classes of PSO(1, d) and specify the structure of its elements.

The hyperbolic space $\mathbb{H}^d$ is defined as the upper sheet of the unit pseudo-sphere of the Minkowski space $\mathbb{R}^{1,d}$. Its boundary $\partial\mathbb{H}^d$ is made of light rays, i.e., the straight lines of the light cone (which are asymptotic to $\mathbb{H}^d$). Physically, a point in $\mathbb{H}^3$ is the velocity of a material point that has unit mass. Poincaré coordinates are defined on $\mathbb{H}^d$ by means of the affine subgroup $\mathbb{A}^d$.

Iwasawa's decomposition of PSO(1, d) is based on the affine subgroup $\mathbb{A}^d$ and the rotation subgroup SO(d), while Cartan's decomposition of PSO(1, d) is related to the hyperbolic metric (and to polar coordinates).

1.1 Lie algebras and groups: introduction

1.1.1 $\mathcal{M}(d)$ and Lie subalgebras of $\mathcal{M}(d)$ We shall consider here only algebras and groups of matrices, that is, subalgebras of the basic Lie algebra

$\mathcal{M}(d)$ (which is the set of all $d \times d$ real square matrices, for some integer $d \geq 2$), and subgroups of the basic Lie group $\mathrm{GL}(d)$ (which is the set of all $d \times d$ real square invertible matrices), known as the **general linear group**.

The real vector space $(\mathcal{M}(d), +, \cdot)$ is made into an algebra, called a **Lie algebra**, by means of the **Lie bracket**

$$[M, M'] := MM' - M'M = -[M', M],$$

MM' being the usual product of the square matrices M and M'. The Lie bracket clearly satisfies the **Jacobi identity**: for any matrices M, M', M'' we have

$$[[M, M'], M''] + [[M', M''], M] + [[M'', M], M'] = 0.$$

The **adjoint action** of $\mathcal{M}(d)$ on itself is defined by

$$\mathrm{ad}(M)(M') := [M, M'], \quad \text{for any } M, M' \in \mathcal{M}(d).$$

The Jacobi identity can be written as follows: for any $M, M' \in \mathcal{G}$,

$$\mathrm{ad}([M, M']) = [\mathrm{ad}(M), \mathrm{ad}(M')] \big(= \mathrm{ad}(M) \circ \mathrm{ad}(M') - \mathrm{ad}(M') \circ \mathrm{ad}(M) \big).$$

The **adjoint action** of the linear group $\mathrm{GL}(d)$ on $\mathcal{M}(d)$ is obtained by conjugation: $\mathrm{GL}(d) \ni g \mapsto \mathrm{Ad}(g)$ is a morphism of groups, defined by

$$\mathrm{Ad}(g)(M) := gMg^{-1}, \quad \text{for any } g \in \mathrm{GL}(d), M \in \mathcal{M}(d).$$

A simple relation between the actions ad and Ad is as follows: for any $g \in \mathrm{GL}(d), M \in \mathcal{M}(d)$, we have

$$\mathrm{ad}\big(\mathrm{Ad}(g)(M)\big) = \mathrm{Ad}(g) \circ \mathrm{ad}(M) \circ \mathrm{Ad}(g)^{-1}. \tag{1.1}$$

Indeed, for any $M' \in \mathcal{G}$, we have

$$\mathrm{ad}\big[\mathrm{Ad}(g)(M)\big](M') = g[M, g^{-1}M'g]g^{-1} = \mathrm{Ad}(g) \circ \mathrm{ad}(M) \circ \mathrm{Ad}(g)^{-1}(M').$$

A **Lie algebra** $\mathcal{G}$ of matrices is a vector subspace of $\mathcal{M}(d)$ which is stable under the Lie bracket $[\cdot, \cdot]$.

The adjoint action of $\mathcal{G}$ then defines a linear map $M \mapsto \mathrm{ad}(M)$ from $\mathcal{G}$ into the vector space of derivations on $\mathcal{G}$. Indeed, the Jacobi identity is equivalent to

$$\mathrm{ad}(M)([M', M'']) = [\mathrm{ad}(M)(M'), M''] + [M', \mathrm{ad}(M)(M'')], \quad \text{for any } M, M', M'' \in \mathcal{G}.$$

The **Killing form** K of a Lie algebra $\mathcal{G}$ is the bilinear form on $\mathcal{G}$ defined (by means of the trace within $\mathcal{G}$) by

$$K(M', M'') := \mathrm{Tr}_{\mathcal{G}}(\mathrm{ad}(M') \circ \mathrm{ad}(M'')), \quad \text{for any } M', M'' \in \mathcal{G}. \tag{1.2}$$

The adjoint action acts skew-symmetrically on K, meaning that

$$K\big(\mathrm{ad}(M)(M'), M''\big) = -K\big(M', \mathrm{ad}(M)(M'')\big), \quad \text{for all } M, M', M'' \in \mathcal{G}.$$

Indeed, this is equivalent to $K\big([M, M'], M''\big) = K\big(M, [M', M'']\big)$ for any M, M', M'' or, using the Jacobi identity in its second formulation above, to

$$\mathrm{Tr}_{\mathcal{G}}\big(\mathrm{ad}(M') \circ \mathrm{ad}(M) \circ \mathrm{ad}(M'')\big) = \mathrm{Tr}_{\mathcal{G}}\big(\mathrm{ad}(M) \circ \mathrm{ad}(M'') \circ \mathrm{ad}(M')\big), \text{which clearly holds.}$$

The Killing form K on $\mathcal{G}$ is necessarily Ad-invariant. Indeed, for any $A, B, C \in \mathcal{G}$ and $g \in G$, we have

$$\mathrm{ad}\big[\mathrm{Ad}(g)(A)\big] \circ \mathrm{ad}\big[\mathrm{Ad}(g)(B)\big](C) = \mathrm{Ad}(g)\Big(\mathrm{ad}(A) \circ \mathrm{ad}(B)\big[\mathrm{Ad}(g^{-1})(C)\big]\Big),$$

that is,

$$\mathrm{ad}\big[\mathrm{Ad}(g)(A)\big] \circ \mathrm{ad}\big[\mathrm{Ad}(g)(B)\big] = \mathrm{Ad}(g) \circ \mathrm{ad}(A) \circ \mathrm{ad}(B) \circ \mathrm{Ad}(g)^{-1},$$

so that

$$K\Big(\mathrm{Ad}(g)(A), \mathrm{Ad}(g)(B)\Big) = \mathrm{Tr}\Big(\mathrm{ad}\big[\mathrm{Ad}(g)(A)\big] \circ \mathrm{ad}\big[\mathrm{Ad}(g)(B)\big]\Big)$$

$$= \mathrm{Tr}\Big(\mathrm{ad}(A) \circ \mathrm{ad}(B)\Big) = K(A, B).$$

1.1.2   Basic examples of Lie algebras

(1) The Lie algebra $\mathrm{sl}(d)$ of traceless matrices.

(2) The Lie algebra $\mathrm{so}(d)$ of antisymmetric matrices.

(3) The Lie algebra $\mathrm{so}(1, d) := \big\{A \in \mathcal{M}(d + 1)\big|{}^t A = -JAJ\big\}$, where $J \in \mathcal{M}(d + 1)$ denotes the diagonal matrix having diagonal entries $(1, -1, \ldots, -1)$ (in that order). Using the Lorentz quadratic form $\langle,\rangle$ defined on $\mathbb{R}^{1+d}$ by $\langle x, x \rangle := x_0^2 - x_1^2 - \cdots - x_d^2$, we also have

$$\mathrm{so}(1, d) = \big\{A \in \mathcal{M}(d + 1)\big|\langle Ax, x \rangle = 0 \ \text{ for all } x \in \mathbb{R}^{1+d}\big\}.$$

Exercise   Verify that in the above cases the Killing form is given by

$$K(M, M') = \mathrm{Tr}\big({}^t M \times M'\big).$$

1.1.3   The exponential map   The **exponential map** is defined by

$$\exp(M) := \sum_{n \in \mathbb{N}} M^n \big/ n!, \text{ for any } M \in \mathcal{M}(d),$$

and is C^∞ from $\mathcal{M}(d)$ into $\mathcal{M}(d)$. The formula $\exp(M + M') = \exp(M)\exp(M')$ is correct for commuting matrices M, M', but does not hold in general.

We denote by $\mathrm{Tr}(M) := \sum_{j=1}^d M_{jj}$ the trace of any matrix $M \in \mathcal{M}(d)$ as usual, and denote by $\det(M)$ its determinant. The formula

$$\det\big(\exp(M)\big) = e^{\mathrm{Tr}(M)} \tag{1.3}$$

holds trivially for diagonalizable matrices M, and hence holds for all $M \in \mathcal{M}(d)$ by density. This shows that the range of exp is included in the general linear group $\mathrm{GL}(d)$.

We denote by $\mathbf{1}$ the unit matrix, that is, the unit element of $\mathrm{GL}(d)$.

Lemma 1.1.3.1 *The differential of* $\mathrm{Ad} \circ \exp$ *at the unit* $\mathbf{1}$ *is* ad. *Furthermore,*

$$\mathrm{Ad}\big(\exp(A)\big) = \exp\big[\mathrm{ad}(A)\big] \ \text{on } \mathcal{M}(d), \text{ for any } A \in \mathcal{M}(d). \qquad (1.4)$$

Proof Consider the analytical map $t \mapsto \Phi(t) := \mathrm{Ad}(\exp(tA))$ from $\mathbb{R}$ into the space of endomorphisms on $\mathcal{M}(d)$. For any real t $(d_0/ds$ denoting the derivative at 0 with respect to $s)$, this satisfies

$$\frac{d}{dt}\Phi(t) = \frac{d_0}{ds}\mathrm{Ad}\Big(\exp\big((s+t)A\big)\Big) = \frac{d_0}{ds}\mathrm{Ad}\big(\exp(sA)\exp(tA)\big)$$

$$= \frac{d_0}{ds}\Phi(s) \circ \Phi(t) = \mathrm{ad}(A) \circ \Phi(t).$$

Hence, Φ solves a linear differential equation, which has the unique solution

$$\Phi(t) = \exp\big[t\,\mathrm{ad}(A)\big] \circ \Phi(0) = \exp\big[t\,\mathrm{ad}(A)\big]. \qquad\qquad \diamond$$

Proposition 1.1.3.2 *The differential* $\mathrm{d}\exp_M$ *of the exponential map at any* $M \in \mathcal{M}(d)$ *is expressed as follows: for any* $B \in \mathcal{M}(d)$,

$$\mathrm{d}\ \exp_M(B) := \frac{d_0}{d\varepsilon}\exp(M+\varepsilon B) = \exp(M)\left(\sum_{k\in\mathbb{N}}\frac{1}{(k+1)!}\mathrm{ad}(-M)^k(B)\right).$$

The exponential map induces a diffeomorphism from a neighbourhood of 0 in $\mathcal{M}(d)$ *onto a neighbourhood of* $\exp(0) = \mathbf{1}$ *in* $\mathrm{GL}(d)$.

Note We denote the derivative at 0 with respect to ε by $d_0/d\varepsilon$, here and henceforth.

Proof We have

$$\mathrm{d}\ \exp_M(B) = \frac{d_0}{d\varepsilon}\exp(M+\varepsilon B) = \sum_{n\in\mathbb{N}}\frac{Y_n}{(n+1)!} = \sum_{n\in\mathbb{N}}\frac{Z_n}{(n+1)!},$$

where $Y_n := \sum_{k=0}^{n}M^{n-k}BM^k$ and $Z_n := \sum_{k=0}^{n}C_{n+1}^{k+1}M^{n-k}\mathrm{ad}(-M)^k(B)$.

Indeed, let us prove by induction that $Y_n = Z_n$ for $n \in \mathbb{N}$. We have $Y_0 = B = Z_0$ and, assuming $Y_{n-1} = Z_{n-1}$, we have

$$Y_n = BM^n + MY_{n-1} = BM^n + MZ_{n-1}$$

$$= BM^n + \sum_{k=0}^{n-1} C_n^{k+1} M^{n-k} \mathrm{ad}(-M)^k(B)$$

$$= BM^n + Z_n - \mathrm{ad}(-M)^n(B) - \sum_{k=0}^{n-1} C_n^k M^{n-k} \mathrm{ad}(-M)^k(B)$$

$$= Z_n + BM^n - \sum_{k=0}^{n} C_n^k M^{n-k} \, \mathrm{ad}(-M)^k(B)$$

$$= Z_n + BM^n - \Big(\mathrm{ad}(-M) + [A \mapsto MA] \Big)^n (B) = Z_n,$$

since $\mathrm{ad}(-M)$ and the left multiplication by M commute.

Finally, the above expression containing Z_n entails the following statement:

$$\mathrm{d}\exp_M(B) = \sum_{0 \le k \le n < \infty} \frac{M^{n-k}}{(n-k)!} \frac{\mathrm{ad}(-M)^k}{(k+1)!} (B)$$

$$= \exp(M) \times \left(\sum_{k \ge 0} \frac{\mathrm{ad}(-M)^k}{(k+1)!} (B) \right).$$

In particular, we see that $\mathrm{d}\exp_0$ is the identity map, and then that $\exp$ induces a diffeomorphism near the origin. $\diamond$

Exercise   Using Proposition 1.1.3.2 and its proof, show that the differential $\mathrm{d}\exp_M$ computed in Proposition 1.1.3.2 can be expressed alternatively as follows:

$$\mathrm{d}\exp_M(B) = \sum_{k \in \mathbb{N}} \frac{\mathrm{ad}(-M)^k}{(k+1)!} \Big(\exp(M)B \Big)$$

$$= \sum_{k \in \mathbb{N}} \frac{\mathrm{ad}(M)^k}{(k+1)!} (B) \exp(M) = \sum_{k \in \mathbb{N}} \frac{\mathrm{ad}(M)^k}{(k+1)!} \Big(B \exp(M) \Big).$$

1.1.4   The group associated with a Lie subalgebra   The groups that we shall consider are the subgroups of the general linear group $\mathrm{GL}(d)$ which are generated by the image (under the exponential map) of a Lie subalgebra of $\mathcal{M}(d)$.

Definition 1.1.4.1 *We call the subgroup of* $\mathrm{GL}(d)$ *generated by* $\exp(\mathcal{G})$ *the group associated with the Lie (sub)algebra* $\mathcal{G}$.

We consider again the action $g \mapsto \mathrm{Ad}(g)$ of any such group G by conjugation, as defined in Section 1.1.1.

Lemma 1.1.4.2 *For any $g \in G$ (associated with $\mathcal{G}$), $\mathrm{Ad}(g)$ maps $\mathcal{G}$ into $\mathcal{G}$.*

Proof For any fixed $A, B \in \mathcal{G}$ and any real t, set

$$\varphi(t) := \exp(tA) \ B \ \exp(-tA) \in \mathcal{M}(d).$$

This defines a C^{∞} map φ from $\mathbb{R}$ into $\mathcal{M}(d)$, solving the order-one linear differential equation: $\varphi' = A\varphi - \varphi A$, and such that $\varphi(0) = B \in \mathcal{G}$, $\varphi'(0) = [A, B] \in \mathcal{G}$. Hence the equation satisfied by φ can be solved in the vector space $\mathcal{G}$, showing by uniqueness that φ must be $\mathcal{G}$-valued. By an immediate induction, this proves that for any $A_1, \ldots, A_n, B \in \mathcal{G}$, we have $\mathrm{Ad}\big(\exp(A_1) \ldots \exp(A_n)\big)(B) \in \mathcal{G}$, that is, $\mathrm{Ad}(g)(B) \in \mathcal{G}$ for any $g \in G$. $\diamond$

The Killing form is invariant under inner automorphisms: for any $g \in G$ and $M, M' \in \mathcal{G}$, we have

$$K\big(\mathrm{Ad}(g)(M), \mathrm{Ad}(g)(M')\big) = K(M, M'). \tag{1.5}$$

Indeed, by the definition of K (eqn (1.2)) and by eqn (1.1), we have

$$K\big(\mathrm{Ad}(g)(M), \mathrm{Ad}(g)(M')\big) = \mathrm{Tr}\big(\mathrm{ad}\big[\mathrm{Ad}(g)(M)\big] \circ \mathrm{ad}\big[\mathrm{Ad}(g)(M')\big]\big)$$
$$= \mathrm{Tr}\big(\mathrm{Ad}(g) \circ \mathrm{ad}(M) \circ \mathrm{ad}(M') \circ \mathrm{Ad}(g)^{-1}\big)$$
$$= \mathrm{Tr}\big(\mathrm{ad}(M) \circ \mathrm{ad}(M')\big) = K(M, M').$$

The group G associated with a Lie subalgebra $\mathcal{G}$ is a **Lie subgroup** of $\mathrm{GL}(d)$ when it satisfies the following condition: there exists a neighbourhood $\mathcal{U}$ of 0 in $\mathcal{M}(d)$ such that $\exp(\mathcal{U}) \cap G = \exp(\mathcal{U} \cap \mathcal{G})$.

This is consistent with the usual definition of a Lie group; see, for example the introductory chapter of [Kn], or [Ho], Chapter II, Section 2. In particular, we have $\mathcal{G} = \{M \in \mathcal{M}(d)| \exp(tM) \in G \text{ for any real } t\}$. Moreover, any Lie subgroup G is necessarily closed in $\mathrm{GL}(d)$: fixing $g \in \overline{G}$, a compact neighbourhood $\mathcal{V}$ of 0 in $\mathcal{M}(d)$ on which the restriction of the exponential map is one-to-one and such that $\exp(\mathcal{V}) \cap G = \exp(\mathcal{V} \cap \mathcal{G})$, another neighbourhood $\mathcal{U}$ of 0 in $\mathcal{M}(d)$ such that $\exp(-\mathcal{U})\exp(\mathcal{U}) \subset \exp(\mathcal{V})$, and a sequence $(g_n) \subset G \cap g\exp(\mathcal{U})$ which converges to g, we have $g_0^{-1}g_n \in \exp(\mathcal{V}) \cap G$, whence $g_0^{-1}g_n = \exp(h_n)$, with some $h_n \in \mathcal{V} \cap \mathcal{G}$; selecting a sub-sequence, we may suppose that $h_n \to h \in \mathcal{V} \cap \mathcal{G}$, whence $g = g_0\exp(h) \in G$.

In contrast, a group G associated with a Lie subalgebra $\mathcal{G}$ does not need to be closed in $\mathrm{GL}(d)$, as the following example shows: the one-parameter subgroup H of $\mathrm{GL}(4)$ defined by $H := \{\exp[t\tilde{A}]| t \in \mathbb{R}\}$, where

$$\tilde{A} := \begin{pmatrix} A & 0 \\ 0 & \pi A \end{pmatrix},$$

with

$$A := \begin{pmatrix} 0 & -1 \\ 1 & 0 \end{pmatrix},$$

does not contain $-\mathbf{1} \in \overline{H}$, as is easily seen by considering a sequence $t_n \in 2\mathbb{N} + 1$ approaching π modulo 2π.

Observe that by continuity of the product and of the inverse maps, the closure in GL(d) of a group associated with a Lie subalgebra $\mathcal{G}$ remains a subgroup of GL(d), which we shall call **the closed group associated with the Lie subalgebra $\mathcal{G}$**. This is a Lie group, since it is known (see [Ho], Chapter VIII) that any closed subgroup of a Lie group is also a Lie group.

In a general and usual way, any left action of any group G on any set S can be extended to any function F on this set by setting

$$L_g F(s) := F(gs), \quad \text{for any } (g, s) \in G \times S.$$

Similarly, any right action of G on S can be extended to functions by setting

$$R_g F(s) := F(sg), \quad \text{for any } (g, s) \in G \times S.$$

We shall say that a real function f on a group G associated with a Lie subalgebra $\mathcal{G}$ is differentiable when its restrictions to all lines $[t \mapsto g \exp(tA)]$ and $[t \mapsto \exp(tA)g]$ are differentiable, for any $g \in G, A \in \mathcal{G}$. We then define the **right Lie derivative $\mathcal{L}_A f$** by

$$\mathcal{L}_A f(g) := \frac{d_0}{d\varepsilon} f\big[g \exp(\varepsilon A)\big], \quad \text{for any } g \in G. \tag{1.6}$$

This defines a **left-invariant** vector field $\mathcal{L}_A$ on G, which means that it commutes with left translations on G:

$$\mathcal{L}_A \circ L_g = L_g \circ \mathcal{L}_A, \quad \text{for all } g \in G \text{ and } A \in \mathcal{G}. \tag{1.7}$$

We also have

$$\mathcal{L}_A \circ R_g = R_g \circ \mathcal{L}_{\mathrm{Ad}(g^{-1})(A)}, \quad \text{for all } g \in G \text{ and } A \in \mathcal{G}. \tag{1.8}$$

Similarly, the **left Lie derivative $\mathcal{L}'_A f$** is defined by

$$\mathcal{L}'_A f(g) := \frac{d_0}{d\varepsilon} f\big[\exp(\varepsilon A)g\big], \tag{1.9}$$

and defines a **right-invariant** vector field $\mathcal{L}'_A$ on G

$$\mathcal{L}'_A \circ R_g = R_g \circ \mathcal{L}'_A, \quad \text{for all } g \in G \text{ and } A \in \mathcal{G},$$

such that

$$\mathcal{L}'_A \circ L_g = L_g \circ \mathcal{L}'_{\mathrm{Ad}(g)(A)}, \quad \text{for all } g \in G \text{ and } A \in \mathcal{G}.$$

Furthermore, we have the following formula:

$$\mathcal{L}_A R_g F(h) = \mathcal{L}'_A L_h F(g), \quad \text{for any } A \in \mathcal{G}, g, h \in G. \tag{1.10}$$

And, in the same spirit, setting $\tilde{f}(g) := f(g^{-1})$, we have

$$\mathcal{L}_A \tilde{f} = -\widetilde{\mathcal{L}'_A f}. \tag{1.11}$$

We say that a function f is C^1 on G when all $\mathcal{L}_A f$ exist and are continuous on G, or, equivalently according to eqn (1.11), when all $\mathcal{L}'_A f$ exist and are continuous on G. Similarly, f is C^2 on G when all $\mathcal{L}_A f$ are C^1.

The map $A \mapsto \mathcal{L}_A$ from $\mathcal{G}$ onto the Lie algebra of left-invariant vector fields on G is an isomorphism of Lie algebras. We indeed have the formula

$$[\mathcal{L}_A, \mathcal{L}_B] = \mathcal{L}_{[A,B]}, \quad \text{for any } A, B \in \mathcal{G}. \tag{1.12}$$

1.1.5   Basic examples of Lie groups

(0) The Lie subgroup $\mathrm{GL}^+(d)$ of matrices that have a positive determinant is the group associated with the Lie algebra $\mathcal{M}(d)$.

In fact, according to Definition 1.1.4.1, the group associated with the Lie algebra $\mathcal{M}(d)$ is the connected component $\mathrm{GL}^1(d)$ of $\mathbf{1}$ in $\mathrm{GL}(d)$, which is included in the subgroup $\mathrm{GL}^+(d)$. Conversely, any element $A \in \mathrm{GL}^+(d)$ is arbitrarily near to a diagonalizable $A' = PDP^{-1} \in \mathrm{GL}^+(d)$, where we can take the matrix D to be diagonal with positive diagonal entries, so that A' is connected to $\mathbf{1}$ within $\mathrm{GL}^+(d)$ by an arc $[0,1] \ni t \mapsto P \exp(t\Delta)P^{-1}$ (the matrix Δ being diagonal and such that $\exp(\Delta) = D$). This shows that $A \in \mathrm{GL}^1(d)$, and then that $\mathrm{GL}^1(d) = \mathrm{GL}^+(d)$.

We consider the examples of Section 1.1.2 again (in the same order).

(1) The Lie group $\mathrm{SL}(d)$ of matrices that have determinant one is the group associated with the Lie subalgebra $\mathrm{sl}(d)$ of traceless matrices.

Indeed, on the one hand, $\mathrm{SL}(d)$ is clearly closed and, by eqn (1.3) we have $\exp[\mathrm{sl}(d)] \subset \mathrm{SL}(d)$. On the other hand, the preceding example shows that any $A \in \mathrm{SL}(d)$ can be written as $A = \exp(M_1) \cdots \exp(M_k) = \exp(M_1 - (\tau_1/d)\mathbf{1}) \cdots \exp(M_k - (\tau_k/d)\mathbf{1})$, with $\tau_j := \mathrm{Tr}(M_j)$, which establishes the claim.

(2) The Lie group $\mathrm{SO}(d)$ of rotation matrices is the group associated with the Lie subalgebra $\mathrm{so}(d)$ of antisymmetric matrices.

Indeed, on the one hand, $\mathrm{SO}(d)$ is obviously closed, and we clearly have $\exp[\mathrm{so}(d)] \subset \mathrm{SO}(d)$. On the other hand, using the well-known fact that any $\varrho \in \mathrm{SO}(d)$ is conjugate in $\mathrm{SO}(d)$ to an element whose non-null entries either are 1's on the diagonal or form diagonal planar rotation blocks

$$\begin{bmatrix} \cos \varphi_j & -\sin \varphi_j \\ \sin \varphi_j & \cos \varphi_j \end{bmatrix},$$

we have directly $\varrho = \exp(PA^tP) \in \mathrm{so}(d)$, where $P \in \mathrm{SO}(d)$ and $A \in \mathrm{so}(d)$, whose non-null entries form diagonal planar blocks

$$\begin{bmatrix} 0 & -\varphi_j \\ \varphi_j & 0 \end{bmatrix}.$$

This establishes that $\exp\big[\mathrm{so}(d)\big] = \mathrm{SO}(d)$, and hence the claim.

(3) The Lorentz–Möbius Lie group PSO$(1, d)$, which will be our main interest, is the group of matrices preserving the Lorentz quadratic form defined on $\mathbb{R}^{1+d}$ by $\langle x, x \rangle := x_0^2 - x_1^2 - \cdots - x_d^2$, the upper sheet of the hyperboloid $\mathbb{H}^d := \{x \in \mathbb{R}^{1+d} | \langle x, x \rangle = 1, x_0 > 0\}$ and the orientation. This is the connected component of the unit element in the group O$(1, d)$ of all matrices which preserve the quadratic form $\langle x, x \rangle$. In other words, it is the group made up of the $g \in \mathrm{SL}(d+1)$ such that ${}^tgJg = J$ and $\big[\langle x, x \rangle > 0, x_0 > 0\big] \Rightarrow (gx)_0 > 0$, where the diagonal matrix J is given by $J = \mathrm{diag}(1, -1, \ldots, -1) \in \mathcal{M}(d+1)$; See Sections 1.2 and 1.3.

The Lorentz–Möbius group PSO$(1, d)$ is the group associated with the Lie algebra so$(1, d)$, as we shall see in Remark 1.5.3, which states more precisely that actually PSO$(1, d) = \exp\big[\mathrm{so}(1, d)\big]$.

Proposition 1.1.5.1 *The groups* PSL$(2) := \mathrm{SL}(2)/\{\pm 1\}$ *and* PSO$(1, 2)$ *are isomorphic. The Lie algebras* sl(2) *and* so$(1, 2)$ *are isomorphic.*

Proof The three matrices

$$Y_0 := \begin{pmatrix} 0 & -1 \\ 1 & 0 \end{pmatrix}, Y_1 := \begin{pmatrix} 0 & 1 \\ 1 & 0 \end{pmatrix}, Y_2 := \begin{pmatrix} 1 & 0 \\ 0 & -1 \end{pmatrix}$$

constitute a basis of sl(2), and we have $\det(x_0 Y_0 + x_1 Y_1 + x_2 Y_2) = x_0^2 - x_1^2 - x_2^2$. Since for any $g \in \mathrm{SL}(2)$ the linear map $\mathrm{Ad}(g)$, acting on sl(2), preserves the determinant, and since SL(2) is connected and the morphism $g \mapsto \mathrm{Ad}(g)$ is continuous, we see that the map $\mathrm{Ad}(g)$ belongs to the group O$(1, 2)$, and even to the connected component of its unit element. Hence the morphism $g \mapsto \mathrm{Ad}(g)$ maps the Lie group SL(2) into PSO$(1, 2)$.

Therefore, by differentiating at the unit element and using Lemma 1.1.3.1, we see that the morphism $A \mapsto \mathrm{ad}(A)$ maps the Lie algebra sl(2) into the Lie algebra so$(1, 2)$. Moreover, if $\mathrm{ad}(A) = 0$, then A commutes with all $M \in \mathrm{sl}(2)$ and then vanishes. Hence $A \mapsto \mathrm{ad}(A)$ is indeed an isomorphism of Lie algebras (by the Jacobi identity).

Hence $\mathrm{Ad}(\mathrm{SL}(2))$ is a neighbourhood of the unit element and a subgroup of the connected Lie group PSO$(1, 2)$, so that it has to be the whole of PSO$(1, 2)$. Finally, if $\mathrm{Ad}(g) = 1$, then $g \in \mathrm{SL}(2)$ must commute with the elements of sl(2), and then be ± 1. $\diamondsuit$

The elements of the group PSL(2) are identified with *homographies* of the Poincaré half plane $\mathbb{R} \times \mathbb{R}_+^* \equiv \{z = x + \sqrt{-1}\, y \in \mathbb{C} | y > 0\}$ by

$$\pm \begin{pmatrix} a & b \\ c & d \end{pmatrix} \longleftrightarrow \left[z \mapsto \frac{az + b}{cz + d} \right].$$

We could have considered matrices with complex entries as well, i.e., $\mathcal{M}_d(\mathbb{C})$ instead of $\mathcal{M}(d) \equiv \mathcal{M}_d(\mathbb{R})$. In this context, considering the group $\mathrm{SL}_2(\mathbb{C})$ of spin matrices, we have the following continuation of Proposition 1.1.5.1.

Proposition 1.1.5.2 *The groups* $\mathrm{PSL}_2(\mathbb{C}) := \mathrm{SL}_2(\mathbb{C})/\{\pm 1\}$ *and* $\mathrm{PSO}(1,3)$ *are isomorphic. The Lie algebras* $\mathrm{sl}_2(\mathbb{C})$ *and* $\mathrm{so}(1,3)$ *are isomorphic.*

Proof We follow the very similar proof of Proposition 1.1.5.1. We identify the Minkowski space $\mathbb{R}^{1,3}$ (see Section 1.2) with the subset $\mathcal{H}_2$ of Hermitian matrices in $\mathcal{M}_2(\mathbb{C})$ by means of the map

$$\xi \mapsto \begin{pmatrix} \xi_0 + \xi_1 & \xi_2 + \sqrt{-1}\xi_3 \\ \xi_2 - \sqrt{-1}\xi_3 & \xi_0 - \xi_1 \end{pmatrix},$$

the determinant giving the Lorentz quadratic form. We define the action $\varphi(g)$ of $g \in \mathrm{SL}_2(\mathbb{C})$ on $\mathcal{H}_2$ by $H \mapsto gH{}^t\bar{g}$. The morphism φ maps the group $\mathrm{SL}_2(\mathbb{C})$ into $\mathrm{PSO}(1,3)$.

Differentiating, we obtain a morphism $d\varphi(1)$ from the Lie algebra $\mathrm{sl}_2(\mathbb{C})$ into the Lie algebra $\mathrm{so}(1,3)$, which are both six-dimensional. The kernel of this morphism is made up of those matrices $A \in \mathcal{M}_2(\mathbb{C})$ which have null trace and which satisfy $AH + H{}^t\bar{A} = 0$ for any $H \in \mathcal{H}_2$; since $A = 0$ is the only solution, the two algebras are indeed isomorphic.

Hence the range $\varphi(\mathrm{SL}_2(\mathbb{C}))$ is the whole of $\mathrm{PSO}(1,3)$. Finally, if $g \in \mathrm{SL}_2(\mathbb{C})$ belongs to the kernel of φ, then $gH{}^t\bar{g} = H$ for any $H \in \mathcal{H}_2$, which implies that $g = \pm 1$. ◇

Exercise The Heisenberg group $\mathcal{H}^3$ is the group associated with the Lie algebra generated by the matrices

$$A_1 := \begin{pmatrix} 0 & 1 & 0 \\ 0 & 0 & 0 \\ 0 & 0 & 0 \end{pmatrix} \quad \text{and} \quad A_2 := \begin{pmatrix} 0 & 0 & 0 \\ 0 & 0 & 1 \\ 0 & 0 & 0 \end{pmatrix}.$$

Describe precisely its Lie algebra, the computation of the exponential map, the product formula in the group, the adjoint actions ad and Ad, and the Killing form K.

1.2   Minkowski space and its pseudo-metric

We fix an integer $d \geq 2$, and consider the **Minkowski space**:

$$\mathbb{R}^{1,d} := \{\xi = (\xi_0, \dots, \xi_d) \in \mathbb{R} \times \mathbb{R}^d\},$$

endowed with the **Minkowski pseudo-metric** (Lorentz quadratic form)

$$\langle \xi, \xi \rangle := \xi_0^2 - \sum_{j=1}^{d} \xi_j^2.$$

We denote by $(e_0, e_1, \dots, e_d)$ the canonical basis of $\mathbb{R}^{1,d}$, and we orientate $\mathbb{R}^{1,d}$ by taking this basis as direct. We have $\langle e_i, e_j \rangle = 1_{\{i=j=0\}} - 1_{\{i=j\neq 0\}}$.

Note that $\{e_0, e_1\}^\perp$ is the subspace generated by $\{e_2, \dots, e_d\}$; we identify it with $\mathbb{R}^{d-1}$. Similarly, we identify the subspace $\{e_0\}^\perp$ generated by $\{e_1, \dots, e_d\}$ with $\mathbb{R}^d$:

$$e_0^\perp \equiv \mathbb{R}^d \quad \text{and} \quad \{e_0, e_1\}^\perp \equiv \mathbb{R}^{d-1}.$$

Note that the negative of the pseudo-metric obviously induces the Euclidean metric on $\mathbb{R}^d$. As usual, we shall denote by $\mathbb{S}^{d-1}, \mathbb{S}^{d-2}$ the corresponding unit Euclidean spheres.

A vector $\xi \in \mathbb{R}^{1,d}$ is called **lightlike** (or isotropic) if $\langle\xi,\xi\rangle = 0$, **timelike** if $\langle\xi,\xi\rangle > 0$, **positive timelike** or **future-directed** if $\langle\xi,\xi\rangle > 0$ and $\xi_0 > 0$, **spacelike** if $\langle\xi,\xi\rangle < 0$, and **non-spacelike** if $\langle\xi,\xi\rangle \geq 0$.

The **light cone** of $\mathbb{R}^{1,d}$ is the upper half-cone of lightlike vectors of Minkowski space,

$$\mathcal{C} := \left\{\xi \in \mathbb{R}^{1,d} \,\middle|\, \langle\xi,\xi\rangle = 0, \xi_0 > 0\right\}.$$

The **solid light cone** of $\mathbb{R}^{1,d}$ is the convex hull of $\mathcal{C}$, i.e., the upper half of the solid cone of non-spacelike vectors,

$$\overline{\mathcal{C}} := \left\{\xi \in \mathbb{R}^{1,d} \,\middle|\, \langle\xi,\xi\rangle \geq 0, \xi_0 > 0\right\}.$$

Lemma 1.2.1 *(i) No plane of $\mathbb{R}^{1,d}$ is included in the solid cone of non-spacelike vectors.*

(ii) For any $\xi, \xi' \in \overline{\mathcal{C}}$, we have $\langle\xi,\xi'\rangle \geq \sqrt{\langle\xi,\xi\rangle\langle\xi',\xi'\rangle}$, with equality if and only if ξ, ξ' are collinear.

Proof Set $|\xi| := \sqrt{\xi_0^2 - \langle\xi,\xi\rangle}$, for any $\xi \in \mathbb{R}^{1,d}$.

(i) For non-collinear $\xi, \xi' \in \mathbb{R}^{1,d}$ such that $\langle\xi,\xi\rangle \geq 0$ and $\langle\xi',\xi'\rangle \geq 0$, we must have $\xi_0 \neq 0$ (since $\xi_0 = 0$ and $\langle\xi,\xi\rangle \geq 0$ imply $\xi = 0$, collinear to ξ'), and then

$$\langle\xi - (\xi_0'/\xi_0)\xi', \xi - (\xi_0'/\xi_0)\xi'\rangle = -\left|\xi - (\xi_0'/\xi_0)\xi'\right|^2 < 0.$$

(ii) For $\xi, \xi' \in \overline{\mathcal{C}}$, we have on the one hand $\xi_0 \geq |\xi|$ and $\xi_0 \xi_0' \geq |\xi| \times |\xi'| \geq \sum_{j=1}^d \xi_j \xi_j'$, whence $\langle\xi,\xi'\rangle \geq 0$, and on the other hand

$$\langle\xi,\xi'\rangle^2 \geq \left(\xi_0 \xi_0' - |\xi||\xi'|\right)^2 = \langle\xi,\xi\rangle\langle\xi',\xi'\rangle + \left(\xi_0|\xi'| - \xi_0'|\xi|\right)^2 \geq \langle\xi,\xi\rangle\langle\xi',\xi'\rangle.$$

Hence $\langle\xi,\xi'\rangle \geq \sqrt{\langle\xi,\xi\rangle\langle\xi',\xi'\rangle}$, and equality occurs if and only if $(\xi_1,\ldots,\xi_d)$ and $(\xi_1',\ldots,\xi_d')$ are collinear $\big($of the same direction, by the case of equality in the Schwarz inequality $|\xi||\xi'| \geq \sum_{j=1}^d \xi_j \xi_j'\big)$ and $\xi_0|\xi'| - \xi_0'|\xi| = 0$. Then $|\xi| = 0$ holds if and only if $|\xi'| = 0$ holds, in which case ξ, ξ' are indeed collinear. Finally, for $|\xi| \neq 0$, we have $(\xi_1',\ldots,\xi_d') = \lambda(\xi_1,\ldots,\xi_d)$ for some positive λ, and then $\xi_0' = \lambda\xi_0$ too. $\diamond$

Definition 1.2.2 *A direct basis $\beta = (\beta_0, \beta_1, \ldots, \beta_d)$ of $\mathbb{R}^{1,d}$ satisfying the pseudo-orthonormality condition $\langle\beta_i, \beta_j\rangle = 1_{\{i=j=0\}} - 1_{\{i=j\neq0\}}$ for $0 \leq i, j \leq d$, and such that the first component of β_0 is positive is called a (direct and future-directed) **Lorentz frame**. We set $\pi_0(\beta) := \beta_0$.*

The set of all (direct and future-directed) Lorentz frames of $\mathbb{R}^{1,d}$ is denoted by $\mathbb{F}^d$.

We shall systematically identify the endomorphisms of $\mathbb{R}^{1,d}$ with their matrices in the canonical basis $(e_0, e_1, \ldots, e_d)$ of $\mathbb{R}^{1,d}$.

1.3 The Lorentz–Möbius group and its Lie algebra

Definition 1.3.1 *The Lorentz group* $\mathrm{O}(1, d)$ *is the group of endomorphisms of* $\mathbb{R}^{1,d}$ *which preserve the Minkowski pseudo-metric* $\langle \cdot, \cdot \rangle$. *The* **Lorentz–Möbius group** $\mathrm{PSO}(1, d)$ *is the connected component of the unit matrix in the Lorentz group* $\mathrm{O}(1, d)$.

We shall call **Lorentzian** any matrix which maps a Lorentz frame (and therefore any Lorentz frame) to another Lorentz frame, and then identify it with an element of $\mathrm{O}(1, d)$.

The special Lorentz group $\mathrm{SO}(1, d)$ is the subgroup of index two in $\mathrm{O}(1, d)$ of endomorphisms which also preserve the orientation, so that $\mathrm{PSO}(1, d)$ is the subgroup of index two in $\mathrm{SO}(1, d)$ of endomorphisms which preserve the light cone $\mathcal{C}$ and the orientation (since the columns of the matrix of such an element g form a Lorentz frame that can be drawn continuously to the canonical basis, as will become clear below, boosts allow one to draw ge_0 continuously to e_0, and then rotations allow one to draw the spacelike vectors $ge_1, \ldots, ge_d$ continuously to $e_1, \ldots, e_d$).

The special orthogonal group $\mathrm{SO}(d)$ (i.e., the rotation group of $\mathbb{R}^d$) is identified with the subgroup of elements that fix the base vector e_0. Similarly, we identify the subgroup of elements that fix both base vectors e_0 and e_1 with $\mathrm{SO}(d - 1)$.

As already seen in Sections 1.1.2 and 1.1.5 (and as will be established in Remark 1.5.3), we have $\mathrm{PSO}(1, d) = \exp\!\big[\mathrm{so}(1, d)\big]$, where

$$\mathrm{so}(1, d) = \big\{ A \in \mathcal{M}(d + 1) \,\big|\, \langle A\xi, \xi \rangle = 0 \ \text{ for all } \xi \in \mathbb{R}^{1,d} \big\},$$

so that the Lorentz–Möbius group $\mathrm{PSO}(1, d)$ is the group associated with the Lie algebra $\mathrm{so}(1, d)$, and is a Lie group (being clearly closed). One says that the Lie algebra of $\mathrm{PSO}(1, d)$ is $\mathrm{so}(1, d)$.

The matrices

$$E_j := \langle e_0, \cdot \rangle e_j - \langle e_j, \cdot \rangle e_0, \qquad \text{for } 1 \le j \le d,$$

belong to the Lie algebra $\mathrm{so}(1, d)$, and generate so-called **boosts** (or **hyperbolic screws**): for any $\xi \in \mathbb{R}^{1,d}, t \in \mathbb{R}, 1 \le j \le d$,

$$e^{tE_j}(\xi_0, \ldots, \xi_j, \ldots) = \big(\xi_0 \mathrm{ch}\, t + \xi_j \mathrm{sh}\, t, \ldots, \xi_0 \mathrm{sh}\, t + \xi_j \mathrm{ch}\, t, \ldots\big).$$

In special relativity, a boost corresponds to the replacement of the reference frame by a frame in uniform translational motion or, equivalently, to an acceleration in the fixed reference frame; see Section 2.4.1.

The matrices

$$E_{kl} := \langle e_k, \cdot \rangle e_l - \langle e_l, \cdot \rangle e_k, \qquad \text{for } 1 \le k, l \le d,$$

belong to the Lie algebra $so(d) \subset so(1, d)$, and generate the subgroup $SO(d)$. In displayed form, we have (with, for example, $d = 4$, $j = 2$, $k = 1$, $l = 3$)

$$E_j = \begin{pmatrix} 0 & 0 & 1 & 0 & 0 \\ 0 & 0 & 0 & 0 & 0 \\ 1 & 0 & 0 & 0 & 0 \\ 0 & 0 & 0 & 0 & 0 \\ 0 & 0 & 0 & 0 & 0 \end{pmatrix}, \quad E_{kl} = \begin{pmatrix} 0 & 0 & 0 & 0 & 0 \\ 0 & 0 & 0 & 1 & 0 \\ 0 & 0 & 0 & 0 & 0 \\ 0 & -1 & 0 & 0 & 0 \\ 0 & 0 & 0 & 0 & 0 \end{pmatrix}.$$

Proposition 1.3.2 *The matrices* $\{E_j, E_{kl} | 1 \leq j \leq d, 1 \leq k < l \leq d\}$ *constitute a pseudo-orthonormal basis of* $so(1, d)$, *endowed with a Killing form* K. *More precisely, they are pairwise orthogonal, and* $K(E_j, E_j) = 2(d - 1) = -K(E_{k\ell}, E_{k\ell})$. *In particular,* $so(1, d)$ *is a* **semisimple** *Lie algebra: its Killing form is non-degenerate.*

Proof It is clear from its definition that $so(1, d)$ has $d(d + 1)/2$ dimensions, and that the E_j, E_{kl} above are linearly independent (this is also an obvious consequence of the computation of K below) and therefore constitute a basis of $so(1, d)$. Recall from Section 1.1.1 that the Killing form K is defined by $K(E, E') := \text{Tr}(\text{ad}(E) \circ \text{ad}(E'))$.

From the definitions of E_j, E_{kl} above, we find easily (for $1 \leq i, j, k, \ell \leq d$) that

$$\text{ad}(E_j)(E_i) = E_{ji}, \text{ad}(E_j)(E_{k\ell}) = -\text{ad}(E_{k\ell})(E_j) = \delta_{jk} E_\ell - \delta_{j\ell} E_k, \tag{1.13}$$

$$\text{ad}(E_{k\ell})(E_{ij}) = \delta_{i\ell} E_{kj} - \delta_{j\ell} E_{ki} - \delta_{ik} E_{\ell j} + \delta_{jk} E_{\ell i}. \tag{1.14}$$

We deduce that (for $1 \leq i, j, j', k, \ell \leq d$)

$$\text{ad}(E_{j'}) \circ \text{ad}(E_j)(E_i) = \delta_{jj'} E_i - \delta_{ij'} E_j, \quad \text{ad}(E_{j'}) \circ \text{ad}(E_j)(E_{k\ell}) = \delta_{jk} E_{j'\ell} + \delta_{j\ell} E_{kj'},$$

so that

$$K(E_j, E_{j'}) = (d - 1)\delta_{jj'} + \sum_{1 \leq k < l \leq d} (\delta_{jk}\delta_{j'k} + \delta_{j\ell}\delta_{j'\ell}) = 2(d - 1)\delta_{jj'}.$$

It is then clear that any $\text{ad}(E_{k\ell}) \circ \text{ad}(E_j)$ maps any infinitesimal boost E_i onto a linear combination of infinitesimal rotations $E_{k'\ell'}$, and any infinitesimal rotation $E_{k'\ell'}$ onto a linear combination of infinitesimal boosts E_i, so that $K(E_j, E_{k\ell}) = 0$ for all j, k, ℓ.

Finally, for $k < \ell, k' < \ell', j, i < j$ in $\{1, \ldots, d\}$, we find

$$\text{ad}(E_{k'\ell'}) \circ \text{ad}(E_{k\ell})(E_j) = (\delta_{j\ell}\delta_{k\ell'} - \delta_{jk}\delta_{\ell\ell'})E_{k'} - (\delta_{j\ell}\delta_{kk'} - \delta_{jk}\delta_{k'\ell})E_{\ell'},$$

and

$$\text{ad}(E_{k'\ell'}) \circ \text{ad}(E_{k\ell})(E_{ij})$$
$$= (\delta_{i\ell}\delta_{k\ell'} - \delta_{ik}\delta_{\ell\ell'})E_{k'j} + (\delta_{ik}\delta_{k'\ell} - \delta_{i\ell}\delta_{kk'})E_{\ell'j}$$
$$+ (\delta_{j\ell}\delta_{k\ell'} - \delta_{jk}\delta_{\ell\ell'})E_{ik'} + (\delta_{jk}\delta_{k'\ell} - \delta_{j\ell}\delta_{kk'})E_{i\ell'} - (\delta_{j\ell}\delta_{i\ell'} - \delta_{i\ell}\delta_{j\ell'})E_{kk'}$$
$$- (\delta_{i\ell}\delta_{jk'} - \delta_{j\ell}\delta_{ik'})E_{k\ell'} - (\delta_{i\ell'}\delta_{jk} - \delta_{j\ell'}\delta_{ik})E_{k'\ell} - (\delta_{ik'}\delta_{jk} - \delta_{jk'}\delta_{ik})E_{\ell\ell'},$$

so that

$$K(E_{k\ell}, E_{k'\ell'}) = (\delta_{k'\ell}\delta_{k\ell'} - \delta_{kk'}\delta_{\ell\ell'}) \times \left(2 + \sum_{j>k'} 1 + \sum_{j>\ell'} 1 + \sum_{i<k'} 1 + \sum_{i<\ell'} 1\right)$$

$$-(\delta_{k'\ell}\delta_{k\ell'} - \delta_{k\ell}\delta_{k'\ell'})(1_{\{k<k'\}} + 1_{\{k>k'\}}) - (\delta_{k\ell}\delta_{k'\ell'} - \delta_{kk'}\delta_{\ell\ell'})(1_{\{k<\ell'\}} + 1_{\{k>\ell'\}})$$

$$-(\delta_{k\ell}\delta_{k'\ell'} - \delta_{kk'}\delta_{\ell\ell'})(1_{\{k'<\ell\}} + 1_{\{k'>\ell\}}) - (\delta_{k'\ell}\delta_{k\ell'} - \delta_{k\ell}\delta_{k'\ell'})(1_{\{\ell<\ell'\}} + 1_{\{\ell>\ell'\}})$$

$$= -\delta_{kk'}\delta_{\ell\ell'} \times (2 + 2(d-1) - 2) = 2(1-d)1_{\{E_{k\ell} = E_{k'\ell'}\}}. \qquad \diamond$$

Remark 1.3.3 Note that $PSO(1,d)$ acts transitively and properly on $\mathbb{F}^d$. Actually, every Lorentz frame $\beta = (\beta_0, \beta_1, \ldots, \beta_d)$ is the image of the canonical Lorentz frame $e = (e_0, e_1, \ldots, e_d)$ by a unique element of $PSO(1,d)$, which we denote henceforth by $\tilde{\beta}$, so that $\tilde{\beta}(e_j) = \beta_j$. Note moreover that we have: $g(\beta) = g\tilde{\beta}$, for any $g \in PSO(1,d)$ and $\beta \in \mathbb{F}^d$. This describes a **left action** of $PSO(1,d)$ on $\mathbb{F}^d$. We have a **right action** as well, namely $(\beta, g) \mapsto \beta g \in \mathbb{F}^d$, defined by $\widetilde{\beta g} := \tilde{\beta}g$.

1.4   Two remarkable subgroups of $PSO(1, d)$

We set

$$\tilde{E}_j := E_j + E_{1j}, \quad \text{for} \ 2 \le j \le d, \qquad (1.15)$$

where the matrices E_j, E_{1j} are as defined in Section 1.3. As $[\tilde{E}_j, \tilde{E}_k]$ vanishes for all $2 \le j, k \le d$, the set $\{\tilde{E}_2, \ldots, \tilde{E}_d\}$ generates a commutative Lie subalgebra of $so(1,d)$, isomorphic to $\mathbb{R}^{d-1}$, which we shall denote by τ_{d-1}. By Proposition 1.3.2, the Killing form K vanishes on the subalgebra τ_{d-1}. For future convenience, we also set $\tilde{E}_1 := E_1$.

Exercise Show that, as a commutative Lie subalgebra of $so(1,d)$, τ_{d-1} is maximal.
    (By first letting the Lie bracket act on e_0, verify that the linear combinations of E_1 and the $E_{k\ell}$ which commute with $\tilde{E}_j$ can use only those $E_{k\ell}$ such that $k \ne j \ne \ell$, and not E_1.)

Definition 1.4.1 (i) *Set, for any* $r \in \mathbb{R}$,

$$\theta_r := \exp[rE_1] = \begin{pmatrix} \operatorname{ch} r & \operatorname{sh} r & 0 & 0 & \cdots & 0 \\ \operatorname{sh} r & \operatorname{ch} r & 0 & 0 & \cdots & 0 \\ 0 & 0 & 1 & 0 & \cdots & 0 \\ \cdots & \cdots & \cdots & \cdots & \cdots & \cdots \\ 0 & 0 & 0 & 0 & \cdots & 1 \end{pmatrix}.$$

(ii) For any $u = (0, 0, u_2, \ldots, u_d) \in \{e_0, e_1\}^\perp \equiv \mathbb{R}^{d-1}$, set $|u|^2 = -\langle u, u \rangle = \sum_{j=2}^d u_j^2$, and

$$\theta_u^+ := \exp\left[\sum_{j=2}^d u_j \tilde{E}_j\right] = \begin{pmatrix} 1 + |u|^2/2 & -|u|^2/2 & u_2 & \cdots & \cdots & u_d \\ |u|^2/2 & 1 - |u|^2/2 & u_2 & \cdots & \cdots & u_d \\ u_2 & -u_2 & 1 & 0 & \cdots & 0 \\ \cdots & \cdots & \cdots & \cdots & \cdots & \cdots \\ u_d & -u_d & 0 & \cdots & 0 & 1 \end{pmatrix}.$$

$\{\theta_r | r \in \mathbb{R}\}$ and $\mathbb{T}_{d-1} := \{\theta_u^+ | u \in \mathbb{R}^{d-1}\}$ are Abelian subgroups of PSO(1, d). We shall call the elements of $\mathbb{T}_{d-1}$ **horizontal translations**.

Note that the above expression for θ_u^+ follows at once from the following observation:

$$\left[\sum_{j=2}^d u_j \tilde{E}_j\right]^2 = \sum_{j=2}^d u_j^2 \tilde{E}_j^2, \text{ and then } \left[\sum_{j=2}^d u_j \tilde{E}_j\right]^3 = 0.$$

Note also that $\theta_{u+v}^+ = \theta_u^+ \theta_v^+$, so that $\mathbb{T}_{d-1}$ is isomorphic to $\mathbb{R}^{d-1}$.

We now denote by $\hat{\tau}_d$ the Lie subalgebra of so(1, d) generated by τ_{d-1} and E_1. Note that $[E_1, E_j] = E_{1,j}$ and $[E_1, E_{1,j}] = E_j$, so that $[E_1, \tilde{E}_j] = \tilde{E}_j$, and $\hat{\tau}_d$ has dimension d.

We denote by $\mathbb{A}^d$ the affine subgroup of PSO(1, d) generated by these matrices, which is also the subgroup of PSO(1, d) associated with the Lie subalgebra $\hat{\tau}_d$ of so(1, d). Note that $\theta_x^+ (e_0 + e_1) = (e_0 + e_1)$ for any $x \in \mathbb{R}^{d-1}$ and that $\theta_t (e_0 + e_1) = e^t (e_0 + e_1)$ for any $t \in \mathbb{R}$. Thus all matrices of $\mathbb{A}^d$ fix a particular half-line $\mathbb{R}_+^* (e_0 + e_1)$ of the light cone $\mathcal{C}$. Note that $\mathbb{A}^d$ is a Lie group, since it is closed, as follows easily from part (i) of Proposition 1.4.3: if $\gamma = \lim_{n \to \infty} \theta_{x_n}^+ \theta_{t_n}$, then

$$\gamma(e_0 + e_1) = \lim_{n \to \infty} \theta_{x_n}^+ \theta_{t_n} (e_0 + e_1) = \lim_{n \to \infty} e^{t_n} (e_0 + e_1),$$

so that $t_n \to t \in \mathbb{R}$, and then x_n must converge to some $x \in \mathbb{R}^{d-1}$, yielding $\gamma = \theta_x^+ \theta_t \in \mathbb{A}^d$.

Lemma 1.4.2 For any $\varrho \in$ SO(d) and $1 \le j \le d$, we have

$$\text{Ad}(\varrho)(E_j) \equiv \varrho E_j \varrho^{-1} = E_{\varrho(e_j)} := -\sum_{k=1}^d \langle \varrho(e_j), e_k \rangle E_k,$$

Proof By the commutation formula in eqn (1.13), we have: $[A, E_j] = E_{Ae_j} = -\sum_{k=1}^d \langle Ae_j, e_k \rangle E_k$ for any $A \in$ so(d) and $1 \le j \le d$, whence, by eqn (1.4),

$$\text{Ad}(\exp(A))(E_j) = E_{\exp(A)(e_j)}. \qquad \diamond$$

Proposition 1.4.3 (*i*) *Any element of the Lie subgroup* $\mathbb{A}^d$ *can be written as* $\theta_x^+ \theta_t$ *in a unique way, for some* $(t, x) \in \mathbb{R} \times \mathbb{R}^{d-1}$. *Moreover, for any* $(t, x) \in \mathbb{R} \times \mathbb{R}^{d-1}$, *we have*

$$\theta_t \theta_x^+ = \theta_{e^t x}^+ \theta_t. \tag{1.16}$$

For any $z = (x, y) \in \mathbb{R}^{d-1} \times \mathbb{R}_+^*$, *set* $T_z \equiv T_{x,y} := \theta_x^+ \theta_{\log y}$. *We thus have* $\mathbb{A}^d = \{T_z | z \in \mathbb{R}^{d-1} \times \mathbb{R}_+^*\}$, *and the product formula*

$$T_{x,y} T_{x',y'} = T_{x+yx', yy'}. \tag{1.17}$$

(*ii*) *For any* $\varrho \in \mathrm{SO}(d-1)$, *we have*

$$\varrho \theta_x^+ \varrho^{-1} = \theta_{\varrho(x)}^+ \quad and \quad \varrho \theta_t \varrho^{-1} = \theta_t.$$

Proof (*i*) The commutation formula in eqn (1.16) follows directly from the expressions for the matrices θ_x^+, θ_t displayed above. The existence of the decomposition $\theta_u^+ \theta_t$ follows at once. The uniqueness is clear, since $\theta_t = \theta_u^+$ implies obviously that $t = 0, u = 0$, looking again at the expressions for these matrices. Equation (1.17) follows at once from eqn (1.16).

(*ii*) follows straightforwardly from Lemma 1.4.2. ◇

Remark 1.4.4 Equation (1.17) shows that the affine subgroup $\mathbb{A}^d$ is isomorphic to a semi-direct product of $\mathbb{R}$ and $\mathbb{R}^{d-1}$, namely the classical group of translations and dilatations of $\mathbb{R}^{d-1}$, or, equivalently, to the group of $d \times d$ triangular matrices

$$\left\{ \begin{pmatrix} 1 & 0 \\ x & y\mathbf{1}_{\mathbf{d-1}} \end{pmatrix} \in \mathrm{GL}(d) \, \middle| \, x \in \mathbb{R}^{d-1} (\text{written as a column}), y > 0, \text{unit } \mathbf{1}_{\mathbf{d-1}} \in \mathrm{GL}(d-1) \right\},$$

associating $\begin{pmatrix} 1 & 0 \\ x & \mathbf{1}_{\mathbf{d-1}} \end{pmatrix}$ with θ_x^+ and $\begin{pmatrix} 1 & 0 \\ 0 & y\mathbf{1}_{\mathbf{d-1}} \end{pmatrix}$ with $\theta_{\log y}$.

1.5   Structure of the elements of PSO(1, d)

Recall that any $\varrho \in \mathrm{SO}(d)$ is conjugate (in $\mathrm{SO}(d)$) to an element whose non-null entries either are 1's on the diagonal or form diagonal planar rotation blocks

$$\begin{bmatrix} \cos \varphi_j & -\sin \varphi_j \\ \sin \varphi_j & \cos \varphi_j \end{bmatrix}.$$

This is, of course, well known. We determine here the structure of the elements of $\mathrm{PSO}(1, d)$, and classify them.

Consider   the   following   three   subgroups   of   PSO$(1,d)$:   SO(d), PSO$(1, 1) \times$ SO$(d - 1) =$

$$\left\{ \begin{pmatrix} \mathrm{ch}\ r & \mathrm{sh}\ r & 0 \\ \mathrm{sh}\ r & \mathrm{ch}\ r & 0 \\ 0 & 0 & R \end{pmatrix} = \theta_r R = R\theta_r \,\middle|\, r \in \mathbb{R}, R \in \mathrm{SO}(d-1) \right\}$$

and $\mathbb{T}_1 \times \mathrm{SO}(d - 2) =$

$$\left\{ \begin{pmatrix} 1 + u^2/2 & -u^2/2 & u & 0 \\ u^2/2 & 1 - u^2/2 & u & 0 \\ u & -u & 1 & 0 \\ 0 & 0 & 0 & \varrho \end{pmatrix} = \theta^+_{ue_2} \varrho = \varrho \theta^+_{ue_2} \,\middle|\, u \in \mathbb{R}, \varrho \in \mathrm{SO}(d-2) \right\}.$$

Note that the eigenvalues of the element

$$\begin{pmatrix} \mathrm{ch}\ r & \mathrm{sh}\ r \\ \mathrm{sh}\ r & \mathrm{ch}\ r \end{pmatrix}$$

are $e^{\pm r}$ (associated with lightlike eigenvectors), and that

$$\begin{pmatrix} 1 + u^2/2 & -u^2/2 & u \\ u^2/2 & 1 - u^2/2 & u \\ u & -u & 1 \end{pmatrix}$$

has the unique eigenvalue 1 (associated with the unique lightlike eigenray $\mathbb{R}_+(1,1,0)$).

Theorem 1.5.1 *The elements of* PSO$(1, d)$ *can be classified into the following three types (and there is no other possibility):*

- *Those with a timelike eigenvector are conjugate to an element of* SO(d), *and will be called **rotations** (or elliptic elements).*
- *Those without any timelike eigenvector and with two lightlike eigenvectors are conjugate to an element of* PSO$(1, 1) \times$ SO$(d - 1)$, *and will be called **boosts** (or loxodromic elements).*
- *Those without any timelike eigenvector and with a unique lightlike eigenvector are conjugate to an element of* $\mathbb{T}_1 \times$ SO$(d - 2)$, *and will be called **parabolic** elements.*

  In other words, using the hyperbolic space $\mathbb{H}^d$ *and its boundary* $\partial\mathbb{H}^d$ *(defined in Section 1.6),*
- *rotations are elements of* PSO$(1, d)$ *which fix at least one point of* $\mathbb{H}^d$;
- *parabolic isometries are elements of* PSO$(1, d)$ *which fix no point of* $\mathbb{H}^d$ *and a unique point of* $\partial\mathbb{H}^d$;
- *boosts are elements of* PSO$(1, d)$ *which fix no point of* $\mathbb{H}^d$ *but two (and not three) points of* $\partial\mathbb{H}^d$.

Specializing to $d = 2$ and using Proposition 1.1.5.1, it can be immediately seen that an element of PSL(2) is a rotation, a parabolic element or a boost, according to whether the absolute value of its trace is $< 2, = 2$ or > 2, respectively.

It appears that all eigenvalues of the elements of PSO$(1, d)$ are real and positive or have modulus 1.

Proof (1) Let us complexify $\mathbb{R}^{1,d}$ in $\mathbb{C}^{1,d}$, endowed with the sesquilinear pseudo-norm $\langle \xi, \xi' \rangle := \xi_0 \overline{\xi'_0} - \sum_{j=1}^d \xi_j \overline{\xi'_j}$, and fix $\gamma \in$ PSO$(1, d)$, linearly extended into an isometry of $\mathbb{C}^{1,d}$, which can be diagonalized. If $\gamma v = \lambda v$, then $\gamma \bar{v} = \bar{\lambda} \bar{v}$ $\langle v, v \rangle = |\lambda|^2 \langle v, v \rangle$ and $\langle v, \bar{v} \rangle = \lambda^2 \langle v, \bar{v} \rangle$. Therefore we have

$$\text{either} \quad |\lambda| = 1 \quad \text{or} \quad \langle v, v \rangle = 0,$$

and

$$\text{either} \quad \lambda = \pm 1 \quad \text{or} \quad \langle v, \bar{v} \rangle = 0.$$

As a consequence, if $\lambda = e^{\sqrt{-1}\,\varphi} \notin \mathbb{R}$ is an eigenvalue, with associated eigenvector v, we must have $\langle v, \bar{v} \rangle = 0$. Therefore, any real eigenvector associated with λ must be lightlike.

(2) Suppose there exists an eigenvalue λ such that $|\lambda| \neq 1$, and let v denote an associated eigenvector. Then $\langle v, v \rangle = \langle v, \bar{v} \rangle = \langle \bar{v}, \bar{v} \rangle = 0$, which by Lemma 1.2.1 forces the two real vectors $(v + \bar{v})$ and $(v - \bar{v})\sqrt{-1}$ to be linearly dependent: there exist real s, t, both non-vanishing, such that $s(v - \bar{v}) = \sqrt{-1}\, t(v + \bar{v})$, meaning that $w := (s - \sqrt{-1}\, t)v$ is real. Thus w is clearly an isotropic eigenvector associated with the eigenvalue λ, which we can choose to belong to the light cone $\mathcal{C}$. Note that this implies also that λ must be real, and even positive, since $\gamma \in$ PSO$(1, d)$ has to preserve $\mathcal{C}$.

(3) In accordance with (2) above, consider a real eigenvalue $\lambda = e^{r'} \neq 1$, associated with an eigenvector $w \in \mathcal{C}$. Since $\det \gamma = 1$, we must have another eigenvalue λ' with modulus $\neq 1$, necessarily also real and associated with some eigenvector $w' \in \mathcal{C}$. By Lemma 1.2.1, we must have $\langle w, w' \rangle \neq 0$, whence $\lambda' = \langle v, \gamma v' \rangle / \langle w, w' \rangle = \langle \gamma^{-1} v, v' \rangle / \langle v, v' \rangle = e^{-r'}$. This implies also that w, w' are not collinear. Moreover, $v_0 := w + \langle w, w' \rangle^{-1} w'$ is in $\mathbb{H}^d$, and $v_1 := w - \langle w, w' \rangle^{-1} w'$ is such that $\langle v_1, v'_1 \rangle = -1$. We obtain at once, for some $\varepsilon = \pm 1$, the result $\gamma v_0 = (\text{ch } r)v_0 + (\text{sh } r)v_1$ and $\gamma v_1 = (\text{sh } r)v_0 + (\text{ch } r)v_1$. The restriction of γ to the spacelike subspace $\{v_0, v_1\}^\perp$ must be a rotation in this subspace, and must then be a conjugate of some $\varrho \in$ SO$(d - 1)$.

Furthermore, the eigenvectors different from $v_0 \pm v_1$ are the (spacelike) eigenvectors of the rotation part (as any other eigenvector could be decomposed into the sum of an eigenvector belonging to the timelike plane $\{v_0, v_1\}$ and an eigenvector in $\{v_0, v_1\}^\perp$ with the same eigenvalue). In particular, γ has no timelike eigenvectors.

(4) Suppose there is a timelike eigenvector v_0. By (1) and (3) above, the corresponding eigenvalue λ has to be ± 1, and in fact 1, since the solid light cone is preserved by γ. The restriction of γ to the spacelike subspace $\{v_0\}^\perp$ must be a rotation in this subspace, and must then be a conjugate of some $\varrho \in$ SO(d), so that γ is a rotation.

(5) We now consider the remaining possibility: the eigenvalues have modulus 1, and there are no timelike eigenvectors.

Consider an eigenvalue $\lambda = e^{\sqrt{-1}\,\varphi} \notin \mathbb{R}$, with associated eigenvector v. By (1) above we must have $\langle v, \bar{v} \rangle = 0$. Consider also the real vectors $u := (v + \bar{v})$ and $u' := (v - \bar{v})\sqrt{-1}$. We have $\langle u, u \rangle = \langle v, v \rangle + \langle \bar{v}, \bar{v} \rangle = \langle u', u' \rangle$, and

$$\gamma u = (\cos\ \varphi)u + (\sin\varphi)u' \quad \text{and} \quad \gamma u' = (\cos\ \varphi)u' - (\sin\ \varphi)u, \qquad (*)$$

whence $\langle u, u'\rangle = \langle \gamma u, \gamma u'\rangle = \cos(2\varphi)\langle u, u'\rangle$. Since $\cos(2\varphi) \neq 1$, this implies $0 = \langle u, u'\rangle$, and then, for all real s, t, $\langle su + tu', su + tu'\rangle = (s^2 + t^2)\langle u, u\rangle$. Hence, by Lemma 1.2.1, u and u' are either spacelike, or collinear and lightlike.

Now, if $u \neq 0$ and $u' = \alpha u$, using the above expressions $(*)$ for $\gamma u, \gamma u'$, we would have $\alpha^2 = -1$, a contradiction. Hence, u and u' span a spacelike plane, and the restriction of γ to this plane is a rotation. And the restriction of γ to $\{u, u'\}^\perp$ is an element of PSO$(1, d - 2)$ which is neither a boost nor a rotation. Hence, by recursion, we are left with the case where $\gamma \in$ PSO$(1, d - 2k)$ can have only ± 1 as eigenvalues. Any such eigenvalue has an associated real eigenvector, which must be non-timelike. We can then restrict γ to the space orthogonal to the space spanned by spacelike eigenvectors.

We are left with the case where all eigenvectors are lightlike. Let us show now that -1 cannot be an eigenvalue: in that case, there would be a $u \in \mathcal{C}$ such that $\gamma u = -u$; choosing some $u' \in \mathcal{C}$ non-collinear to u, we would have $\langle u, u'\rangle > 0$ and $\langle u, \gamma u'\rangle = \langle \gamma\ u, u'\rangle < 0$, a contradiction, since $\gamma u' \in \mathcal{C}$.

It follows that the eigenvector is unique, since the sum of two non-collinear eigenvectors in $\mathcal{C}$ would be timelike. Note at this stage that the multiplicity of -1 in the decomposition of γ must be even, so that the restriction of γ to the subspace spanned by spacelike eigenvectors is a rotation.

(6) We are left finally with $\gamma \in$ PSO$(1, d')$ (with $d' = d - 2m > 0$) possessing a unique eigenvector $v \in \mathcal{C}$ (up to a scalar), which is associated with the eigenvalue 1. We conclude the proof by showing that $d' = 2$, and that γ is conjugate to an element of $\mathbb{T}_1$.

Choose some $v' \in \mathcal{C}$ such that $\langle v, v'\rangle = \frac{1}{2}$, and set $v'' := \gamma v'$, $v_0 := v + v'$, $v_1 :=$ v-v', $u := \sqrt{2\langle v', v''\rangle}$. Note that $v'' \in \mathcal{C}$, $\langle v, v''\rangle = \frac{1}{2}$, $\langle v_1, v_1\rangle = -\langle v_0, v_0\rangle = -1$, $\langle v_0, v_1\rangle = 0$, and consider $v_2 := (v'' - u^2 v - v')/u$. We have $\langle v_2, v\rangle = \langle v_2, v'\rangle = 0$, $\langle v_2, v_2\rangle = -1$, so that we can complete (v_0, v_1, v_2) to form a Lorentz basis $(v_0, v_1, \ldots, v_{d'})$. Let $\tilde\gamma$ denote the element of PSO$(1, d')$ which has the matrix $\theta^+_{ue_2}$ in this basis. We then have

$$\gamma v' = v'' = u^2 v + v' + u v_2 = \tilde\gamma v' \text{ and } \gamma v = v = \tilde\gamma v,$$

whence $\tilde\gamma^{-1}\gamma v_0 = v_0$ and $\tilde\gamma^{-1}\gamma v_1 = v_1$. Hence $P := \tilde\gamma^{-1}\gamma$ must belong to SO$(d' - 1)$, i.e., in the basis $(v_0, v_1, \ldots, v_{d'})$, it must have a matrix of the form

$$\begin{pmatrix} 1 & 0 & 0 \\ 0 & 1 & 0 \\ 0 & 0 & Q \end{pmatrix},$$

with $Q \in$ SO$(d' - 1)$. We must also have, for any λ, $(\lambda - 1)^{d'+1} = \det(\lambda\mathbf{1} - \gamma) = \det(\lambda\theta^+_{-ue_2} - P)$. Now, this last determinant is easily computed by adding the first column to the second and then subtracting the second line from the first. We obtain $\det(\lambda\theta^+_{-ue_2} - P) = (\lambda - 1)^2 \det(\lambda\mathbf{1} - Q)$, which entails $\det(\lambda\mathbf{1} - Q) = (\lambda - 1)^{d'-1}$, whence $Q = \mathbf{1}$ and then $\gamma = \tilde\gamma$. Finally, since γ cannot have any spacelike eigenvector, this forces $d' = 2$, and $\tilde\gamma$ is conjugate to $\theta^+_{ue_2} \in \mathbb{T}_1$. $\diamond$

Remark 1.5.2 The non-trivial boosts form a dense open set. The complement of this set is negligible for the Haar measure of PSO$(1, d)$ (defined in eqn (3.7)).

Proof By Theorem 1.5.1, all rotations and parabolic elements have the eigenvalue 1, and hence are included in the algebraic hypersurface having equation $\det(T_{x,y}\ \varrho - 1) = 0$ (in the Iwasawa coordinates). The statement is now clear, by eqn (3.7) and Theorem 3.3.5. $\diamond$

Remark 1.5.3 The Lorentz–Möbius group is the image of its Lie algebra under the exponential map: $\mathrm{PSO}(1, d) = \exp\big[\mathrm{so}(1, d)\big]$.

Proof The property of $\gamma \in \mathrm{PSO}(1, d)$ of belonging to the range of the exponential map is clearly stable under conjugation, and is obviously true for $\mathrm{SO}(d), \mathrm{PSO}(1, 1)$ and $\mathbb{T}_1$, by Definition 1.4.1. $\diamond$

1.6   The hyperbolic space $\mathbb{H}^d$ and its boundary $\partial\mathbb{H}^d$

The set of vectors that have pseudo-norm 1 and a positive first coordinate is of particular interest, and constitutes the basic model for hyperbolic space, which thus appears most naturally in the framework of Minkowski space.

Notation 1.6.1 We denote by $\mathbb{H}^d$ the d-dimensional hyperbolic space, defined as the positive half of the unit pseudo-sphere of $\mathbb{R}^{1,d}$, that is to say the hypersurface of $\mathbb{R}^{1,d}$ made up of all vectors that have pseudo-norm 1 and a positive first coordinate:

$$\mathbb{H}^d := \big\{\xi = (\xi_0, \dots, \xi_d) \in \mathbb{R}^{1,d} \,\big|\, \langle \xi, \xi \rangle = 1, \xi_0 > 0\big\}$$
$$= \big\{(\mathrm{ch}\ r)e_0 + (\mathrm{sh}\ r)u \,\big|\, r \in \mathbb{R}_+, u \in \mathbb{S}^{d-1}\big\}.$$

This is, of course, a sheet of a hyperboloid. Observe that any element of $\mathrm{PSO}(1, d)$ maps $\mathbb{H}^d$ onto $\mathbb{H}^d$, as is clear if we notice that the elements $\xi \in \mathbb{H}^d$ are, among the vectors that have pseudo-norm 1, those which satisfy $\langle \xi, v \rangle > 0$ for any $v \in \mathcal{C}$.

In special relativity, an element of $\mathbb{H}^d$ represents the $(d + 1)$-velocity of a particle of unit mass. Boosts can be interpreted as an acceleration of the particle in a fixed reference frame; see Section 2.4.

Proposition 1.6.2 *The Lie subgroup $\mathbb{A}^d$ acts transitively and properly on $\mathbb{H}^d$: for any $p \in \mathbb{H}^d$, there exists a unique $T_{x,y} \in \mathbb{A}^d$ such that $T_{x,y}e_0 = p$.*

*x, y are called the **Poincaré coordinates** of p.*

In other words, in the canonical basis $e = (e_0, \dots, e_d)$, for a unique $(x, y) \in \mathbb{R}^{d-1} \times \mathbb{R}_+^*$, we have

$$p = \left(\frac{y^2 + |x|^2 + 1}{2y}\right)e_0 + \left(\frac{y^2 + |x|^2 - 1}{2y}\right)e_1 + \sum_{j=2}^{d} \frac{x^j}{y}e_j. \qquad (1.18)$$

Equivalently, the Poincaré coordinates (x, y) of a point $p \in \mathbb{H}^d$ that has coordinates $(p^0, \dots, p^d)$ in the canonical basis $(e_0, \dots, e_d)$ are given by

$$y = \frac{1}{p^0 - p^1} = \frac{1}{\langle p, e_0 + e_1 \rangle} \quad and \quad x^j = \frac{p^j}{p^0 - p^1} = \frac{-\langle p, e_j \rangle}{\langle p, e_0 + e_1 \rangle} \qquad (1.19)$$

for $2 \leq j \leq d$.

The Poincaré coordinates (x, y) can also be used to parametrize the subgroup $\mathbb{A}^d$ by $\mathbb{R}^{d-1} \times \mathbb{R}_+^*$, which is usually called the **Poincaré (upper) half-space**.

Proof Equation (1.18) is given simply by the first column of the matrix $T_{x,y}$ (recall Definition 1.4.1 and Proposition 1.4.3). It is easily solved in a unique way if we notice that we must have $|x|^2 = y^2\big((p^0)^2 - (p^1)^2 - 1\big)$, which yields eqn (1.19). $\diamondsuit$

A **light ray** is a future-oriented lightlike direction, i.e., an element of

$$\partial \mathbb{H}^d := \mathcal{C}/\mathbb{R}_+ = \{\mathbb{R}_+\xi | \xi \in \mathcal{C}\}.$$

In the projective space of $\mathbb{R}^{1,d}$, the set of light rays $\partial \mathbb{H}^d$ is identified with the **boundary** of $\mathbb{H}^d$. Note that the Lorentz–Möbius group $PSO(1, d)$ acts on $\partial \mathbb{H}^d$. Note also that for any light ray $\eta \in \partial \mathbb{H}^d$, $\eta^\perp$ is the hyperplane tangent to the light cone $\mathcal{C}$ at η (in particular, $\eta^\perp$ contains η).

The pseudo-metric of $\mathbb{R}^{1,d}$ induces canonically a Euclidean structure on each d-space $p^\perp$, for any $p \in \mathbb{H}^d$; the inner product here is simply the negative $-\langle \cdot, \cdot \rangle$ of the restriction of the pseudo-metric.

1.7 The Cartan and Iwasawa decompositions of PSO(1, d)

The Cartan decomposition will be used to define the hyperbolic metric, and then (in Section 3.5) polar coordinates on $\mathbb{H}^d$.

Theorem 1.7.1 (**Cartan decomposition** of $PSO(1, d)$) Any γ in $PSO(1, d)$ can be written as $\gamma = \varrho \theta_r \varrho'$, with $r \in \mathbb{R}_+$ and $\varrho, \varrho' \in SO(d)$. Moreover, we have (i) $r = \log \|\gamma\| = \log \|\gamma^{-1}\|$, where $\|\gamma\| := \max\{|\gamma v|/v \in \mathbb{S}^d\}$ denotes the operator norm of γ acting on the Euclidean space $\mathbb{R}^{d+1}$, identified as a vector space with $\mathbb{R}^{1,d}$; and (ii) $\gamma = \varrho_1 \theta_r \varrho_1'$ holds if and only if $\varrho_1 = \varrho \tilde{\varrho}$ and $\varrho' = \tilde{\varrho} \varrho_1'$, with $\tilde{\varrho} \in SO(d-1)$ if $r > 0$.

Proof We must obviously have $\gamma e_0 = \varrho \theta_r e_0 \in \mathbb{H}^d$. For some $r \in \mathbb{R}_+$ and $u \in \mathbb{S}^{d-1}$, we can write $\gamma e_0 = (\operatorname{ch} r)e_0 + (\operatorname{sh} r)u$. Denote by $\varrho \in SO(d)$ the rotation that acts trivially in $\{e_1, u\}^\perp$ and maps e_1 to u: we thus have $\varrho^{-1}\gamma e_0 = (\operatorname{ch} r)e_0 + (\operatorname{sh} r)e_1 = \theta_r e_0$, meaning that $\varrho' := \theta_r^{-1}\varrho^{-1}\gamma \in SO(d)$.

(i) We clearly have $\|\gamma\| = \|\varrho \theta_r \varrho'\| = \|\theta_r\|$, and, using the matrix expression for θ_r, we easily obtain $\|\theta_r\| = e^r$. Considering then the rotation $\varrho_0 \in SO(d)$ that fixes $(e_0, e_3, \ldots, e_d)$ and maps (e_1, e_2) to $(-e_1, -e_2)$, we have $\theta_{-r} = \varrho_0 \theta_r \varrho_0$, whence $\|\gamma^{-1}\| = \|\gamma\|$.

(ii) Suppose now that $\gamma = \varrho \theta_r \varrho' = \varrho_1 \theta_r \varrho_1'$, and set $\tilde{\varrho} := \varrho^{-1}\varrho_1$. We have:

$$(\operatorname{ch} r)e_0 + (\operatorname{sh} r)\tilde{\varrho} e_1 = \tilde{\varrho}\theta_r e_0 = \tilde{\varrho}\theta_r \varrho_1' e_0$$

$$= \theta_r \varrho' e_0 = \theta_r e_0 = (\operatorname{ch} r)e_0 + (\operatorname{sh} r)e_1,$$

and hence if $r > 0$, then $\tilde{\varrho} e_1 = e_1$, meaning that $\tilde{\varrho} \in SO(d-1)$. The remainder of (ii) is obvious, since θ_r commutes with any $\tilde{\varrho} \in SO(d-1)$. $\diamondsuit$

Example The Cartan decomposition of $\theta_t \in \mathbb{A}^2$ is given for any $t < 0$ by

$$\theta_t = \varrho\theta_{|t|}\varrho, \quad \text{where} \quad \varrho e_1 + e_1 = \varrho e_2 + e_2 = 0 \quad \text{and} \quad \varrho = 1 \text{ on } \{e_1, e_2\}^\perp.$$

The Cartan decomposition of $\theta_u^+ \in \mathbb{A}^2$ is given for any $u \in \mathbb{R}$ by

$$\begin{bmatrix} 1 + \frac{u^2}{2} & -\frac{u^2}{2} & u \\ \frac{u^2}{2} & 1 - \frac{u^2}{2} & u \\ u & -u & 1 \end{bmatrix} = \begin{bmatrix} 1 & 0 & 0 \\ 0 & \frac{u}{\sqrt{u^2+4}} & \frac{-2}{\sqrt{u^2+4}} \\ 0 & \frac{2}{\sqrt{u^2+4}} & \frac{u}{\sqrt{u^2+4}} \end{bmatrix} \begin{bmatrix} 1 + \frac{u^2}{2} & \frac{u}{2}\sqrt{u^2+4} & 0 \\ \frac{u^2}{2}\sqrt{u^2+4} & 1 + \frac{u^2}{2} & 0 \\ 0 & 0 & 1 \end{bmatrix} \begin{bmatrix} 1 & 0 & 0 \\ 0 & \frac{-u}{\sqrt{u^2+4}} & \frac{2}{\sqrt{u^2+4}} \\ 0 & \frac{-2}{\sqrt{u^2+4}} & \frac{-u}{\sqrt{u^2+4}} \end{bmatrix}.$$

We shall actually use mainly the following Iwasawa decomposition of $\mathrm{PSO}(1, d)$, which asserts that any element of $\mathrm{PSO}(1, d)$ is in a unique way the product of an element of $\mathbb{A}^d$ with a rotation.

Theorem 1.7.2 (***Iwasawa decomposition*** *of* $\mathrm{PSO}(1, d)$) *Any* $\gamma \in \mathrm{PSO}(1, d)$ *can be written in a unique way:*

$$\gamma = \theta_u^+ \theta_t \varrho, \quad \text{with } \theta_u^+ \theta_t \in \mathbb{A}^d \text{ and } \varrho \in \mathrm{SO}(d).$$

We denote by Iw and I^{w} the canonical projections from $\mathrm{PSO}(1, d)$ onto $\mathbb{A}^d$ and $\mathrm{SO}(d)$, respectively:

$$\mathrm{Iw}(\theta_u^+ \theta_t \varrho) := \theta_u^+ \theta_t, \quad \mathrm{I}^{\mathrm{w}}(\theta_u^+ \theta_t \varrho) := \varrho.$$

Proof We must have: $\gamma e_0 = \theta_u^+ \theta_t e_0 \in \mathbb{H}^d$. By Proposition 1.6.2, this determines a unique $\theta_u^+ \theta_t \in \mathbb{A}^d$, proving the uniqueness. As to the existence, fixing $\gamma \in \mathrm{PSO}(1, d)$ and using Proposition 1.6.2 to obtain $\theta_u^+ \theta_t \in \mathbb{A}^d$ such that $\theta_u^+ \theta_t e_0 = \gamma e_0$, we have $\varrho := (\theta_u^+ \theta_t)^{-1}\gamma \in \mathrm{SO}(d)$. $\diamond$

1.8 Notes and comments

The reference [Kn] contains a historical account of Lie theory, as well as a lot of references. See in particular [Bk], [He], [Ho] and also [DFN].

The Cartan and Iwasawa decompositions are usually presented in the more general context of semisimple Lie groups; see for example [He] and [KN]. The Lorentz group and Minkowski space are fundamental in relativistic mechanics (at least for $d = 3$); see for example [LL], [HE] and [W].

There are various presentations of hyperbolic space; see, in particular, [Rac]. We have chosen to focus on the pseudo-sphere model, in particular to benefit from the physical interpretation. Other models will appear in the remainder of the book as coordinate systems (see Section 2.5).

In general relativity, Minkowski space is replaced by manifolds locally modelled on it. The bundle of frames becomes a principal $\mathrm{PSO}(1, d)$ bundle. The fibres of the unit tangent bundle (which describes particle positions and $(d + 1)$-velocities) are hyperbolic spaces. See, for example, [HE] and [W].

Chapter Two
Hyperbolic geometry

Celui-ci se croyait l'hyperbole permise:
J'ai vu dit-il un chou plus grand qu'une maison. (Jean de La Fontaine)

This chapter presents the basic notions of hyperbolic geometry, systematically derived from the properties of Minkowski space and the Lorentz–Möbius group. Geodesics are given by planes intersecting the light cone in two rays, and horospheres by affine hyperplanes parallel to a light ray and intersecting the hyperbolic space. We use only elementary linear algebra within the Minkowski space $\mathbb{R}^{1,d}$ to calculate intrinsic formulae for the hyperbolic distance and the distance from a point to a geodesic.

We show that any tangent vector to the hyperbolic space $\mathbb{H}^d$ (viewed as a subspace of Minkowski space) can be parametrized by its base point and a point at infinity (i.e., on the boundary $\partial \mathbb{H}^d$). Poincaré coordinates are extended to $\partial \mathbb{H}^d$. We also discuss harmonic conjugation in this framework.

Then we define the geodesic and horocyclic flows by the right action of the affine subgroup $\mathbb{A}^d$ on frames, and we give physical interpretations. We present the classical ball and upper-half-space models, relate the latter to Poincaré coordinates and prove a result which illustrates the instability of the geodesic flow.

In this same chapter, we also establish a commutation relation in the Lorentz–Möbius group, and specify two particular cases, which will be crucial in the proofs of the mixing theorem and the central limit (Sinai's) theorem (in Sections 5.3 and 8.7, respectively). Finally, we introduce stable leaves and the Busemann function.

2.1 The hyperbolic metric

The hyperbolic space $\mathbb{H}^d$ carries a natural metric, induced by $\mathbb{R}^{1,d}$, as follows. Recall that the notation $\tilde{\beta}$ was introduced in Remark 1.3.3.

Proposition 2.1.1 *Given two elements p, p' of $\mathbb{H}^d$, let $\beta, \beta' \in \mathbb{F}^d$ be two Lorentz frames such that $\beta_0 = p$ and $\beta'_0 = p'$. Then the size $r = \log \left\| \tilde{\beta}^{-1}\widetilde{\beta'} \right\| =: \mathrm{dist}\,(p, p')$ of $\tilde{\beta}^{-1}\widetilde{\beta'}$ in its Cartan decomposition (recall Theorem 1.7.1) depends only on (p, p'). This defines a metric on $\mathbb{H}^d$, called the **hyperbolic metric**. Moreover, we have $\mathrm{dist}(p, p') = \mathrm{argch}\left[\langle p, p'\rangle\right]$.*

Proof According to Theorem 1.7.1, write $\tilde{\beta}^{-1}\widetilde{\beta'} = \varrho\,\theta_r\,\varrho'$. Now, if $\gamma, \gamma' \in \mathbb{F}^d$ are such that $\gamma_0 = p$ and $\gamma'_0 = p'$ too, then we have $\tilde{\gamma}^{-1}\tilde{\beta} \in \mathrm{SO}(d)$ and $\widetilde{\gamma'}^{-1}\widetilde{\beta'} \in$

SO(d), so that $\tilde\gamma^{-1}\widetilde{\gamma'} = (\tilde\gamma^{-1}\tilde\beta\varrho)\theta_r(\varrho'\widetilde{\beta'}^{-1}\widetilde{\gamma'})$ determines the same $r \in \mathbb{R}_+$ as $\tilde\beta^{-1}\widetilde{\beta'}$. By Theorem 1.7.1($i$), we have $r = \log\|\tilde\beta^{-1}\widetilde{\beta'}\|$, and then the triangle inequality is clear. And $r = 0$ occurs if and only if $\tilde\beta^{-1}\widetilde{\beta'} \in$ SO(d), which means precisely that $p = p'$. Finally, we have:

$$\mathrm{argch}[\langle p, p'\rangle] = \mathrm{argch}[\langle\beta_0, \beta_0'\rangle] = \mathrm{argch}\Big[\langle e_0, \tilde\beta^{-1}\widetilde{\beta'}(e_0)\rangle_{\mathbb{R}^{1,d}}\Big]$$

$$= \mathrm{argch}[\langle e_0, \theta_r e_0\rangle] = r. \qquad \diamondsuit$$

It is obvious from Proposition 2.1.1 that elements of PSO($1, d$) define isometries of $\mathbb{H}^d$. In fact, all orientation-preserving hyperbolic isometries can be obtained in this way.

Proposition 2.1.2 *The group of orientation-preserving hyperbolic isometries, i.e., orientation-preserving isometries of the hyperbolic space $\mathbb{H}^d$, is canonically isomorphic to the Lorentz–Möbius group* PSO($1, d$).

Proof Consider first $g, g' \in$ PSO($1, d$), which induce the same isometry of $\mathbb{H}^d$. Then $g'g^{-1} \in$ PSO($1, d$) fixes all points of $\mathbb{H}^d$, and hence all points of the vector space generated by $\mathbb{H}^d$, meaning that this is the identity map.

Now consider an orientation-preserving isometry f of $\mathbb{H}^d$. Since PSO($1, d$) acts transitively on $\mathbb{H}^d$, we can suppose that $f(e_0) = e_0$. This implies that $\langle f(\xi), e_0\rangle = \langle\xi, e_0\rangle$ for any $\xi \in \mathbb{H}^d$. The projection $P := (\xi \mapsto \xi - \langle\xi, e_0\rangle e_0)$ is a bijection from $\mathbb{H}^d$ onto $e_0^\perp \equiv \mathbb{R}^d$. Consider $\tilde f := P \circ f \circ P^{-1} : \mathbb{R}^d \to \mathbb{R}^d$. For any $v = P(\xi), v' = P(\xi') \in \mathbb{R}^d$, we have $\langle\tilde f(v), \tilde f(v')\rangle = \langle f(\xi) - \langle f(\xi), e_0\rangle e_0, f(\xi') - \langle f(\xi'), e_0\rangle e_0\rangle = \langle f(\xi), f(\xi')\rangle - \langle f(\xi), e_0\rangle\langle f(\xi'), e_0\rangle = \langle\xi, \xi'\rangle - \langle\xi, e_0\rangle\langle\xi', e_0\rangle = \langle v, v'\rangle$. Thus $\tilde f \in$ SO($\mathbb{R}^d$), and hence it must be linear. We extend it by linearity to the whole of $\mathbb{R}^{1,d}$ by setting $\tilde f(e_0) := e_0$. We thus have an element $\tilde f \in$ SO(d) $\subset$ PSO($1, d$), which agrees with f on $\mathbb{H}^d$, since for any $\xi \in \mathbb{H}^d$,

$$\tilde f(\xi) = \tilde f\big(\langle\xi, e_0\rangle e_0 + P(\xi)\big) = \langle\xi, e_0\rangle e_0 + \tilde f\big(P(\xi)\big) = \langle f(\xi), e_0\rangle e_0 + P\big(f(\xi)\big) = f(\xi).$$

Finally, $f \in$ PSO($1, d$) since, by linearity, the preservation of orientation and of the pseudo-metric must extend from $\mathbb{H}^d$ to $\mathbb{R}^{1,d}$, and the preservation of $\mathbb{H}^d$ entails that of the light cone. $\qquad \diamondsuit$

Proposition 2.1.3 *The hyperbolic distance between points $q_1, q_2 \in \mathbb{H}^d$, which have Poincaré coordinates $(x_1, y_1), (x_2, y_2)$, respectively, can be expressed as*

$$\mathrm{dist}(q_1, q_2) = \mathrm{argch}[\langle q_1, q_2\rangle] = \mathrm{argch}\left[\frac{|x_1 - x_2|^2 + y_1^2 + y_2^2}{2y_1 y_2}\right]. \qquad (2.1)$$

Proof We apply Propositions 2.1.1 and 1.6.2, and observe that by eqn (1.17) we have

$$\langle q_1, q_2\rangle = \langle T_{x_1,y_1} e_0, T_{x_2,y_2} e_0\rangle = \langle e_0, T_{x_1,y_1}^{-1} T_{x_2,y_2} e_0\rangle$$

$$= \langle e_0, T_{x_2 - x_1/y_1, y_2/y_1} e_0\rangle.$$

The result follows at once, using eqn (1.18). $\qquad \diamondsuit$

Remark 2.1.4 The hyperbolic length of the line element in the upper half-space $\mathbb{R}^{d-1} \times \mathbb{R}^*_+$ of Poincaré coordinates is the Euclidean length divided by the height y: $ds = \sqrt{|dx|^2 + dy^2}/y$. This results directly from eqn (2.1), since for small $|\delta|^2 + \varepsilon^2$ we have

$$\operatorname{dist}\left(T_{x,y}e_0, T_{x+\delta,y+\varepsilon}e_0\right) = \operatorname{argch}\left[1 + \frac{|\delta|^2 + \varepsilon^2}{2y(y+\varepsilon)}\right] \sim \sqrt{\frac{|\delta|^2 + \varepsilon^2}{y(y+\varepsilon)}} \sim \frac{\sqrt{|\delta|^2 + \varepsilon^2}}{y}.$$

2.2 Geodesics and light rays

2.2.1 Hyperbolic geodesics

Definition 2.2.1.1 *We call any non-empty intersection of $\mathbb{H}^d$ with a vector plane of $\mathbb{R}^{1,d}$ a geodesic of the hyperbolic space $\mathbb{H}^d$. Thus, the set of geodesics of $\mathbb{H}^d$ is identified with the set of vector planes of $\mathbb{R}^{1,d}$ which intersect $\mathbb{H}^d$.*

The following remark justifies the above definition, and yields a natural identification between the set of geodesics of $\mathbb{H}^d$ and the set of pairs $\{\eta, \eta'\}$ of distinct $\eta, \eta' \in \partial\mathbb{H}^d$, or between the set of **oriented geodesics** of $\mathbb{H}^d$ and the set of ordered pairs (η, η') of distinct $\eta, \eta' \in \partial\mathbb{H}^d$.

In fact, for $\eta \neq \eta' \in \partial\mathbb{H}^d$ and for non-null $\xi \in \eta, \xi \in \eta'$, we have $\langle \xi, \xi' \rangle > 0$ by Lemma 1.2.1, and then $(2\langle \xi, \xi' \rangle)^{-1}(\xi + \xi') \in \mathbb{H}^d$: the plane cone generated by η, η' does indeed intersect $\mathbb{H}^d$.

Remark 2.2.1.2 *There exists a unique geodesic containing any two given distinct points of $\mathbb{H}^d \cup \partial\mathbb{H}^d$.*

The geodesic containing the two points of $\partial\mathbb{H}^d$ fixed by a boost is called its axis.

Proposition 2.2.1.3 *In two dimensions, consider two oriented distinct geodesics γ, γ' of $\mathbb{H}^2$, determined (up to orientation) by distinct planes P, P', respectively.*

- *If $P \cap P'$ is timelike, then there exists a unique isometry $\psi \in \mathrm{PSO}(1,2)$ mapping γ to γ' and such that $\psi(P \cap P') = P \cap P'$, and ψ is a rotation.*
- *If $P \cap P'$ is spacelike, then there exists a unique isometry $\psi \in \mathrm{PSO}(1,2)$ mapping γ to γ' and such that $\psi(P \cap P') = P \cap P'$, and ψ is a boost. Moreover, there exists a unique geodesic γ'' intersecting γ and γ' and perpendicular to both of them, and γ'' is the axis of the boost ψ.*
- *If $P \cap P'$ is lightlike, then there exists a unique parabolic isometry $\psi \in \mathrm{PSO}(1,2)$ mapping the non-oriented γ to the non-oriented γ' and such that $\psi(P \cap P') = P \cap P'$.*

  There is no longer any such uniqueness in larger dimensions.

Proof If $P \cap P'$ is spacelike, its orthogonal is a plane intersecting $\mathbb{H}^2$ (it must contain a timelike vector) and hence is a geodesic, clearly perpendicular to both P and P'. Consider a Lorentz frame β such that $\beta_2 \in P \cap P'$ and $\beta_0 \in P$. P' contains $p \in \mathbb{H}^2 \cap (P \cap P')^\perp$, necessarily of the form $p = \operatorname{ch} r\, \beta_0 + \operatorname{sh} r\, \beta_1$ with $r \in \mathbb{R}^*$. Any solution ψ must map β_0 to p, and β_2 to $\pm\beta_2$, and hence it is the unique possibility, depending on the orientations of γ, γ'. Obviously, the boost with matrix θ_r in the frame β is the unique solution, and its axis is $(P \cap P')^\perp$, as claimed.

If $P \cap P'$ is timelike, any solution is a rotation according to Theorem 1.5.1, which in any Lorentz frame β such that $\beta_0 \in P \cap P'$, must have its matrix in SO(2). Obviously, a unique solution satisfies this condition, taking the orientations of γ, γ' into account.

If $P \cap P'$ is lightlike, consider a Lorentz frame β such that $\beta_0 \in P$ and $\beta_0 + \beta_1 \in P \cap P'$. We can write the ends of γ, γ' as $\{\mathbb{R}_+(\beta_0 + \beta_1), \mathbb{R}_+(\beta_0 - \beta_1)\}$ and $\{\mathbb{R}_+(\beta_0 + \beta_1), \mathbb{R}_+(\beta_0 - \cos \varphi \, \beta_1 + \sin \varphi \, \beta_2)\}$, respectively, with $\sin \varphi \neq 0$. Then the parabolic isometry ψ with matrix $\theta^+_{\text{tg}(\varphi/2)}$ in the frame β satisfies this condition. If ψ' is another solution, then $\psi^{-1} \circ \psi'$ fixes $\mathbb{R}_+(\beta_0 + \beta_1)$ and $\mathbb{R}_+(\beta_0 - \beta_1)$ and then (by Theorem 1.5.1) is a boost, which must have matrix θ_r in the frame β. Thus ψ' has, in the frame β, a matrix $\theta^+_{\text{tg}(\varphi/2)}\theta_r$, which happens to have eigenvalues $1, e^r, (1 + \text{tg}^2(\varphi/2))e^{-r}$. By Theorem 1.5.1, since ψ' must be parabolic, this forces $r = 0$, and hence $\psi' = \psi$.   ◇

2.2.2 Projection onto a light ray, and tangent bundle

We have the following useful projection from hyperbolic space onto a light ray, illustrated in Figure 2.1 (note that the location of p in this figure is generic; analogous figures with other locations of p can be deduced merely by using an isometry).

Proposition 2.2.2.1 *For any $(p, \eta) \in \mathbb{H}^d \times \partial\mathbb{H}^d$, there exists a unique $(p_\eta, \eta_p) \in \mathbb{R}^{1,d} \times \eta$ such that*

$$\langle p, p_\eta \rangle = 0, \quad \langle p_\eta, p_\eta \rangle = -1 \quad and \quad p + p_\eta = \eta_p \ (\neq 0).$$

Proof Consider $p' := p - \alpha\eta_0$, for any given $\eta_0 \in \eta \cap \mathcal{C}$. Then $\langle p, \eta_0 \rangle > 0$, and $\langle p, p' \rangle = 0 \Leftrightarrow \langle p', p' \rangle = -1 \Leftrightarrow \alpha\langle p, \eta_0 \rangle = 1$ shows that there is indeed a unique solution.   ◇

Conversely if q belongs to the unit sphere of $p^\perp$, i.e., if $\langle p, q \rangle = 0$ and $\langle q, q \rangle = -1$, then $\eta := \mathbb{R}_+(p + q) \in \partial\mathbb{H}^d$, and $p_\eta = q$.

Notation For any $p \in \mathbb{H}^d$, we denote by J_p the one-to-one map from $\partial\mathbb{H}^d$ into $p^\perp$ defined by $J_p(\eta) = p_\eta$. Its range is the unit sphere of the Euclidean d-space $p^\perp$.

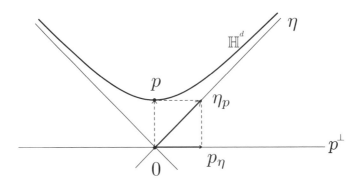

Fig. 2.1 Projection from $\mathbb{H}^d$ onto a light ray η

We denote by $T^1\mathbb{H}^d$ the **unit tangent bundle** of $\mathbb{H}^d$, defined by:

$$T^1\mathbb{H}^d := \left\{ (p,q) \in \mathbb{H}^d \times \mathbb{R}^{1,d} \,\middle|\, q \in p^\perp, \langle q,q \rangle = -1 \right\}.$$

Accordingly, $\mathbb{F}^d$ is known as the **frame bundle** of $\mathbb{H}^d$. Note that $p^\perp$ is the vector hyperplane directing the affine hyperplane tangent to $\mathbb{H}^d$ at p.

We identify $\mathbb{H}^d \times \partial\mathbb{H}^d$ with the unit tangent bundle $T^1\mathbb{H}^d$, by means of the bijection J defined by $J(p,\eta) := (p, p_\eta) = (p, J_p(\eta))$.

The projection π_0 (recall Definition 1.2.2) goes from $\mathbb{F}^d$ onto $\mathbb{H}^d$. We also define a projection π_1 from $\mathbb{F}^d$ onto $\mathbb{H}^d \times \partial\mathbb{H}^d$ by the following: for any $\beta \in \mathbb{F}^d$,

$$\pi_1(\beta) := \left(\beta_0, \mathbb{R}_+(\beta_0 + \beta_1) \right) = \left(\pi_0(\beta), J_{\beta_0}^{-1}(\beta_1) \right) = J^{-1}(\beta_0, \beta_1) \in \mathbb{H}^d \times \partial\mathbb{H}^d.$$

Note that by the above definitions, using the right action of $\mathrm{PSO}(1,d)$ on $\mathbb{F}^d$ (recall Remark 1.3.3), we have at once the following identifications:

$$\mathbb{F}^d/\mathrm{SO}(d) \equiv \pi_0(\mathbb{F}^d) = \mathbb{H}^d \quad \text{and} \quad \mathbb{F}^d/\mathrm{SO}(d-1) \equiv \pi_1(\mathbb{F}^d) = \mathbb{H}^d \times \partial\mathbb{H}^d.$$

Remark 2.2.2.2 As we have identified $\mathrm{SO}(d)$ (and $\mathrm{SO}(d-1)$) with the elements of $\mathrm{PSO}(1,d)$ that fix the first canonical vector e_0 (and the vectors e_0, e_1, respectively), we can identify $\mathrm{PSO}(1, d-1)$ with the elements of $\mathrm{PSO}(1,d)$ that fix the second canonical vector e_1. Considering the projection π_1' defined by $\pi_1'(\beta) := \beta_1 = \tilde{\beta}(e_1)$, we then have the following identifications, analogous to the above ones:

$$\mathbb{F}^d/\mathrm{PSO}(1,d-1) \equiv \pi_1'(\mathbb{F}^d) = \mathrm{Sitt}^d \quad \text{and} \quad \mathbb{F}^d/\mathrm{SO}(d-1) \equiv \pi_1(\mathbb{F}^d) \equiv T^1\mathrm{Sitt}^d,$$

where $\mathrm{Sitt}^d := \left\{ \xi \in \mathbb{R}^{1,d} \,\middle|\, \langle \xi, \xi \rangle = -1 \right\}$ is the one-sheeted unit hyperboloid (as $\mathbb{H}^d$ is the upper sheet of the two-sheeted unit hyperboloid), known as the **de Sitter space**, and $T^1\mathrm{Sitt}^d$ is its unit tangent bundle. Note that $T^1\mathbb{H}^d$ and $T^1\mathrm{Sitt}^d$ differ only by their canonical projections onto $\mathbb{H}^d$ and Sitt^d respectively, which are the first projection π_0 and the second projection π_1', respectively. In other words, the first two canonical vectors e_0, e_1 have their roles exchanged between the two structures. Moreover, Proposition 2.1.2 has the following analogue: the Minkowski pseudo-metric canonically induces a Lorentzian structure on Sitt^d, with $\mathrm{PSO}(1,d)$ and $(\mathbb{F}^d, \pi_1')$ as the isometry group and frame bundle, respectively.

Proposition 2.2.2.3 *A sequence $(q_n) \subset \mathbb{H}^d$ converges to $\eta \in \partial\mathbb{H}^d$ in the projective space of $\mathbb{R}^{1,d}$ (identifying q_n with the line $\mathbb{R}q_n$) if and only if $\langle q_n, p_\eta \rangle / \langle q_n, p \rangle$ goes to -1 for any $p \in \mathbb{H}^d$ or, equivalently, for one $p \in \mathbb{H}^d$. This implies that $\mathrm{dist}(p, q_n)$ goes to infinity.*

Proof Take a Lorentz frame $\beta \in \mathbb{F}^d$ such that $\beta_0 = p$ and $\beta_1 = p_\eta$, and write $q_n = \sum_{j=0}^d q_n^j \beta_j$. Then convergence in the projective space of $\mathbb{R}^{1,d}$ holds if and only if $q_n^0 \to +\infty$ and $q_n^j/q_n^0 \to 1_{\{j=1\}}$. Now, since $|q_n^0|^2 - \sum_{j=1}^d |q_n^j|^2 = 1$, this holds if and only if $q_n^1/q_n^0 \to 1$. As $q_n^0 = \langle q_n, p \rangle = \mathrm{ch}\left[\mathrm{dist}(p, q_n)\right]$ and $q_n^1 = -\langle q_n, p_\eta \rangle$, the proof is complete. $\diamond$

Proposition 2.2.2.4 *Let $\{\eta', \eta\}$ be a geodesic of $\mathbb{H}^d$, and let $q \in \mathbb{H}^d \smallsetminus \{\eta', \eta\}$. We have the following expression for the hyperbolic distance from the point q to the geodesic $\{\eta', \eta\}$:*

$$\mathrm{ch}^2 \left[\mathrm{dist}\,(q, \{\eta', \eta\}) \right] = \frac{2}{\langle \eta_q, \eta'_q \rangle}. \tag{2.2}$$

*Moreover, the minimizing geodesic from q to $\{\eta', \eta\}$ intersects the plane $\{\eta', \eta\}$ orthogonally. The intersection point is the **orthogonal projection** of q on the geodesic $\{\eta', \eta\}$ in $\mathbb{H}^d$.*

Proof Given $q \in \mathbb{H}^d \smallsetminus \{\eta', \eta\}$, it can be straightforwardly verified that $q' := \langle \eta_q, \eta'_q \rangle^{-1} (\eta_q + \eta'_q)$ is the pseudo-orthogonal projection of q onto the vector plane generated by $\{\eta_q, \eta'_q\}$, and that $\bar{q} = (2\langle \eta_q, \eta'_q \rangle)^{-1/2} (\eta_q + \eta'_q)$ belongs to the geodesic $\{\eta', \eta\}$. Recall that we have $\langle \eta_q, \eta'_q \rangle > 0$ by Lemma 1.2.1. By Proposition 2.1.1, we have $\mathrm{ch}^2 [\mathrm{dist}(q, \bar{q})] = \langle q, \bar{q} \rangle^2 = 2/\langle \eta_q, \eta'_q \rangle$.

Let p be any point on the geodesic $\{\eta', \eta\}$. This implies $p = x\eta_q + x'\eta'_q$, with $x, x' > 0$ and $2xx'\langle \eta_q, \eta'_q \rangle = 1$. Hence, we have

$$\mathrm{ch}[\mathrm{dist}(q, p)] - \mathrm{ch}[\mathrm{dist}(q, \bar{q})] = \langle p, q \rangle - \langle \bar{q}, q \rangle = x + x' - \sqrt{2/\langle \eta_q, \eta'_q \rangle}$$

$$= \left[\sqrt{x} - \sqrt{x'} \right]^2 \geq 0,$$

which shows that $\mathrm{dist}(q, \bar{q})$ does indeed realize $\mathrm{dist}\left[q, \{\eta', \eta\} \right]$.

Applying the preceding procedure to p instead of q, we obtain $2 = \langle \eta_p, \eta'_p \rangle$. As we have $\eta_p = \langle q, \eta_p \rangle \eta_q$ and $\eta'_p = \langle q, \eta'_p \rangle \eta'_q$, we find indeed that $2 = \langle \eta_p, \eta'_p \rangle = \langle q, \eta_p \rangle \langle q, \eta'_p \rangle \langle \eta_q, \eta'_q \rangle$.

Finally,

$$q_\varepsilon := \left[1 + 2\varepsilon \sqrt{2/\langle \eta_q, \eta'_q \rangle} + \varepsilon^2 \right]^{-1/2} (\bar{q} + \varepsilon q)$$

runs along the geodesic $\{\bar{q}, q\}$, so that a tangent vector at $\bar{q}$ to this geodesic is

$$\frac{d_0}{d\varepsilon} q_\varepsilon = q - \sqrt{2/\langle \eta_q, \eta'_q \rangle}\, \bar{q} = q - q',$$

and the proof is complete. $\diamond$

Remark 2.2.2.5 (*i*) As shown in the above proof, we also have an alternative expression for eqn (2.2), valid for any point p on the geodesic $\{\eta', \eta\}$:

$$\frac{2}{\langle \eta_q, \eta'_q \rangle} = \langle q, \eta_p \rangle \times \langle q, \eta'_p \rangle. \tag{2.3}$$

(*ii*) From the above proof, we can verify that the orthogonal projection $\bar{q}$ of q onto the geodesic $\{\eta', \eta\}$ is closer than q to any point p on the geodesic $\{\eta', \eta\}$, and strictly closer if q does not belong to the geodesic $\{\eta', \eta\}$.

We have in fact

$$\mathrm{ch}\big[\mathrm{dist}(p,\bar{q})\big] = \langle p,\bar{q}\rangle = \left\langle x\eta_q + x'\eta'_q, \frac{\eta_q + \eta'_q}{\sqrt{2\langle\eta_q,\eta'_q\rangle}}\right\rangle$$

$$= \sqrt{\frac{\langle\eta_q,\eta'_q\rangle}{2}}\,(x + x') = \frac{\mathrm{ch}\big[\mathrm{dist}(p,q)\big]}{\mathrm{ch}\big[\mathrm{dist}(q,\{\eta',\eta\})\big]}.$$

We then have the following statement, similar to Proposition 1.6.1, about the action of the subgroup $\mathbb{T}_{d-1}$ on the boundary $\partial\mathbb{H}^d$, which extends the Poincaré coordinates to the boundary $\partial\mathbb{H}^d$. Recall that $\theta_u^+(e_0 + e_1) = (e_0 + e_1)$, and then that $T_{x,y}(e_0 + e_1) = y(e_0 + e_1)$.

Proposition 2.2.2.6 *The Lie subgroup $\mathbb{T}_{d-1}$ of horizontal translations acts transitively and properly on $\partial\mathbb{H}^d \smallsetminus \mathbb{R}_+(e_0 + e_1)$: for any light ray $\eta \in \partial\mathbb{H}^d \smallsetminus \mathbb{R}_+(e_0 + e_1)$, there exists a unique $\theta_u^+ \in \mathbb{T}_{d-1}$ such that $\theta_u^+(e_0 - e_1) \in \eta$. u is called the **Poincaré coordinate** of η. By convention, the Poincaré coordinate of $\mathbb{R}_+(e_0 + e_1)$ is ∞.*

In other words, in the canonical basis $e = (e_0, \ldots, e_d)$, for a unique $u \in \mathbb{R}^{d-1}$, we have

$$\eta_{e_0} = e_0 + \left(\frac{|u|^2 - 1}{|u|^2 + 1}\right)e_1 + \frac{2}{|u|^2 + 1}\sum_{j=2}^{d} u^j e_j. \tag{2.4}$$

Equivalently, the Poincaré coordinate u of the light ray η in $\partial\mathbb{H}^d \smallsetminus \mathbb{R}_+(e_0 + e_1)$ having coordinates proportional to $(\eta^0, \ldots, \eta^d)$ in the canonical basis $(e_0, \ldots, e_d)$ is given by:

$$u^j = \frac{\eta^j}{\eta^0 - \eta^1} = \frac{-\langle\eta_{e_0}, e_j\rangle}{\langle\eta_{e_0}, e_0 + e_1\rangle} \quad \text{for } 2 \le j \le d. \tag{2.5}$$

Proof Equation (2.4) can be obtained simply by subtracting the first two columns of the matrix θ_u^+ (recall Definition 1.4.1) and using Proposition 2.2.2.1. It is then easily solved in a unique way, which yields eqn (2.5). ◇

Remark 2.2.2.7 Consider a sequence $(q_n) \subset \mathbb{H}^d$ with Poincaré coordinates (x_n, y_n) and a light ray $\eta \in \partial\mathbb{H}^d$, with Poincaré coordinate u (as in Propositions 1.6.1 and 2.2.2.6). Then the sequence (q_n) goes to the boundary point η (recall Proposition 2.2.2.3) if and only if its Poincaré coordinates $(x_n, y_n) \in \mathbb{R}^{d-1} \times \mathbb{R}_+^*$ go to $(u, 0)$ in the Euclidean topology of $\overline{\mathbb{R}^{d-1} \times \mathbb{R}_+^*}$. Check this, as an exercise.

2.2.3   Harmonic conjugation

Proposition 2.2.3.1 *Consider two pairs $\{\eta, \eta'\}$ and $\{\eta'', \eta'''\}$ of distinct light rays in $\partial\mathbb{H}^d$. The following three statements are equivalent.*

(i) The intersection of the two vector planes defined by $\{\eta, \eta'\}$ and $\{\eta'', \eta'''\}$ is a line, and these planes are perpendicular.

(ii) The two geodesics defined by the vector planes $\{\eta, \eta'\}$ and $\{\eta'', \eta'''\}$ intersect orthogonally in $\mathbb{H}^d$.

(iii)

$$\frac{\langle \eta, \eta' \rangle \langle \eta'', \eta''' \rangle}{\langle \eta, \eta''' \rangle \langle \eta', \eta'' \rangle} = 4 \quad and \quad \frac{\langle \eta, \eta'' \rangle \langle \eta', \eta''' \rangle}{\langle \eta, \eta''' \rangle \langle \eta', \eta'' \rangle} = 1.$$

(Note that these fractions make sense, since by homogeneity we can think of any spanning vector for each of these light rays.)

Moreover, if these conditions are fulfilled, then (iv) the intersection of the geodesics $\{\eta, \eta'\}$ and $\{\eta'', \eta'''\}$ is the point of $\mathbb{H}^d$ which belongs to $\sqrt{\langle \eta, \eta'' \rangle \langle \eta, \eta''' \rangle} \eta' + \sqrt{\langle \eta', \eta'' \rangle \langle \eta', \eta''' \rangle} \eta$ (note that this is a well-defined timelike direction; in particular, it does not depend on the vectors that we can choose in $\mathcal{C}$ to represent the light rays $\eta, \eta', \eta'', \eta'''$); and (v) there exists a $g \in \mathrm{PSO}(1, d)$ which exchanges at the same time η and η', and η'' and η''': $g(\eta) = \eta'$, $g(\eta') = \eta$, $g(\eta'') = \eta'''$, $g(\eta''') = \eta''$.

Remark 2.2.3.2 *(i)* The conditions of item *(iii)* in Proposition 2.2.3.1 are not redundant for $d \geq 3$, as the following two examples show. Taking (in the canonical frame of $\mathbb{R}^{1,3}$) $\eta = \mathbb{R}_+(1, 1, 0, 0)$, $\eta' = \mathbb{R}_+(1, -1, 0, 0)$, $\eta'' = \mathbb{R}_+(1, 0, 1, 0)$, and

- either $\eta''' = \mathbb{R}_+(1, 0, 0, 1)$, where we get 2 for the first cross-ratio and 1 for the second,
- or $\eta''' = (2, 1, 0, \sqrt{3})$, where we get 4 for the first cross-ratio and 3 for the second,

and the intersection of the two planes reduces to $\{0\}$ in both cases.

(ii) However, if $d = 2$, then the second condition entails the first one, but the converse does not hold. Indeed, we can use a Lorentz frame such that $\eta = \mathbb{R}_+(1, 1, 0)$, $\eta' = \mathbb{R}_+(1, -1, 0)$, and for $\eta'' = \mathbb{R}_+(1, \cos\alpha, \sin\alpha)$, $\eta''' = \mathbb{R}_+(1, \cos\varphi, \sin\varphi)$, the second cross-ratio equals 1 if and only if $\mathrm{tg}^2(\alpha/2) = \mathrm{tg}^2(\varphi/2)$, i.e., if and only if $\alpha + \varphi \in 2\pi\mathbb{Z}$, while the first cross-ratio equals 4 if and only if

$$[\mathrm{tg}(\alpha/2) - 3\mathrm{tg}(\varphi/2)][\mathrm{tg}(\alpha/2) + \mathrm{tg}(\varphi/2)]\mathrm{cotg}(\varphi/2) = 0.$$

(iii) Note that by the definition of the pseudo-norm on the exterior algebra,

$$\langle \eta_p \wedge \eta'_p, \eta''_p \wedge \eta'''_p \rangle = \det \begin{pmatrix} \langle \eta_p, \eta''_p \rangle & \langle \eta_p, \eta'''_p \rangle \\ \langle \eta'_p, \eta''_p \rangle & \langle \eta'_p, \eta'''_p \rangle \end{pmatrix} = \langle \eta_p, \eta''_p \rangle \langle \eta'_p, \eta'''_p \rangle - \langle \eta_p, \eta'''_p \rangle \langle \eta'_p, \eta''_p \rangle,$$

so that we have

$$\frac{\langle \eta_p, \eta''_p \rangle \langle \eta'_p, \eta'''_p \rangle}{\langle \eta_p, \eta'''_p \rangle \langle \eta'_p, \eta''_p \rangle} = 1 \iff \langle \eta_p \wedge \eta'_p, \eta''_p \wedge \eta'''_p \rangle = 0.$$

Definition 2.2.3.3 *Two pairs of distinct light rays $\{\eta, \eta'\}$ and $\{\eta'', \eta'''\}$ which satisfy the conditions of Proposition 2.2.3.1, are called **harmonically conjugate**. In this case, the ideal quadrangle $\{\eta, \eta'', \eta', \eta'''\}$ is called a **harmonic quadrangle**.*

Proof of Proposition 2.2.3.1.   We use Proposition 2.2.2.1.

Suppose first that condition *(i)* holds. We pick a non-null vector q in the intersection of the vector planes $\{\eta, \eta'\}$ and $\{\eta'', \eta'''\}$, and pick also some reference point $p \in \mathbb{H}^d$. We have $q = \alpha\eta_p + \beta\eta'_p = \gamma\eta''_p + \delta\eta'''_p$, and there exists some non-null vector $a\eta_p +$

$b\eta'_p$ orthogonal to η''_p, η'''_p. This implies $0 = \langle \alpha\eta_p + \beta\eta'_p, a\eta_p + b\eta'_p \rangle = (\alpha b + a\beta)\langle \eta_p, \eta'_p \rangle$, and then $\alpha b + a\beta = 0$, so that we can suppose that $a = \alpha, b = -\beta$. Then we have $0 = \langle \alpha\eta_p - \beta\eta'_p, \eta'_p \rangle$ or, equivalently, $\alpha\langle \eta_p, \eta'_p \rangle = \beta\langle \eta'_p, \eta'_p \rangle$, which implies $\alpha\beta > 0$. Now, since $\langle q, q \rangle = 2\alpha\beta\langle \eta_p, \eta'_p \rangle > 0$, we can suppose that $q \in \mathbb{H}^d$ (up to multiplying it by a scalar), so that it must belong to both geodesics of $\mathbb{H}^d$ defined by $\{\eta, \eta'\}$ and $\{\eta'', \eta'''\}$, and we can then take $p = q$.

The tangent at p to the line $\{\eta, \eta'\} \cap \mathbb{H}^d$ is the limit of chords joining p to $(1 + 2\varepsilon)^{-1/2}(p + \varepsilon\eta_p)$, so that $\lim_{\varepsilon \to 0} \varepsilon^{-1}\left[(1 + 2\varepsilon)^{-1/2}(p + \varepsilon\eta_p) - p\right] = \eta_p - p = p_\eta$ spans this tangent. Since it is also orthogonal to p, it must be collinear with $\alpha\eta_p - \beta\eta'_p$, and therefore orthogonal to η''_p, η'''_p. Thus $(i) \Rightarrow (ii)$ is proved.

Suppose now that condition (ii) holds, and denote by p the intersection of the geodesic lines defined by $\{\eta, \eta'\}$ and $\{\eta'', \eta'''\}$. As we have just seen in the proof of $(i) \Rightarrow (ii)$ above, the non-null vector p_η is tangent to the line $\{\eta, \eta'\} \cap \mathbb{H}^d$ and orthogonal to p, and similarly $p_{\eta''}$ is tangent to the line $\{\eta'', \eta'''\} \cap \mathbb{H}^d$. Hence p_η belongs the plane $\{\eta, \eta'\}$ and is orthogonal to $\{p, p_{\eta''}\}$, and hence to η'', η'''. This proves $(ii) \Rightarrow (i)$, whence $(i) \Leftrightarrow (ii)$.

Moreover (still under hypothesis (ii)), $p = \alpha\eta_p + \beta\eta'_p = (\alpha + \beta)p + (\alpha p_\eta + \beta p_{\eta'})$ implies $\alpha + \beta = 1$ and $\alpha p_\eta + \beta p_{\eta'} = 0$, and then $\alpha = \beta = \frac{1}{2}$ and $p_{\eta'} = -p_\eta$. Similarly, $p_{\eta'''} = -p_{\eta''}$. Hence $\eta_p + \eta'_p = 2p$, and $\langle \eta_p, \eta'_p \rangle = \langle \eta''_p, \eta'''_p \rangle = 2$.

We must also have $\langle p_\eta, p_{\eta''} \rangle = 0$. Hence $\langle \eta_p, \eta''_p \rangle = \langle \eta_p, \eta'''_p \rangle = \langle \eta'_p, \eta''_p \rangle = \langle \eta'_p, \eta'''_p \rangle = 1$.

These values obviously satisfy condition (iii), so that $(ii) \Rightarrow (iii)$ and (iv) is proved.

Furthermore, we can complete $(p, p_\eta, p_{\eta''})$ into some Lorentz frame $(p, p_\eta, p_{\eta''}, p_3, \dots, p_d)$, and consider the isomorphism g which fixes $p, p_3, \dots, p_d$ and maps $(p_\eta, p_{\eta''})$ onto $(-p_\eta, -p_{\eta''})$: it belongs to $\mathrm{PSO}(1, d)$ and is as in (v). This proves $(ii) \Rightarrow (v)$.

Suppose, conversely, that condition (iii) holds, we fix some reference point $p \in \mathbb{H}^d$, and consider $q_0 := \sqrt{\langle \eta_p, \eta''_p \rangle \langle \eta_p, \eta'''_p \rangle}\, \eta'_p + \sqrt{\langle \eta'_p, \eta''_p \rangle \langle \eta'_p, \eta'''_p \rangle}\, \eta_p$ and

$$q := \frac{q_0}{\sqrt{\langle q_0, q_0 \rangle}} = \left(2\langle \eta_p, \eta'_p \rangle\right)^{-1/2}\left[\left(\frac{\langle \eta_p, \eta''_p \rangle \langle \eta_p, \eta'''_p \rangle}{\langle \eta'_p, \eta''_p \rangle \langle \eta'_p, \eta'''_p \rangle}\right)^{1/4}\eta'_p\right.$$
$$\left. + \left(\frac{\langle \eta'_p, \eta''_p \rangle \langle \eta'_p, \eta'''_p \rangle}{\langle \eta_p, \eta''_p \rangle \langle \eta_p, \eta'''_p \rangle}\right)^{1/4}\eta_p\right] \in \mathbb{H}^d.$$

Similarly, we set

$$q' := \left(2\langle \eta''_p, \eta'''_p \rangle\right)^{-1/2}\left[\left(\frac{\langle \eta_p, \eta''_p \rangle \langle \eta'_p, \eta''_p \rangle}{\langle \eta_p, \eta'''_p \rangle \langle \eta'_p, \eta'''_p \rangle}\right)^{1/4}\eta'''_p + \left(\frac{\langle \eta_p, \eta'''_p \rangle \langle \eta'_p, \eta'''_p \rangle}{\langle \eta_p, \eta''_p \rangle \langle \eta'_p, \eta''_p \rangle}\right)^{1/4}\eta''_p\right] \in \mathbb{H}^d.$$

We then have

$$\langle q, q' \rangle = \sqrt{\frac{\langle \eta_p, \eta''_p \rangle \langle \eta'_p, \eta'''_p \rangle}{\langle \eta_p, \eta'_p \rangle \langle \eta''_p, \eta'''_p \rangle}} + \sqrt{\frac{\langle \eta_p, \eta'''_p \rangle \langle \eta'_p, \eta''_p \rangle}{\langle \eta_p, \eta'_p \rangle \langle \eta''_p, \eta'''_p \rangle}} = 1$$

by (iii). Now, this means that $q = q'$, so that q_0 is indeed a non-null vector belonging to both of the planes $\{\eta, \eta'\}$ and $\{\eta'', \eta'''\}$.

Now consider $u := \sqrt{\langle \eta'_p, \eta''_p \rangle \langle \eta'_p, \eta'''_p \rangle}\, \eta_p - \sqrt{\langle \eta_p, \eta''_p \rangle \langle \eta_p, \eta'''_p \rangle}\, \eta'_p$, which is clearly orthogonal to q. By (iii), we have

$$\langle u, \eta''_p \rangle = \sqrt{\langle \eta'_p, \eta''_p \rangle \langle \eta'_p, \eta'''_p \rangle}\, \langle \eta_p, \eta''_p \rangle - \sqrt{\langle \eta_p, \eta''_p \rangle \langle \eta_p, \eta'''_p \rangle}\, \langle \eta'_p, \eta''_p \rangle = 0,$$

which means that the non-null vector u of the plane $\{\eta, \eta'\}$ is orthogonal to the plane $\{\eta'', \eta'''\}$. This proves that $(iii) \Rightarrow (i)$, thereby concluding the proof. $\diamond$

Proposition 2.2.3.4 *Consider two pairs $\{\eta, \eta'\}$ and $\{\eta'', \eta'''\}$ of distinct light rays in $\partial\mathbb{H}^d$, and their Poincaré coordinates (recall Proposition 2.2.2.6) u, u', u'', u''', respectively. Then $\{\eta, \eta'\}$ and $\{\eta'', \eta'''\}$ are harmonically conjugate if and only if*

$$\frac{|u - u'| \times |u'' - u'''|}{|u - u'''| \times |u' - u''|} = 2 \quad and \quad \frac{|u - u''| \times |u' - u'''|}{|u - u'''| \times |u' - u''|} = 1.$$

If $d = 2$ (and then $u, u', u'', u''' \in \mathbb{R} \cup \{\infty\}$), this is equivalent to the more usual cross-ratio condition

$$[u, u', u'', u'''] := \frac{u'' - u}{u'' - u'} \times \frac{u''' - u'}{u''' - u} = -1.$$

Proof The first claim is merely a transcription of condition (iii) of Proposition 2.2.3.1 in terms of the Poincaré coordinates, since by Definition 1.4.1(ii) we have simply

$$\langle \theta^+_{u_1}(e_0 - e_1), \theta^+_{u_2}(e_0 - e_1) \rangle = \langle \theta^+_{u_1 - u_2}(e_0 - e_1), (e_0 - e_1) \rangle = 2|u_1 - u_2|^2.$$

For $d = 2$, this yields $\varepsilon, \varepsilon' \in \{\pm 1\}$ such that

$$[u, u', u'', u'''] = \varepsilon \quad and \quad \frac{(u' - u) \times (u''' - u'')}{(u'' - u') \times (u''' - u)} = 2\varepsilon'.$$

But writing $(u' - u) = (u' - u''') + (u''' - u)$ and $(u'' - u'') = (u''' - u') + (u' - u'')$, we obtain at once

$$\frac{(u' - u) \times (u''' - u'')}{(u'' - u') \times (u''' - u)} = [u, u', u'', u'''] - 1,$$

so that the condition reduces to $2\varepsilon' = \varepsilon - 1$, which is clearly equivalent to $\varepsilon = \varepsilon' = -1$, and to $\varepsilon = -1$ as well. $\diamond$

Remark 2.2.3.5 *The Lorentz–Möbius group* $\mathrm{PSO}(1, 2)$ *acts transitively on the set of harmonic quadrangles of $\mathbb{H}^2$. Any harmonic quadrangle is isometric to the quadrangle $\{-1, 0, 1, \infty\}$ of the Poincaré half-plane $\mathbb{R} \times \mathbb{R}^*_+$.*

Proof Of course, any isometry maps a harmonic quadrangle onto a harmonic quadrangle. Conversely by Remark 2.3.4, a change of Lorentz frame, and hence an isometry, maps a given harmonic quadrangle onto another one.

Let us give, however, an alternative proof that all harmonic quadrangles are isometric, by considering the half-plane $\mathbb{R} \times \mathbb{R}^*_+$ and identifying an ideal vertex (i.e., a boundary point or light ray) with its Poincaré coordinate by means of Proposition 2.2.2.6. By using a first homography (seen as an element of $\mathrm{SL}(2)$; recall Proposition

1.1.5.1), we move one vertex to ∞. We next move the leftmost vertex (on the real line) to -1 by a horizontal translation, and then the neighbouring vertex on the right of -1 to 0 by a dilatation (centred at -1). So far, we have obtained a new quadrangle $\{-1, 0, \alpha, \infty\}$, which is harmonic too, so that the geodesics $[-1, \alpha]$ and $[0, \infty]$ are orthogonal. Finally, this forces α to be equal to 1: any harmonic quadrangle is indeed isometric to $\{-1, 0, 1, \infty\}$. $\diamondsuit$

2.3 Flows and leaves

Recall that $PSO(1, d)$ has a right action on the set $\mathbb{F}^d$ of Lorentz frames (recall Definition 1.2.2 and Remark 1.3.3). In particular, the right action of the subgroups (θ_t) and (θ_u^+) introduced in Definition 1.4.1 defines the two fundamental flows acting on Lorentz frames.

Definition 2.3.1 The **geodesic flow** is the one-parameter group defined on $\mathbb{F}^d$ by

$$\beta \mapsto \beta\theta_t, \quad \text{for any } \beta \in \mathbb{F}^d \text{ and } t \in \mathbb{R}. \tag{2.6}$$

The **horocyclic flow** is the $(d-1)$-parameter group defined on $\mathbb{F}^d$ by

$$\beta \mapsto \beta\theta_u^+, \quad \text{for any } \beta \in \mathbb{F}^d \text{ and } u \in \{e_0, e_1\}^\perp \equiv \mathbb{R}^{d-1}. \tag{2.7}$$

Proposition 2.3.2 The projection $\pi_0(\beta\theta_t)$ of the orbit of a Lorentz frame β under the action of the geodesic flow is a geodesic of $\mathbb{H}^d$. More precisely, this is the geodesic determined by the plane $\{\beta_0, \beta_1\}$, and we have:

$$\frac{d}{dt}(\beta\theta_t)_0 = (\beta\theta_t)_1 \quad \text{and} \quad \text{dist}(\beta_0, (\beta\theta_t)_0) = |t|.$$

Proof The expression for θ_t yields at once $(\beta\theta_t)_0 = (\text{ch }t)\beta_0 + (\text{sh }t)\beta_1$ and $(\beta\theta_t)_1 = (\text{sh }t)\beta_0 + (\text{ch }t)\beta_1 = d/dt(\beta\theta_t)_0$. Finally, by Proposition 2.1.1, we have

$$\text{dist}(\beta_0, (\beta\theta_t)_0) = \text{argch}\big[\langle\beta_0, (\beta\theta_t)_0\rangle\big] = \text{argch}[\text{ch }t] = |t|. \qquad \diamondsuit$$

Corollary 2.3.3 The geodesic segment $[p, p']$ joining $p, p' \in \mathbb{H}^d$ has a length $\text{dist}(p, p')$, and is the unique minimizing curve joining p to p'.

Proof The first claim follows at once from Proposition 2.3.2: we necessarily have $[p, p'] = \{(\beta\theta_t)_0 | 0 \leq t \leq \text{dist}(p, p')\}$, for some Lorentz frame β. Then, if q belongs to some minimizing curve joining p to p' in $\mathbb{H}^d$, by Remark 2.2.2.5(ii) we must have

$$\text{dist}(p, p') = \text{dist}(p, q) + \text{dist}(q, p') \geq \text{dist}(p, \bar{q}) + \text{dist}(\bar{q}, p') \geq \text{dist}(p, p'),$$

which forces the equality $\text{dist}(p, q) = \text{dist}(p, \bar{q})$ and then the fact that q belongs to the geodesic $\{\eta', \eta\}$, by Remark 2.2.2.5(ii) again. $\diamondsuit$

Note that the geodesic flow makes sense also at the level of line elements of $\mathbb{H}^d$: recalling the identification $\left(\mathbb{H}^d \times \partial\mathbb{H}^d \equiv T^1\mathbb{H}^d\right)$ of Section 2.2.2, we can set $\pi_1(\beta)\theta_t := \pi_1(\beta\theta_t)$.

Indeed, if $\pi_1(\beta) = \pi_1(\beta')$, then $\varrho := \tilde{\beta}^{-1}\tilde{\beta}' \in SO(d-1)$, so that $\pi_1(\beta'\theta_t) = \pi_1(\beta\varrho\theta_t) = \pi_1(\beta\theta_t\varrho) = \pi_1(\beta\theta_t)$. Moreover, we have $(\beta\theta_t)_0 + (\beta\theta_t)_1 = e^t(\beta_0 + \beta_1)$, for any real t.

Thus the action of the geodesic flow (θ_t) on the boundary component of $\pi_1(\beta) \in \mathbb{H}^d \times \partial\mathbb{H}^d$ is trivial, and θ_t moves the generic line element along the geodesic that it generates, by an algebraic hyperbolic distance t.

In contrast, the horocyclic flow does not make sense at the level of line elements (for $d \geq 3$).

The geodesic determined by a given $(p, p_\eta) = J(p, \eta) = J \circ \pi_1(\beta) \in T^1\mathbb{H}^d$ (using the notation of Section 2.2.2), with $\beta \in \mathbb{F}^d$ (determined up to the right action of $SO(d-1)$), is parametrized by its real arc length s as follows: $s \mapsto (\text{ch } s)\beta_0 + (\text{sh } s)\beta_1$. In other words, any geodesic of $\mathbb{H}^d$ is the isometric image (under $\tilde{\beta}$) of the geodesic run along by the point having Poincaré coordinates $(0, e^s)$ (s being the arc length). We can thus say that it has Poincaré coordinates $(0, e^s)$ *in the frame* β (instead of the canonical frame).

Remark 2.3.4 Example. For an appropriate $\beta \in \mathbb{F}^d$, the image under $\tilde{\beta}^{-1}$ of a given harmonic quadrangle Q (recall Definition 2.2.3.3) has Poincaré coordinates $\{-1, 0, 1, \infty\}$. More precisely, this means (recall Proposition 2.2.2.6) that

$$\tilde{\beta}^{-1}(Q) = \{\mathbb{R}_+\theta^+_{-e_2}(e_0 - e_1), \mathbb{R}_+(e_0 - e_1), \mathbb{R}_+\theta^+_{e_2}(e_0 - e_1), \mathbb{R}_+(e_0 + e_1)\}.$$

Indeed, denoting the harmonic quadrangle by $\{\eta, \eta'', \eta', \eta'''\}$, we can restrict our considerations to the 3-subspace containing it, and take β_0 to be the intersection of the geodesics $\{\eta, \eta'\}$ and $\{\eta'', \eta'''\}$, and $\beta_1 := (\beta_0)_{\eta'''}$ and $\beta_2 := (\beta_0)_\eta$. (Note that the '$-1, 0, 1$' actually means $(-1, 0)$, $(0, 0)$, $(1, 0)$; this is a commonly used notation, since no ambiguity can occur as long as it is clear that ideal points (i.e., light rays) are being considered.) This extends Remark 2.2.3.5 to $d \geq 2$.

Definition 2.3.5 (i) *Given* $\eta \in \partial\mathbb{H}^d$, *a* **horosphere based at** η *is the intersection of* $\mathbb{H}^d$ *with an affine hyperplane (of* $\mathbb{R}^{1,d}$*) orthogonal to* η *(Figure 2.2). Given a Lorentz frame* β, $\mathcal{H}(\beta)$ *denotes the horosphere based at* $\mathbb{R}_+(\beta_0 + \beta_1)$, *defined by the hyperplane* $(\beta_0 + \beta_1)^\perp$ *containing* β_0:

$$\mathcal{H}(\beta) := \mathbb{H}^d \cap \big(\beta_0 + (\beta_0 + \beta_1)^\perp\big).$$

(ii) *A horosphere* $\mathcal{H}$ *based at* η *determines the* **horoball** $\mathcal{H}^+$, *which is the intersection of* $\mathbb{H}^d$ *with the closed affine half-space of* $\mathbb{R}^{1,d}$ *delimited by* $\mathcal{H}$ *and containing* η.

Proposition 2.3.2 states in particular, that any geodesic is the projection of an orbit of the geodesic flow. Analogously, the following proposition states in particular that any horosphere is the projection of an orbit of the horocyclic flow.

Proposition 2.3.6 *The horosphere through a Lorentz frame* β *is the projection of its orbit under the action of the horocyclic flow:* $\mathcal{H}(\beta) = \pi_0\big(\beta\theta^+_{\mathbb{R}^{d-1}}\big)$. *For any* $u \in \mathbb{S}^{d-2}$, $\left(\beta_0, \dfrac{d_0}{d\varepsilon}(\beta\theta^+_{\varepsilon u})_0\right)$ *belongs to* $T^1\mathbb{H}^d$. *Moreover, we have*

$$\mathcal{H}(\beta) = \mathcal{H}(\beta') \iff \big(\exists x \in \mathbb{R}^{d-1}, \varrho \in SO(d-1)\big) \quad \beta = \beta'\theta^+_x\varrho.$$

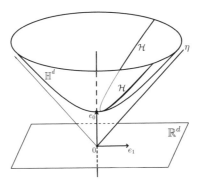

Fig. 2.2 A horosphere $\mathcal{H}$ based at $\eta = \mathbb{R}_+(e_0 + e_1)$

Proof By the Iwasawa decomposition, any point of $\mathbb{H}^d$ can be written as $\left(\beta\theta_u^+\theta_y\right)_0$ for some real y. Now, for any real y and $u \in \mathbb{R}^{d-1}$, we have

$$\left\langle \left(\beta\theta_u^+\theta_y\right)_0 - \beta_0, \beta_0 + \beta_1 \right\rangle = \left\langle \theta_u^+\theta_y e_0 - e_0, e_0 + e_1 \right\rangle = \left\langle \theta_y e_0, e_0 + e_1 \right\rangle - 1$$
$$= e^{-y} - 1.$$

By Definition 2.3.5(i), this shows that the current point $\left(\beta\theta_u^+\theta_y\right)_0 \in \mathbb{H}^d$ belongs to $\mathcal{H}(\beta)$ if and only if $y = 0$, i.e., if and only if it belongs to $\pi_0\left(\beta\theta_{\mathbb{R}^{d-1}}^+\right)$. Then, for $u \in \mathbb{S}^{d-2}$, we have

$$\left\langle \beta_0, \frac{d_0}{d\varepsilon}(\beta\theta_{\varepsilon u}^+)_0 \right\rangle = \frac{d_0}{d\varepsilon}\langle e_0, \theta_{\varepsilon u}^+ e_0 \rangle = \frac{d_0}{d\varepsilon}(1 + \varepsilon^2|u|^2/2) = 0$$

and

$$\left\langle \frac{d_0}{d\varepsilon}(\beta\theta_{\varepsilon u}^+)_0, \frac{d_0}{d\varepsilon}(\beta\theta_{\varepsilon u}^+)_0 \right\rangle = \left\langle \frac{d_0}{d\varepsilon}(\theta_{\varepsilon u}^+ e_0), \frac{d_0}{d\varepsilon}(\theta_{\varepsilon u}^+ e_0) \right\rangle = \langle u, u \rangle = -1.$$

Suppose that $\mathcal{H}(\beta) = \mathcal{H}(\beta')$. According to Definition 2.3.5, we have $\beta_0 = (\beta'\theta_x^+)_0$ for some $x \in \mathbb{R}^{d-1}$. Hence, up to a change of β' into $\beta'\theta_{-x}^+$, we can suppose also that $\beta_0 = \beta_0'$. Set $\beta_1' =: \Sigma_{j=1}^d \lambda_j\beta_j$, with $\Sigma_{j=1}^d \lambda_j^2 = 1$. We must then have, for any $u \in \mathbb{R}^{d-1}$, $(\beta\theta_u^+)_0 \in \beta_0 + (\beta_0 + \beta_1')^\perp$, i.e.,

$$0 = \left\langle \tfrac{1}{2}|u|^2(\beta_0 + \beta_1) + \beta(u), \beta_0 + \beta_1' \right\rangle = \tfrac{1}{2}|u|^2(1 - \lambda_1) - \sum_{j=2}^d \lambda_j u_j,$$

whence $\lambda_1 = 1$ and $\lambda_2 = \ldots = \lambda_d = 0$. Hence, $\beta_1' = \beta_1$, and then $\tilde{\beta}^{-1}\tilde{\beta}' \in$ SO(d-1). The converse is obvious, by Definition 2.3.5. $\diamond$

By mixing the actions of the two flows, we get the notion of a stable leaf, as follows; see Figure 2.3 (for a two-dimensional picture in the half-space of Poincaré coordinates, according to Section 2.5.2).

Definition 2.3.7 *For any light ray η, we denote by $\mathbb{F}^d(\eta)$ the set of all frames $\beta \in \mathbb{F}^d$ **pointing at** η, i.e., such that $\beta_0 + \beta_1 \in \eta$, and call it the **stable leaf** associated with the light ray η. We also write $\mathbb{F}^d\big(\mathbb{R}_+(\beta_0 + \beta_1)\big) =: \mathbb{F}^d(\beta)$ for the stable leaf containing β.*

Proposition 2.3.8 *The flows act on every stable leaf $\mathbb{F}^d(\eta)$. More precisely, $\mathbb{F}^d(\beta)$ is the orbit of β under the right action of the subgroup of $\mathrm{PSO}(1, d)$ generated by $\mathbb{A}^d \cup \mathrm{SO}(d-1)$,*

$$\mathbb{F}^d(\beta) = \Big\{\beta \varrho \theta_t \theta_u^+ \Big| \varrho \in \mathrm{SO}(d-1), t \in \mathbb{R}, u \in \mathbb{R}^{d-1}\Big\}$$

$$= \Big\{\beta T_z \varrho \Big| T_z \in \mathbb{A}^d, \varrho \in \mathrm{SO}(d-1)\Big\}.$$

Proof We have already noticed in Section 1.4 that the lightlike vector $(e_0 + e_1) \in \mathbb{R}^{1,d}$ is an eigenvector for every matrix $\theta_t \theta_u^+ \in \mathbb{A}^d$. Owing to Definition 2.3.1, this means exactly that the flows act on every stable leaf $\mathbb{F}^d(\eta)$. The Iwasawa decomposition, on applying Theorem 1.7.2 to γ^{-1}, yields $\mathrm{PSO}(1, d) = \{\varrho \theta_t \theta_u^+ | \varrho \in \mathrm{SO}(d), t \in \mathbb{R}, u \in \mathbb{R}^{d-1}\}$.

Since $\varrho \in \mathrm{SO}(d)$ fixes $(e_0 + e_1)$ if and only if it belongs to $\mathrm{SO}(d-1)$, this implies at once that the subgroup of $\mathrm{PSO}(1, d)$ that fixes the light ray $\mathbb{R}_+(e_0 + e_1) \in \partial \mathbb{H}^d$ is precisely $\{\varrho \theta_t \theta_u^+ | \varrho \in \mathrm{SO}(d-1), t \in \mathbb{R}, u \in \mathbb{R}^{d-1}\}$. Hence the first characterization. The second characterization follows at once, by eqn (1.17) and Proposition 1.4.3. $\diamond$

The following statement illustrates the instability of the geodesic flow in a simple way. This instability is actually a consequence of the commutation relations in the Lorentz–Möbius group.

Proposition 2.3.9 *For any $\beta \in \mathbb{F}^d$ and $\varepsilon, t \in \mathbb{R}$, we have*

$$\mathrm{dist}\big((\beta\theta_t)_0, (\beta\theta^+_{\varepsilon e_2}\theta_t)_0\big) = \mathrm{argch}\big(1 + \varepsilon^2/2e^{-2t}\big),$$

$$\mathrm{dist}\big((\beta\theta_t)_0, (\beta\mathcal{R}_\varepsilon\theta_t)_0\big) = \mathrm{argch}\big(1 + (1 - \cos\varepsilon)\,\mathrm{sh}^2 t\big),$$

where $\mathcal{R}_\varepsilon$ denotes the planar rotation mapping e_1 to $(\cos\varepsilon)e_1 - (\sin\varepsilon)e_2$.

Note that the first formula in Proposition 2.3.9 shows the stability of horospheres for positive t and their instability for negative t, while the second shows the systematic rotational instability of the geodesic flow, see Figures 2.3 and 2.4 (for two-dimensional pictures in the half-space of Poincaré coordinates, according to Section 2.5.2).

Proof We use Definition 1.4.1 and eqn (2.1), according to which we can clearly replace β by the canonical frame, so that we easily obtain

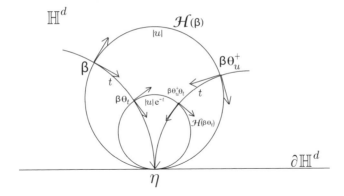

Fig. 2.3 Four frames of the stable leaf $\mathbb{F}^d(\eta) \equiv \mathbb{F}^d(\beta)$ and two horospheres of $\mathcal{H}_\eta$

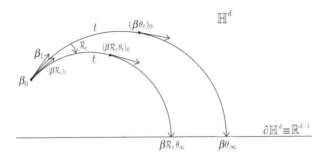

Fig. 2.4 Rotational instability of the geodesic flow

$$\mathrm{dist}\Big((\beta\theta_t)_0, (\beta\theta_{\varepsilon e_2}^+\theta_t)_0\Big) = \mathrm{dist}\big(\theta_t e_0, \theta_{\varepsilon e_2}^+\theta_t e_0\big)$$

$$= \mathrm{dist}\Big((\mathrm{ch}\,t)e_0 + (\mathrm{sh}\,t)e_1, (\mathrm{ch}\,t + \varepsilon^2/2e^{-t})e_0 + (\mathrm{sh}\,t + \varepsilon^2/2e^{-t})e_1 + e^{-t}e_2\Big)$$

$$= \mathrm{argch}\big(1 + \varepsilon^2/2e^{-2t}\big)$$

and

$$\mathrm{dist}\Big((\beta\theta_t)_0, (\beta\mathcal{R}_\varepsilon\theta_t)_0\Big) = \mathrm{dist}\big(\theta_t e_0, \mathcal{R}_\varepsilon\theta_t e_0\big)$$

$$= \mathrm{dist}\Big((\mathrm{ch}\,t)e_0 + (\mathrm{sh}\,t)e_1, (\mathrm{ch}\,t)e_0 + (\mathrm{sh}\,t\cos\varepsilon)e_1 - (\mathrm{sh}\,t\sin\varepsilon)e_2\Big)$$

$$= \mathrm{argch}\big(1 + (1 - \cos\varepsilon)\,\mathrm{sh}^2 t\big). \qquad \diamondsuit$$

2.4   Physical interpretations

2.4.1   Change of frame and relative velocities   In special relativity, space–time is represented by Minkowski space, seen as an affine space, no frame being canonically given (and thus the canonical frame $(e_0, \ldots, e_d)$ that we introduced for notational convenience has no particular physical meaning). Changes of frames are obtained by use of elements of the Lorentz–Möbius group $\mathrm{PSO}(1, d)$.

For any $V \in \mathbb{R}^d$ such that $|V| < 1$, consider the boost in the direction of V defined by $B_V = \varrho_V \theta_{\mathrm{argth}\,|V|} \varrho_V^{-1}$, $\varrho_V \in \mathrm{SO}(d)$ being such that $\varrho_V e_1 = V/|V|$. In particular, we have $\theta_t = B_{(\mathrm{th}\,t)e_1}$. Note that B_V does not depend on the choice of ϱ_V, since θ_t commutes with $\mathrm{SO}(d-1)$. Note also that for any $\varrho' \in \mathrm{SO}(d)$, we have $\varrho' B_V = B_{\varrho' V} \varrho'$.

The Cartan decomposition (recall Theorem 1.7.1) can be written in the following slightly different way: $g = B_V \varrho$, with uniquely determined $\varrho \in \mathrm{SO}(d)$ and $V \in \mathbb{R}^d$ such that $|V| < 1$. We have

$$g e_0 = \frac{e_0 + V}{\sqrt{1 - |V|^2}}.$$

Now, if two frames $\beta', \beta'' \in \mathbb{F}^d$ are such that $\beta'' = \beta' B_V \varrho$, we say that V is the **relative velocity** of β'' with respect to β'. We then have the addition rule for velocities (in which $U \cdot V = -\langle U, V \rangle$ denotes the usual Euclidean inner product in $\mathbb{R}^d$).

Proposition 2.4.1.1 *For any $U, V \in \mathbb{R}^d$ such that $|U|, |V| < 1$, we have $B_U B_V = B_W \varrho$, with $\varrho \in \mathrm{SO}(d)$ and*

$$W = \frac{\left(1 + U \cdot V/|U|^2\right) U + \sqrt{1 - |U|^2}\left(V - (U \cdot V/|U|^2) U\right)}{1 + (U \cdot V)}.$$

If two frames $\beta, \beta' \in \mathbb{F}^d$ are such that $\beta' = \beta B_U \varrho'$ and if V is the relative velocity of $\beta'' \in \mathbb{F}^d$ with respect to β', then the relative velocity of β'' with respect to β is the above W, in which V is replaced by $\varrho' V$.

Proof Setting $r := \mathrm{argth}\,|U|$ and $\lambda := \mathrm{argth}\,|V|$, we first have $B_U e_0 = \mathrm{ch}\,r\,(e_0 + U)$, $B_U U = \mathrm{ch}\,r\,(|U|^2 e_0 + U)$, $B_V e_0 = \mathrm{ch}\,\lambda\,(e_0 + V)$. Then

$$B_U B_V e_0 = \mathrm{ch}\,\lambda\;\mathrm{ch}\,r\,(e_0 + U) + \mathrm{ch}\,\lambda\; B_U \Big((U \cdot V/|U|^2) U + (V - (U \cdot V/|U|^2) U)\Big)$$

$$= \mathrm{ch}\,\lambda\;\mathrm{ch}\,r\,\big((1 + U \cdot V) e_0 + U\big) + \mathrm{ch}\,\lambda\;\mathrm{ch}\,r\,(U \cdot V/|U|^2) U + \mathrm{ch}\,\lambda\,(V - (U \cdot V/|U|^2) U)$$

$$= \mathrm{ch}\,\lambda\;\mathrm{ch}\,r\,\Big((1 + U \cdot V) e_0 + U + (U \cdot V/|U|^2) U + \sqrt{1 - |U|^2}(V - (U \cdot V/|U|^2) U)\Big)$$

$$= \mathrm{ch}\,\lambda\;\mathrm{ch}\,r\,(1 + U \cdot V)(e_0 + W).$$

This shows that $\varrho := B_W^{-1} B_U B_V \in \mathrm{PSO}(1, d)$ preserves the direction of e_0, hence e_0, hence is a rotation.

Finally for $\beta, \beta', \beta'' \in \mathbb{F}^d$ such that $\beta'' = \beta' B_V \varrho$ and $\beta' = \beta B_U \varrho'$, we merely have $\beta'' = \beta B_U \varrho' B_V \varrho = \beta B_U B_{\varrho' V} \varrho' \varrho = \beta B_{W'} \varrho''$. $\diamondsuit$

2.4.2   Motion of particles The motion of a particle of unit mass is represented by a C^1 timelike path $\xi_s \in \mathbb{R}^{1,d}$ parametrized by its proper time, i.e., its arc length, so that its $(d+1)$-**velocity** $\dot{\xi}_s$ belongs to $\mathbb{H}^d$. If ξ_s is C^2, $\ddot{\xi}_s$ is its acceleration.

Given a frame $\beta \in \mathbb{F}^d$, ξ_s can be represented by its coordinates $(t(s), X_s)$:

$$\xi_s = t(s)\beta_0 + \sum_{j=1}^{d} X_s^j \beta_j.$$

$\tau \equiv t(s)$ can be seen as an absolute time in the frame β (note that this is an increasing function of s), and then $Z_\tau^j := X_{t^{-1}(\tau)}^j \in \beta_0^\perp$ describes the path of the particle in the frame β. The β-**velocity** $V_\tau = (V_\tau^1, \ldots, V_\tau^d)$ is defined by

$$V_\tau^j := \frac{d}{d\tau} Z_\tau^j = \frac{\dot{X}_s^j}{\dot{t}(s)} = \frac{\dot{X}_s^j}{\sqrt{1 + |\dot{X}_s|^2}}, \quad \text{where } \tau = t(s).$$

$\big($Thus the relative velocity V of $\beta'' \in \mathbb{F}^d$ with respect to $\beta' \in \mathbb{F}^d$ (defined in Section 2.4.1) is the β'-velocity of β_0''.$\big)$

Equivalently,

$$\dot{X}_s^j = \frac{V_\tau^j}{\sqrt{1 - |V_\tau|^2}}.$$

The β-**energy** is

$$\dot{t}(s) = \frac{1}{\sqrt{1 - |V_\tau|^2}}.$$

Thus we have

$$\dot{\xi}_s = \frac{1}{\sqrt{1 - |V_\tau|^2}} \left(\beta_0 + \sum_{j=1}^{d} V_\tau^j \beta_j \right).$$

Note that the magnitude

$$|V_\tau| = \sqrt{\sum_{j=1}^{d} \left(V_\tau^j \right)^2} = \text{th}\big[\text{dist}\,(\beta_0, \dot{\xi}_s)\big]$$

is always less than 1, 1 being the velocity of light in the present normalization.

Note also that V_τ is constant if and only if $\dot{\xi}_s$ is.

2.4.3   Geodesics Recall from Section 2.2.2 that $J_{\dot{\xi}_s}(\eta)$ is the unit spacelike vector (i.e., of pseudo-norm -1) orthogonal to $\dot{\xi}_s$ such that $J_{\dot{\xi}_s}(\eta) + \dot{\xi}_s \in \eta$. Thus the equation $\ddot{\xi}_s = J_{\dot{\xi}_s}(\eta)$ can be interpreted by saying that the motion ξ_s is accelerated uniformly in the direction defined by η. On the other hand, the equation $(\dot{\xi}_s, \eta) = (\dot{\xi}_0, \eta)\theta_s$ means that the $(d+1)$-velocity $\dot{\xi}_s$ runs along the half-geodesic determined by $\{\dot{\xi}_0, \eta\}$ (recall Proposition 2.3.2).

Actually, the $(d+1)$-velocity $\dot{\xi}_s$ runs along a geodesic line ending at $\eta \in \partial \mathbb{H}^d$ if and only if the motion ξ_s is accelerated uniformly in the direction defined by η. More precisely, we have the following statement.

Proposition 2.4.3.1 *Given $\eta \in \partial \mathbb{H}^d$, a timelike C^2 path (ξ_s) parametrized by its proper time satisfies the differential equation $\ddot{\xi}_s = J_{\dot{\xi}_s}(\eta)$ for all $s \geq 0$ if and only if it satisfies $(\dot{\xi}_s, \eta) = (\dot{\xi}_0, \eta)\theta_s$ for all $s \geq 0$.*

Proof Since $J_{e_0}(\mathbb{R}_+(e_0 + e_1)) = e_1$ (see Figure 2.1), for any $\beta \in \mathbb{F}^d$ and $z \in \mathbb{R}^{d-1} \times \mathbb{R}_+^*$, by applying $\tilde{\beta}T_z$ we have $J_{(\beta T_z)_0}(\mathbb{R}_+(\beta_0 + \beta_1)) = (\beta T_z)_1$.

Choose $\beta \in \mathbb{F}^d$ such that $\beta_0 = \dot{\xi}_0$ and $\beta_0 + \beta_1 \in \eta$, so that $(\dot{\xi}_0, \eta) = \pi_1(\beta)$. We can write $\dot{\xi}_s = (\beta T_{z_s})_0$ for any $s \geq 0$ and some $z_s \in \mathbb{R}^{d-1} \times \mathbb{R}_+^*$, so that we have $J_{\dot{\xi}_s}(\eta) = (\beta T_{z_s})_1$.

Hence the equation $\ddot{\xi}_s = J_{\dot{\xi}_s}(\eta)$ is equivalent to $d/ds(\beta T_{z_s})_0 = (\beta T_{z_s})_1$, and then to the first-order equation $d/ds T_{z_s} e_0 = T_{z_s} e_1$, which (owing to eqn 1.18 and the initial condition $z_0 = (0,1)$) has a unique solution z_s. Now, by Proposition 2.3.2, this unique solution is indeed $T_{z_s} = \theta_s = T_{(0,e^s)}$ as wanted. $\diamond$

2.4.4   Horospheres

We have the following physical characterization of horospheres, which makes the embedding of the hyperbolic space in $\mathbb{R}^{1,d}$ appear rather natural.

Proposition 2.4.4.1 *Consider two particles ξ_s and ξ_s' parametrized by their proper times and accelerated uniformly in the direction defined by the same light ray η. Then the vector $\xi_s - \xi_s'$ converges in $\mathbb{R}^{1,d}$ as $s \to \infty$ if and only if there is a horosphere $\mathcal{H}$ (unique if $d = 2$) based at η containing the initial velocities $\dot{\xi}_0$ and $\dot{\xi}_0'$. (Then $\dot{\xi}_s$ and $\dot{\xi}_s'$ belong to the horosphere $\mathcal{H}\theta_s$ for any $s > 0$.)*

Proof According to Proposition 2.4.3.1, we have $(\dot{\xi}_s, \eta) = (\dot{\xi}_0, \eta)\theta_s$ and $(\dot{\xi}_s', \eta) = (\dot{\xi}_0', \eta)\theta_s$ for all $s \geq 0$. According to Proposition 2.3.6, the initial velocities $\dot{\xi}_0$ and $\dot{\xi}_0'$ belong to the same horosphere based at η if and only if they can be written as $\pi_0(\beta)$ and $\pi_0(\beta\theta_u^+)$ for some $\beta \in \mathbb{F}^d(\eta)$ and $u \in \mathbb{R}^{d-1}$, so that $\dot{\xi}_s = \pi_0(\beta\theta_s)$ and $\dot{\xi}_s' = \pi_0(\beta\theta_u^+\theta_s)$. Then we have

$$\tilde{\beta}^{-1}(\xi_s - \xi_s') - \tilde{\beta}^{-1}(\xi_0 - \xi_0') = \int_0^s (\theta_t e_0 - \theta_u^+ \theta_t e_0) \, dt$$

$$= (1 - \theta_u^+)\big((\mathrm{sh}\, s)e_0 + (\mathrm{ch}\, s - 1)e_1\big) = (e^{-s} - 1)\Big(\frac{|u|^2}{2}(e_0 + e_1) + u\Big),$$

which shows that $(\xi_s - \xi_s')$ converges to $\xi_0 - \xi_0' - |u|^2/2(\beta_0 + \beta_1) - \tilde{\beta}(u) \in \mathbb{R}^{1,d}$ as $s \to \infty$.

Conversely if $\dot{\xi}_s = \pi_0(\beta\theta_{r+s})$ and $\dot{\xi}'_s = \pi_0(\beta\theta_u^+\theta_s)$ belong to two distinct horospheres of the stable leaf $\mathbb{F}^d(\eta)$, computing as above leads to

$$\tilde{\beta}^{-1}(\xi_s - \xi'_s) - \tilde{\beta}^{-1}(\xi_0 - \xi'_0) = (\theta_r - \theta_u^+)\big((\operatorname{sh} s)e_0 + (\operatorname{ch} s - 1)e_1\big)$$

$$= (e^{-s} - 1)\Big(\frac{|u|^2}{2}(e_0 + e_1) + u\Big)$$

$$+ \big(\operatorname{sh}(s+r) - \operatorname{sh} s - \operatorname{sh} r\big)e_0 + \big(\operatorname{ch}(s+r) - \operatorname{ch} s - \operatorname{ch} r\big)e_1$$

$$= (\frac{e^r - 1}{2})e^s(e_0 + e_1) + \mathcal{O}(1),$$

which shows that $(\xi_s - \xi'_s)$ diverges in the light direction η.   ◇

Finally, recall that the case of $\dot{\xi}_s$ and $\dot{\xi}'_s$ evolving in two distinct stable leaves has been considered in Proposition 2.3.9.

2.5 Poincaré ball and half-space models

We present here the classical ball and upper-half-space models for the hyperbolic space $\mathbb{H}^d$. We shall use them essentially to draw pictures, namely Figures 2.3, 2.4, 2.6, 2.7, 2.8, 3.2, 4.1–4.8, 5.1, 5.2, 8.1, 8.2 and B.1.

2.5.1 Stereographic projection and the ball model   The ball model is obtained by projecting $\mathbb{H}^d$ stereographically onto $\mathbb{B}^d$ (the unit open ball of $\mathbb{R}^d$) from the vector $-e_0$ (see Figure 2.5). More precisely, denoting this stereographic projection by ψ, we have

$$\psi\big((\operatorname{ch} r)e_0 + (\operatorname{sh} r)u\big) = \frac{\operatorname{sh} r}{1+\operatorname{ch} r}u = \operatorname{th}(r/2)u \quad \text{for any } r \geq 0, u \in \mathbb{S}^{d-1}.$$

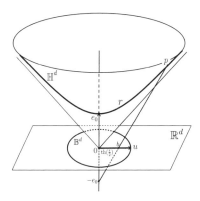

Fig. 2.5 Stereographic projection $\psi(p) = b$ from $\mathbb{H}^d$ onto $\mathbb{B}^d$

Note that ψ extends continuously to $\partial \mathbb{H}^d$, by $\psi\big(\mathbb{R}_+(e_0 + u)\big) := u$.

Let us see how a geodesic (that is to say, the image under ψ of a hyperbolic geodesic in $\mathbb{H}^d$) looks in this model.

Proposition 2.5.1.1 *In the ball model $\mathbb{B}^d$, geodesics appear either as diagonal segments or as circles (intersecting with $\mathbb{B}^d$) orthogonal to $\partial \mathbb{B}^d = \mathbb{S}^{d-1}$.*

Proof As ψ is clearly invariant under the action of $SO(d)$, it is enough to consider the geodesic ending at e_1 and $u := \big((\cos\alpha)e_1 + (\sin\alpha)e_2\big)$ (with $\cos\alpha \neq 1$; recall Remark 2.2.1.2). In $\mathbb{H}^d$, this geodesic is determined by the plane

$$\{(\xi_0 = \lambda + \mu, \xi_1 = \lambda + \mu\cos\alpha, \xi_2 = \mu\sin\alpha, 0, \ldots, 0)\,|\,\lambda, \mu \in \mathbb{R}\},$$

and then is the image of $\mathbb{R}_+^*$ under the map:

$$\lambda \longmapsto \left(\lambda + \frac{1}{4\lambda\,\sin^2(\alpha/2)}, \lambda + \frac{\cos\alpha}{4\lambda\,\sin^2(\alpha/2)}, \frac{\sin\alpha}{4\lambda\,\sin^2(\alpha/2)}, 0, \ldots, 0\right) \in \mathbb{H}^d.$$

Note that the trivial case $\sin\alpha = 0$ yields a diagonal segment, so that we can restrict our considernations to $\sin\alpha \neq 0$. The analogue in $\mathbb{B}^d$ is thus obtained as the image of $\mathbb{R}_+^*$ under the map

$$\lambda \longmapsto \left(0, \xi_1' = \frac{\cos\alpha + 4\lambda^2\,\sin^2(\alpha/2)}{1 + 4\lambda(\lambda+1)\sin^2(\alpha/2)}, \xi_2' = \frac{\sin\alpha}{1 + 4\lambda(\lambda+1)\sin^2(\alpha/2)}, 0, \ldots, 0\right) \in \mathbb{B}^d,$$

for which the equation $(\xi_1' - 1)^2 + (\xi_2' - \mathrm{tg}^2(\alpha/2))^2 = \mathrm{tg}^2(\alpha/2)$ can be directly verified. This equation is indeed that of the circle orthogonal to $\mathbb{S}^{d-1}$ and containing e_1, u.   ◇

Let us now then see how a horosphere (that is to say, the image under ψ of a horosphere in $\mathbb{H}^d$) looks in this model.

Proposition 2.5.1.2 *In the ball model $\mathbb{B}^d$, horospheres appear as hyperspheres included in $\mathbb{B}^d$ and tangent to $\partial \mathbb{B}^d = \mathbb{S}^{d-1}$.*

Proof We proceed as for Proposition 2.5.1.1, using Definition 2.3.5(i). Owing to the action of $SO(d)$, it is enough to consider the horosphere based at $\eta = (e_0 + e_1)$ and containing $p = (\mathrm{ch}\,r)e_0 + (\mathrm{sh}\,r)u$. In $\mathbb{H}^d$ this horosphere is determined by the hyperplane

$$\{(\mu + \mathrm{ch}\,r, \mu + \mathrm{sh}\,r, \lambda_2, \ldots, \lambda_d)\,|\,\mu \in \mathbb{R}, \lambda = (\lambda_2, \ldots, \lambda_d) \in \mathbb{R}^{d-1}\},$$

and then is the image of $\mathbb{R}^{d-1}$ under the map

$$\lambda \longmapsto \big(\mathrm{ch}\,r + e^r|\lambda|^2/2,\ \mathrm{sh}\,r + e^r|\lambda|^2/2, \lambda_2, \ldots, \lambda_d\big) \in \mathbb{H}^d.$$

The analogue in $\mathbb{B}^d$ is thus obtained as the image of $\mathbb{R}^{d-1}$ under the map

$$\lambda \longmapsto \big(1 + \mathrm{ch}\,r + e^r|\lambda|^2/2\big)^{-1}\big(\mathrm{sh}\,r + e^r|\lambda|^2/2, \lambda_2, \ldots, \lambda_d\big) \in \mathbb{B}^d,$$

for which the equation

$$\left(\xi_1' - \tfrac{1+e^r}{2(1+\operatorname{ch} r)}\right)^2 + (\xi_2')^2 + \cdots + (\xi_d')^2 = \left(\tfrac{1+e^{-r}}{2(1+\operatorname{ch} r)}\right)^2$$

can be directly verified. This equation is indeed that of a hypersphere tangent to $\mathbb{S}^{d-1}$ at e_1. $\diamond$

2.5.2   Upper-half-space model and the Poincaré coordinates As
already mentioned just after Proposition 1.6.1, the upper-half-space model model is $\mathbb{R}^{d-1} \times \mathbb{R}_+^*$. It is obtained by performing the following inversion about e_1:

$$\mathbb{B}^d \ni b \longmapsto e_1 + 2|b - e_1|^{-2}(b - e_1) =: z \equiv y e_1 + x \equiv (x, y) \in \mathbb{R}^{d-1} \times \mathbb{R}_+^*.$$

We then have $|b - e_1| \times |z - e_1| = 2$, and the above is equivalent to $b = e_1 + 2|z - e_1|^{-2}(z - e_1)$ or, equivalently $\left(\text{writing } b = \sum_{j=1}^d b^j e_j, x = \sum_{j=2}^d x^j e_j\right)$,

$$y = \frac{1 - |b|^2}{|b - e_1|^2} \quad \text{and} \quad x^j = \frac{2b^j}{|b - e_1|^2} \quad \text{for } 2 \le j \le d.$$

Proposition 2.5.2.1 (i) *The above coordinates* (x, y) *are precisely the Poincaré coordinates of Proposition 1.6.1.*

(ii) *In the upper-half-space model* $\mathbb{R}^{d-1} \times \mathbb{R}_+^*$, *geodesics appear either as vertical half-lines or as half-circles, both orthogonal to* $\mathbb{R}^{d-1}$, *and horospheres appear either as horizontal hyperplanes or as hyperspheres tangent to* $\mathbb{R}^{d-1}$.

Proof (i) Given $p = \sum_{j=0}^d p^j e_j \in \mathbb{H}^d$, we have the corresponding $b = \sum_{j=1}^d b^j e_j = \psi(p) = (1 + p^0)^{-1} \sum_{j=1}^d p^j e_j \in \mathbb{B}^d$, whence $1 - |b|^2 = 2/(1 + p^0)$ and

$$|b - e_1|^2 = (1 + p^0)^{-2}[(p^1 - p^0 - 1)^2 + (p^0)^2 - 1 - (p^1)^2] = 2(p^0 - p^1)/(1 + p^0).$$

Therefore $y = 1/p^0 - p^1$ and $x^j = p^j/p^0 - p^1$: we recover eqn (1.19).

(ii) It is enough to use Propositions 2.5.1.1 and 2.5.1.2 and to recall that the above inversion about e_1 is a conformal map, which maps orthogonal circles/lines to orthogonal circles/lines, and tangent hyperspheres to tangent hyperspheres or horizontal hyperplanes. $\diamond$

Remark 2.5.2.2 Hyperbolic space can be seen as an abstract metric space. The above models are then parametrizations which depend on some non-canonical choices. Actually, they depend on a reference point in the space and also (in the case of the half-space model) on a reference point on the boundary, which can be changed by an isometry.

We saw in Proposition 2.3.2 that the geodesic flow drives any $\beta \in \mathbb{F}^d$ by a distance $|t|$. For any $u \in \mathbb{R}^{d-1}$, we have the following horocyclic analogue: $\operatorname{dist}_{\mathcal{H}}(\beta, \beta\theta_u^+) \equiv \operatorname{dist}_{\mathcal{H}(\beta)}\big(\beta_0, (\beta\theta_u^+)_0\big) = |u|$, where $\operatorname{dist}_{\mathcal{H}(\beta)}\big(\beta_0, (\beta\theta_u^+)_0\big)$ denotes the **horocyclic distance**, i.e., the hyperbolic length of the minimal curve which links β_0 to $(\beta\theta_u^+)_0$

within the horocycle $\mathcal{H}(\beta)$ (as depicted in Figure 2.3). In contrast $\mathrm{dist}\big(\beta_0,(\beta\theta_u^+)_0\big) = \mathrm{argch}\big[|u|^2/2 + 1\big]$.

This can be seen at once by applying the isometry $\tilde{\beta}^{-1}$, which maps β_0 to e_0 and $\mathcal{H}(\beta)$ to the horizontal hyperplane having equation $\{y = 1\}$ in Poincaré coordinates.

Hence, given any $p, q \in \mathbb{H}^d$, the **horocyclic distance** $\mathrm{dist}_{\mathcal{H}}(p,q) = 4\,\mathrm{sh}^2\big[\mathrm{dist}(p,q)/2\big]$ does not depend on the horosphere $\mathcal{H}$ containing p,q (which can also be seen by mapping isometrically the geodesic segment $[p,q]$ to a vertical one). See Lemma B.3.2 for the horocyclic distance from a point to a geodesic.

2.6   A commutation relation

We establish here the part that we shall need, in several places, of the commutation relation between an element $T_{x,y} \in \mathbb{A}^d$ and a rotation $\varrho \in \mathrm{SO}(d)$. We postpone the full commutation formula (which we do not need) to the appendices, see Section A.2. According to the Iwasawa decomposition (Theorem 1.7.2, applied to γ^{-1} instead of γ), there exist unique $T_{x',y'} \in \mathbb{A}^d$ and $\varrho' \in \mathrm{SO}(d)$ such that

$$T_{x,y}\varrho = \varrho' T_{x',y'}. \tag{2.8}$$

We then have $T_{x,y}e_0 = \varrho' T_{x',y'}e_0$, and the following theorem.

Theorem 2.6.1 *Denote by $u(\varrho)$ the Poincaré coordinate of $\mathbb{R}_+\varrho(e_0 + e_1)$ $\big(u(\varrho) = \infty$ if and only if $\varrho \in \mathrm{SO}(d-1)\big)$. We have*

(i) $u(\varrho') = yu(\varrho) + x$ or, equivalently,

(ii) $u(\varrho^{-1}) = \frac{u(\varrho'^{-1}) - x'}{y'}$.

Proof (ii) By the definition of $u(\varrho)$ and eqn (2.5), we have

$$\langle u(\varrho), e_j \rangle = \frac{\langle \varrho(e_0 + e_1), e_j \rangle}{\langle \varrho(e_0 + e_1), e_0 + e_1 \rangle} = \frac{\langle \varrho e_1, e_j \rangle}{1 + \langle \varrho e_1, e_1 \rangle}.$$

On the other hand, since $T_{x,y}(e_0 + e_1) = y(e_0 + e_1)$ (recall Section 1.4), for any $v \in \mathbb{R}^{1,d}$, we have:

$$\langle \varrho^{-1}(e_0 + e_1), v \rangle = y\langle T_{x,y}^{-1}(e_0 + e_1), \varrho v \rangle = y\langle e_0 + e_1, T_{x,y}\varrho v \rangle$$

$$= y\langle e_0 + e_1, \varrho' T_{x',y'}v \rangle = y\langle \varrho'^{-1}(e_0 + e_1), T_{x',y'}v \rangle.$$

Hence we obtain

$$u(\varrho^{-1}) = -\sum_{j=2}^{d} \langle u(\varrho^{-1}), e_j \rangle e_j = -\sum_{j=2}^{d} \frac{\langle \varrho^{-1}(e_0 + e_1), e_j \rangle}{\langle \varrho^{-1}(e_0 + e_1), e_0 + e_1 \rangle} e_j$$

$$= -\sum_{j=2}^{d} \frac{\langle \varrho'^{-1}(e_0 + e_1), T_{x',y'} e_j \rangle}{\langle \varrho'^{-1}(e_0 + e_1), T_{x',y'}(e_0 + e_1) \rangle} e_j$$

$$= -\sum_{j=2}^{d} \frac{\langle \varrho'^{-1}(e_0 + e_1), x'_j(e_0 + e_1) + e_j \rangle}{\langle \varrho'^{-1}(e_0 + e_1), y'(e_0 + e_1) \rangle} e_j = \frac{u(\varrho'^{-1}) - x'}{y'}.$$

(i) can be deduced at once from (ii), applied to $T_{x',y'} \varrho^{-1} = \varrho'^{-1} T_{x,y}$. ◇

A rotation $\varrho \in \mathrm{SO}(d)$ such that $\varrho e_1 \neq -e_1$ can plainly be decomposed in a unique way as $\varrho = R_{\alpha,\sigma} \tilde{\varrho}$, with $\tilde{\varrho} \in \mathrm{SO}(d-1)$, $\alpha \in [0, \pi[$ and $\sigma \in \mathbb{S}^{d-2} \subset \mathbb{R}^{d-1} \equiv \{e_0, e_1\}^{\perp}$, where $R_{\alpha,\sigma} \in \mathrm{SO}(d)$ denotes the planar rotation by an angle α in the plane oriented by (e_1, σ) (for $\alpha \neq 0$). Thus $R_{\alpha,\sigma} e_1 = \varrho e_1 = (\cos \alpha) e_1 + (\sin \alpha)\sigma$.

Proposition 2.6.2 *For any $\varrho \in \mathrm{SO}(d)$, using the unique decomposition $\varrho = R_{\alpha,\sigma} \tilde{\varrho}$ with $\tilde{\varrho} \in \mathrm{SO}(d-1)$, $\alpha \in [0, \pi]$ and $\sigma \in \mathbb{S}^{d-2}$, we have*

$$u(\varrho) = \mathrm{cotg}\left(\tfrac{\alpha}{2}\right)\sigma = \frac{\varrho e_1 + \langle \varrho e_1, e_1 \rangle e_1}{1 + \langle \varrho e_1, e_1 \rangle}.$$

Then

$$|e_1 - \varrho e_1| = 2\sin(\tfrac{\alpha}{2}) = \frac{2}{\sqrt{|u(\varrho)|^2 + 1}}$$

or, alternatively,

$$|u(\varrho)| = \mathrm{cotg}\left(\tfrac{\alpha}{2}\right) = |u(\varrho^{-1})|.$$

This yields the interpretation

$$u(\varrho) \to \infty \iff \varrho e_1 \to e_1 \iff \alpha \to 0 \iff \varrho \to \mathrm{SO}(d-1).$$

Proof By the Definition of $u(\varrho)$ and eqn (2.5), we have

$$\langle u(\varrho), e_j \rangle = \frac{\langle \varrho(e_0 + e_1), e_j \rangle}{\langle \varrho(e_0 + e_1), e_0 + e_1 \rangle} = \frac{\langle \varrho e_1, e_j \rangle}{1 + \langle \varrho e_1, e_1 \rangle} = \frac{(\sin \alpha)\langle \sigma, e_j \rangle}{1 - \cos \alpha},$$

whence

$$u(\varrho) = \mathrm{cotg}\left(\tfrac{\alpha}{2}\right)\sigma = \frac{\varrho e_1 + \langle \varrho e_1, e_1 \rangle e_1}{1 + \langle \varrho e_1, e_1 \rangle}.$$

Therefore $|u(\varrho)| = \mathrm{cotg}(\alpha/2)$, and otherwise

$$|e_1 - \varrho e_1|^2 = (1 - \cos \alpha)^2 + (\sin \alpha)^2 = 4\sin^2(\tfrac{\alpha}{2}) = \frac{4}{\mathrm{cotg}^2(\alpha/2) + 1}.$$

Finally, replacing ϱ by ϱ^{-1} does not change α. ◇

In Section 5.3, we shall need two particular cases of the commutation relation of Theorem 2.6.1, for which it will be sufficient to restrict ourselves to $d = 2$, but with some more specification of the elements ϱ', x', y'. We now gather together what we shall use.

Proposition 2.6.3 *We restrict our considerations here to the case $d = 2$, and use $\mathcal{R}_\alpha$ to denote $\mathcal{R}_{\alpha,-e_2}$. Then we have the following commutation relations.*

(i) For any $\alpha \in [0, \pi]$ and $x \in \mathbb{R}$, there exists a unique $\alpha' \in [0, \pi]$ such that $\cotg(\alpha'/2) = \cotg(\alpha/2) - x$ and

$$\theta_x^+ \mathcal{R}_\alpha = \mathcal{R}_{\alpha'} \theta_{x'}^+ \theta_{\log y'}, \quad \text{with } y' := \frac{1 - \cos\alpha}{1 - \cos\alpha'}, \quad x' := \frac{\sin\alpha' - \sin\alpha}{1 - \cos\alpha'}.$$

(ii) For any real r, we have a unique $\alpha_r \in [0, \pi]$ such that

$$\theta_r \mathcal{R}_{\pi/2} = \mathcal{R}_{\alpha_r} \theta_{\operatorname{sh} r}^+ \theta_{\log \operatorname{ch} r}, \quad \text{with } \cotg\alpha_r = \operatorname{sh} r.$$

Proof *(i)* By Theorem 2.6.1, applied with $y = 1$, and Proposition 2.6.2, we have

$$\theta_x^+ \mathcal{R}_\alpha = \mathcal{R}_{\alpha'} T_{x',y'}, \quad \text{with } \cotg(\alpha'/2) = \cotg(\alpha/2) - x.$$

Then we have $T_{x',y'} e_0 = \mathcal{R}_{\alpha'}^{-1} \theta_x^+ e_0$, and hence by eqn (1.19)

$$y' = \langle T_{x',y'} e_0, e_0 + e_1 \rangle^{-1} = \langle \mathcal{R}_{\alpha'}^{-1} \theta_x^+ e_0, e_0 + e_1 \rangle^{-1}$$

$$= \langle \theta_x^+ e_0, e_0 + \mathcal{R}_{\alpha'} e_1 \rangle^{-1} = \langle \theta_x^+ e_0, e_0 + (\cos\alpha')e_1 - (\sin\alpha')e_2 \rangle^{-1}$$

$$= \left(1 + \frac{x^2}{2} - (\cos\alpha')\frac{x^2}{2} + (\sin\alpha')x\right)^{-1} = \left(1 + x^2 \sin^2(\tfrac{\alpha'}{2}) + x\sin\alpha'\right)^{-1}$$

$$= \left(1 + \left(\cotg(\tfrac{\alpha'}{2}) - \cotg(\tfrac{\alpha}{2})\right)^2 \sin^2(\tfrac{\alpha'}{2}) - \left(\cotg(\tfrac{\alpha'}{2}) - \cotg(\tfrac{\alpha}{2})\right)\sin\alpha'\right)^{-1}$$

$$= \frac{\sin^2(\tfrac{\alpha}{2})}{\sin^2(\tfrac{\alpha'}{2})} = \frac{1 - \cos\alpha}{1 - \cos\alpha'}.$$

By eqn (1.19) again, we have similarly

$$x' = -y' \langle T_{x',y'} e_0, e_2 \rangle = -y' \langle \mathcal{R}_{\alpha'}^{-1} \theta_x^+ e_0, e_2 \rangle = -y' \langle \theta_x^+ e_0, \mathcal{R}_{\alpha'} e_2 \rangle$$

$$= -y' \langle \theta_x^+ e_0, (\sin\alpha')e_1 + (\cos\alpha')e_2 \rangle = y'x\left(\tfrac{x}{2}\sin\alpha' + \cos\alpha'\right)$$

$$= y'x\left(\sin(\tfrac{\alpha'}{2})\cos(\tfrac{\alpha'}{2})\cotg(\tfrac{\alpha}{2}) - \cos^2(\tfrac{\alpha'}{2}) + \cos\alpha'\right)$$

$$= y'\big(\mathrm{cotg}(\tfrac{\alpha}{2}) - \mathrm{cotg}(\tfrac{\alpha'}{2})\big)\sin(\tfrac{\alpha'}{2})\big(\cos(\tfrac{\alpha'}{2})\mathrm{cotg}(\tfrac{\alpha}{2}) - \sin(\tfrac{\alpha'}{2})\big)$$

$$= \sin^{-2}(\tfrac{\alpha'}{2})\big(\cos\tfrac{\alpha}{2}\sin\tfrac{\alpha'}{2} - \cos\tfrac{\alpha'}{2}\sin\tfrac{\alpha}{2}\big)\big(\cos\tfrac{\alpha'}{2}\cos\tfrac{\alpha}{2} - \sin\tfrac{\alpha'}{2}\sin\tfrac{\alpha}{2}\big)$$

$$= \sin^{-2}(\tfrac{\alpha'}{2})\sin(\tfrac{\alpha'-\alpha}{2})\cos(\tfrac{\alpha+\alpha'}{2}) = \frac{\sin\alpha' - \sin\alpha}{1 - \cos\alpha'}.$$

(ii) By Theorem 2.6.1, applied with $(x = 0, y = e^r)$, and Proposition 2.6.2, we have

$$\theta_r \mathcal{R}_{\pi/2} = \mathcal{R}_{\alpha_r} T_{x',y'}, \quad \text{with } \mathrm{cotg}\,(\tfrac{\alpha_r}{2}) = e^r \mathrm{cotg}\,(\tfrac{\pi}{4}) = e^r,$$

whence $\mathrm{cotg}\,\alpha_r = (e^{2r} - 1)/2e^r = \mathrm{sh}\,r$. Then we have $T_{x',y'}e_0 = \mathcal{R}_{\alpha_r}^{-1}\theta_r e_0$, whence by eqn (1.19)

$$y' = \langle T_{x',y'}e_0, e_0 + e_1 \rangle^{-1} = \langle \theta_r e_0, e_0 + \mathcal{R}_{\alpha_r}e_1 \rangle^{-1}$$

$$= \langle \theta_r e_0, e_0 + (\cos\alpha_r)e_1 - (\sin\alpha_r)e_2 \rangle^{-1} = \big(\mathrm{ch}\,r - (\cos\alpha_r)\mathrm{sh}\,r\big)^{-1}$$

$$= \big(\mathrm{ch}\,r - \mathrm{th}\,r\,\mathrm{sh}\,r\big)^{-1} = \mathrm{ch}\,r.$$

And, by eqn (1.19) again, we have similarly

$$x' = -y'\langle T_{x',y'}e_0, e_2 \rangle = -y'\langle \mathcal{R}_{\alpha_r}^{-1}\theta_r e_0, e_2 \rangle = -y'\langle \theta_r e_0, \mathcal{R}_{\alpha_r}e_2 \rangle$$

$$= -y'\langle \theta_r e_0, (\sin\alpha_r)e_1 + (\cos\alpha_r)e_2 \rangle = y'(\sin\alpha_r)\mathrm{sh}\,r = \mathrm{sh}\,r. \qquad \diamondsuit$$

Figure 2.6 provides a visual proof of Proposition 2.6.3(i): it is enough to verify that the two isometries $\theta_x^+\mathcal{R}_\alpha$ and $\mathcal{R}_{\alpha'}\theta_{x'}^+\theta_{\log y'}$ map (under their right action) a given line element $\xi \in T^1\mathbb{H}^2$ to some common line element ξ'. We choose ξ to

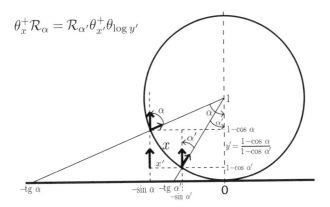

Fig. 2.6 Proof of the commutation relation of Proposition 2.6.3(i)

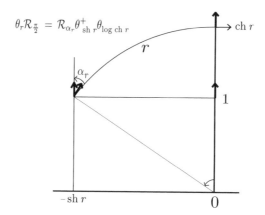

Fig. 2.7 Proof of the commutation relation of Proposition 2.6.3(*ii*)

be based at the point on the half-plane having coordinates $(-\sin\alpha', 1-\cos\alpha')$ and pointing to the centre $(0, 1)$ of the horocycle based at 0. It is moved by $\theta^+_x\mathcal{R}_\alpha$ first to the line element based at $(-\sin\alpha, 1-\cos\alpha)$ and pointing to the centre of the horocycle and then to the vertical line element ξ' based at the same point. Similarly, it is moved by $\mathcal{R}_{\alpha'}\theta^+_{x'}\theta_{\log y'}$ to ξ' as well.

Similarly, Figure 2.7 provides a visual proof of Proposition 2.6.3(*ii*): we now choose the line element ξ based at the point having coordinates $(-\operatorname{sh} r, 1)$ and tangent to the (geodesic) circle centred at 0, which is moved by $\theta_r\mathcal{R}_{\pi/2}$ to the vertical line element ξ' based at $(0, \operatorname{ch} r)$, and by $\mathcal{R}_{\alpha_r}\theta^+_{\operatorname{sh} r}\theta_{\log \operatorname{ch} r}$ as well.

2.7 The Busemann function

We denote by $\mathcal{H}_\eta$ the family of all horospheres based at $\eta \in \partial\mathbb{H}^d$ (recall Definition 2.3.5 and Figure 2.3). For any $p \in \mathbb{H}^d$, there exists a unique $\mathcal{H}_\eta(p) \in \mathcal{H}_\eta$ such that $p \in \mathcal{H}_\eta(p)$. Note that $\mathcal{H}_\eta(p) = (p+\eta^\perp)\cap \mathbb{H}^d$, so that $\mathcal{H}_\eta(p) = \mathcal{H}_\eta(p') \Leftrightarrow p-p' \in \eta^\perp$. For any $\beta \in \mathbb{F}^d$, we have $\mathcal{H}(\beta) = \mathcal{H}_{\mathbb{R}_+(\beta_0+\beta_1)}(\beta_0)$.

Theorem 2.7.1 (*i*) *For any* $\eta \in \partial\mathbb{H}^d$, $H \in \mathcal{H}_\eta$, *and any real* t, *the set* $H\theta_t :=$ $\{(\beta\theta_t)_0 \mid \beta \in \mathbb{F}^d, \beta_0 \in H, \beta_0+\beta_1 \in \eta\}$ *belongs to* $\mathcal{H}_\eta$.
(*ii*) *For any* $H, H' \in \mathcal{H}_\eta$, *there exists a unique real* $B_{H,H'}$ *such that* $H\theta_{B_{H,H'}} = H'$.

Proof (*i*) We fix $\bar\beta \in \mathbb{F}^d$ such that $\bar\beta_0 \in H$ and $\bar\beta_0 + \bar\beta_1 \in \eta$, so that $H = \mathcal{H}(\bar\beta)$, and we verify that $H\theta_t = \mathcal{H}(\bar\beta\theta_t)$. For any $\beta \in \mathbb{F}^d$ such that $\beta_0 \in H$ and $\beta_0 + \beta_1 \in \eta$, we have also $H = \mathcal{H}(\beta)$, whence by Proposition 2.3.6: $\beta = \bar\beta\theta^+_x\varrho$ for some $x \in \mathbb{R}^{d-1}$ and $\varrho \in SO(d-1)$. Hence $\beta\theta_t = \bar\beta\theta_t\theta^+_{xe^{-t}}\varrho$, and then $(\beta\theta_t)_0 = (\bar\beta\theta_t\theta^+_{xe^{-t}})_0 \in \mathcal{H}(\bar\beta\theta_t)$. Conversely, if $p \in \mathcal{H}(\bar\beta\theta_t)$, then $p = (\bar\beta\theta_t\theta^+_u)_0 = (\bar\beta\theta^+_{ue^t}\theta_t)_0$ for some $u \in \mathbb{R}^{d-1}$, while $(\bar\beta\theta^+_{ue^t})_0 \in \mathcal{H}(\bar\beta) = H$ and $(\bar\beta\theta^+_{ue^t})_0 + (\bar\beta\theta^+_{ue^t})_1 = \bar\beta\left(\theta^+_{ue^t}(e_0+e_1)\right) = \bar\beta(e_0+e_1) \in \eta$.

(*ii*) We fix $\beta, \beta' \in \mathbb{F}^d$ such that $\beta_0 \in H, \beta'_0 \in H', \beta_0 + \beta_1 \in \eta$ and $\beta'_0 + \beta'_1 \in \eta$, so that $H = \mathcal{H}(\beta), H' = \mathcal{H}(\beta')$. By Definition 2.3.7, we have $\mathbb{F}^d(\beta) = \mathbb{F}^d(\beta')$, so that Proposition 2.3.8 yields $t \in \mathbb{R}$, $u \in \mathbb{R}^{d-1}$ and $\varrho \in \mathrm{SO}(d-1)$ such that $\beta' = \beta(\varrho\theta_t\theta_u^+) = \beta\theta_t\theta_u^+\varrho$, showing that $H' = \mathcal{H}(\beta') = \mathcal{H}(\beta\theta_t) = H\theta_t$. Moreover, if $H\theta_s = H\theta_t$, then as in the proof of (*i*) above, we can fix $\beta \in \mathbb{F}^d$ such that $\mathcal{H}(\beta\theta_s) = \mathcal{H}(\beta\theta_t)$, whence by Proposition 2.3.6, $\beta\theta_s = \beta\theta_t\theta_x^+\varrho$, for some $x \in \mathbb{R}^{d-1}$ and $\varrho \in \mathrm{SO}(d-1)$, and then $\theta_{s-t} = \theta_x^+\varrho$.

By the uniqueness of the Iwasawa decomposition (recall Theorem 1.7.2), this implies $s = t$, i.e., uniqueness of the real t such that $H' = H\theta_t$. $\diamond$

Remark 2.7.2 *For any $\eta \in \partial\mathbb{H}^d$, $H \in \mathcal{H}_\eta$ and $p \in \mathbb{H}^d$, let $p^H \in \mathbb{H}^d$ be defined by $(p^H, \eta) := (p, \eta)\theta_{B_{\mathcal{H}_\eta(p),H}}$. Then p^H is the orthogonal projection (recall Proposition 2.2.2.4) in $\mathbb{H}^d$ of p on H, and $\mathbb{R}p^H$ is the orthogonal projection in $\mathbb{R}^{1,d}$ of $\mathbb{R}p$ on the affine hyperplane $p^H + \eta^\perp$ defining H.*

Proof We fix $\beta \in \mathbb{F}^d(\eta)$ such that $\pi_1(\beta) = (p, \eta)$, and then $\pi_1(\beta\theta_t) = (p^H, \eta)$ for $t := B_{\mathcal{H}_\eta(p),H}$. We also have $\mathcal{H}_\eta(p) = \mathcal{H}(\beta)$ and $\mathcal{H}_\eta(p^H) = \mathcal{H}(\beta\theta_t) = H$. Then, for any $q = (\beta\theta_t\theta_u^+)_0 \in H$, we have

$$\mathrm{ch}\big[\mathrm{dist}(p,q)\big] = \big\langle \tilde\beta(e_0), \tilde\beta(\theta_t\theta_u^+ e_0)\big\rangle = \big\langle \theta_{-t}e_0, \theta_u^+ e_0\big\rangle = \mathrm{ch}\,t + \frac{|u|^2}{2}e^t \geq \mathrm{ch}\,t$$

$$= \mathrm{ch}\big[\mathrm{dist}(p, p^H)\big].$$

Finally, for any real λ, we have

$$\langle \lambda p - \lambda e^t p^H, \eta\rangle = \lambda\langle \tilde\beta(e_0 - e^t\theta_t e_0), \tilde\beta(e_0 + e_1)\rangle$$

$$= \lambda - \lambda e^t\langle (\mathrm{ch}\,t)e_0 + (\mathrm{sh}\,t)e_1, e_0 + e_1\rangle = 0,$$

whence $\lambda p - (\lambda e^t - 1)p^H \in p^H + \eta^\perp \subset H + \eta^\perp$: the orthogonal projection of λp on $H + \eta^\perp$ is $(\lambda e^t - 1)p^H \in \mathbb{R}p^H$. $\diamond$

Using the notation introduced in Theorem 2.7.1(*ii*), we can now define the Busemann function; see Figure 2.8 (depicted in the Poincaré upper-half-space model of Section 2.5.2).

Definition 2.7.3 *The **Busemann function** $B.(\cdot, \cdot)$ is defined by the following: for any $\eta \in \partial\mathbb{H}^d$ and any $p, q \in \mathbb{H}^d$, $B_\eta(p, q) := B_{\mathcal{H}_\eta(p),\mathcal{H}_\eta(q)}$.*

Proposition 2.7.4 *For any $\eta \in \partial\mathbb{H}^d$ and any $p, q \in \mathbb{H}^d$, we have*

$$B_\eta(p, q) = \log\langle p, \eta_q\rangle = -\log\langle \eta_p, q\rangle.$$

Proof We fix $\beta, \beta' \in \mathbb{F}^d(\eta)$ such that $\beta_0 = p$ and $\beta'_0 = q$. We then have $\mathcal{H}(\beta) = \mathcal{H}_\eta(p)$ and $\mathcal{H}(\beta') = \mathcal{H}_\eta(q)$. Set $t := B_\eta(p, q)$, so that $\mathcal{H}(\beta') = \mathcal{H}(\beta)\theta_t = \mathcal{H}(\beta\theta_t)$, by Theorem 2.7.1. By Proposition 2.3.6, there exist $u \in \mathbb{R}^{d-1}$ and $\varrho \in \mathrm{SO}(d-1)$ such that $\beta' = \beta\theta_t\theta_u^+\varrho$, whence $q = \tilde\beta(\theta_t\theta_u^+ e_0)$, and then

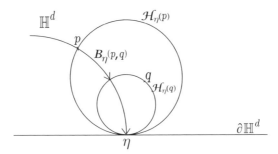

Fig. 2.8 The Busemann function $B_\eta(p, q)$

$$\langle q, \eta_p \rangle = \langle \tilde\beta(\theta_t \theta_u^+ e_0), \tilde\beta(e_0 + e_1) \rangle = \langle \theta_u^+ e_0, \theta_{-t}(e_0 + e_1) \rangle$$
$$= e^{-t} \langle e_0, \theta_{-u}^+(e_0 + e_1) \rangle = e^{-t}.$$

Finally, we have $\eta_q = \langle p, \eta_q \rangle \eta_p$, and hence $1 = \langle p, \eta_q \rangle \langle q, \eta_p \rangle$.   ◇

Remark 2.7.5 For any $\eta \in \partial\mathbb{H}^d$, $p \in \mathbb{H}^d$ and $(q_k) \subset \mathbb{H}^d$, we have $B_\eta(p, q_k) \to \infty$ if and only if the points of the horosphere $\mathcal{H}_\eta(q_k)$ go uniformly to η in the projective space of $\mathbb{R}^{1,d}$ (in the sense of Proposition 2.2.2.3).

Proof On the one hand, by Theorem 2.7.1(ii) and Remark 2.7.2, we have $B_\eta(p, q_k) = \mathrm{dist}(p, \mathcal{H}_\eta(q_k))$ (as appears clearly in Figure 2.8), and on the other hand, by Proposition 2.7.4, we have $B_\eta(p, q_k) \to \infty$ if and only if $\langle p + p_\eta, q_k \rangle \to 0$. Hence $B_\eta(p, q_k) \to \infty$ implies $\langle p_\eta, q_k \rangle / \langle p, q_k \rangle \to -1$, and then the statement follows directly from Proposition 2.2.2.3.   ◇

2.8   Notes and comments

Hyperbolic geometry can be presented in many ways. See [Gh], [Mr] and [Rac] for a historical account, starting with Riemann and Lobachevsky.

The point of view that we have adopted, of deriving all notions from the geometry of Minkowski space, is not so common, although it gives some nice intrinsic formulae, such as eqn (2.2); see also [Rac].

Classical hyperbolic geometry has been developed in two directions:

- in Lie group theory, with more general groups (of rank one and beyond) and associated homogeneous spaces; see in particular [He];
- in differential geometry, with the study of manifolds with (generally non-constant) negative curvature and their generalizations, such as CAT(-1) spaces and Gromov hyperbolicity; see [Bn], [CG], [Gr] and [Pa].

Chapter Three
Operators and measures

The Casimir operator Ξ on $\mathrm{PSO}(1,d)$, i.e., the second-order differential operator associated with the Killing form, is the fundamental operator of the theory. It commutes with Lie derivatives, and induces the Laplace operator $\mathcal{D}$ on the affine group $\mathbb{A}^d$ and the hyperbolic and spherical Laplacians Δ and $\Delta^{\mathbb{S}^{d-1}}$, for which we specify some decomposition formulae and analytical expressions.

After proving some fundamental properties of Haar measures on groups, we determine the Haar measure of $\mathrm{PSO}(1,d)$ and show that it is indeed bi-invariant.

This chapter ends with a presentation of harmonic, Liouville and volume measures. We derive all of them from the Haar measure, and also derive analytical expressions for them and the fundamental properties of the hyperbolic Laplacian, namely covariance with isometries and self-adjointness.

3.1 The Casimir operator on $\mathrm{PSO}(1,d)$

We define the Casimir operator Ξ by means of right Lie derivatives (recall eqn (1.6)) and the pseudo-orthonormal basis of $\mathrm{so}(1,d)$ presented in Proposition 1.3.2.

Definition 3.1.1 *The **Casimir operator** on $\mathrm{PSO}(1,d)$ is the second-order differential operator Ξ defined on C^2 functions on $\mathrm{PSO}(1,d)$ by*

$$\Xi := \sum_{j=1}^{d}(\mathcal{L}_{E_j})^2 - \sum_{1\le k<l\le d}(\mathcal{L}_{E_{kl}})^2, \tag{3.1}$$

where $\{E_j, E_{kl}|\, 1\le j\le d, 1\le k<l\le d\}$ is the basis of $\mathrm{so}(1,d)$ encountered in Proposition 1.3.2.

We show the invariance of the expression of Ξ under a change of basis in the proof of Proposition 3.1.2. Using the subalgebra τ_{d-1} of Section 1.4, i.e., writing $\mathcal{L}_{E_j} = \mathcal{L}_{\tilde{E}_j} - \mathcal{L}_{E_{1j}}$ for $2\le j\le d$, we immediately have the following alternative expression for Ξ:

$$\Xi = (\mathcal{L}_{E_1})^2 + \sum_{j=2}^{d}(\mathcal{L}_{\tilde{E}_j})^2 - \sum_{j=2}^{d}(\mathcal{L}_{\tilde{E}_j}\mathcal{L}_{E_{1j}} + \mathcal{L}_{E_{1j}}\mathcal{L}_{\tilde{E}_j}) - \sum_{2\le k<l\le d}(\mathcal{L}_{E_{kl}})^2. \tag{3.2}$$

Note that $\sum_{1\le k<l\le d}(\mathcal{L}_{E_{kl}})^2$ and $\sum_{2\le k<l\le d}(\mathcal{L}_{E_{kl}})^2$ are the Casimir operators on $\mathrm{SO}(d)$ and $\mathrm{SO}(d-1)$, respectively.

Proposition 3.1.2 *The Casimir operator Ξ commutes with any right deriva-*
tive $\mathcal{L}_A$, $A \in \mathrm{so}(1,d)$: $\Xi \circ \mathcal{L}_A = \mathcal{L}_A \circ \Xi$. It also commutes with the left and
right actions of $\mathrm{PSO}(1,d)$: $\Xi \circ L_g = L_g \circ \Xi$ and $\Xi \circ R_g = R_g \circ \Xi$, for any $g \in$
$\mathrm{PSO}(1,d)$.

Proof We show in (1) below the invariance of the expression for Ξ under a change
of basis. In (2), we apply this to deduce the $\mathrm{Ad}(g)$-invariance of Ξ. The latter then
implies directly the commutation with $\mathcal{L}_A$ and the commutation with R_g as well, since
the commutation with L_g is obvious from eqns (1.7) and (3.1).

(1) We verify first that for any basis $(A_\ell)_{1 \le \ell \le d(d+1)/2}$ of $\mathrm{so}(1,d)$ and its dual basis
(A_ℓ^*) defined by $K(A_\ell^*, A_{\ell'}) = 1_{\{\ell=\ell'\}}$, we have

$$\Xi = \sum_{1 \le \ell, \ell' \le d(d+1)/2} K(A_\ell, A_{\ell'}) \mathcal{L}_{A_\ell^*} \mathcal{L}_{A_{\ell'}^*}.$$

Note that by Proposition 1.3.2, this expression is correct for the particular basis
$\{E_j, E_{kl}\}$.

Then, if $(\tilde{A}_\ell)$ is another basis of $\mathrm{so}(1,d)$, we have $\tilde{A}_\ell = \sum_m P_{m,\ell} A_m$ for some
$P \in \mathrm{GL}(d(d+1)/2)$, whence $\tilde{A}_\ell^* = \sum_m P_{\ell,m}^{-1} A_m^*$, since

$$K\left(\sum_m P_{\ell,m}^{-1} A_m^*, \tilde{A}_{\ell'} \right) = \sum_m P_{\ell,m}^{-1} \sum_{m'} P_{m',\ell'} K(A_m^*, A_{m'}) = \sum_m P_{\ell,m}^{-1} P_{m,\ell'} = 1_{\{\ell=\ell'\}}.$$

Hence we have

$$\sum_{\ell,\ell'} K(\tilde{A}_\ell, \tilde{A}_{\ell'}) \mathcal{L}_{\tilde{A}_\ell^*} \mathcal{L}_{\tilde{A}_{\ell'}^*} = \sum_{\ell,\ell'} \sum_{m,m'} \sum_{n,n'} P_{m,\ell} P_{m',\ell'} P_{\ell,n}^{-1} P_{\ell',n'}^{-1} K(A_m, A_{m'}) \mathcal{L}_{A_n^*} \mathcal{L}_{A_{n'}^*}$$

$$= \sum_{m,m'} \sum_{n,n'} 1_{\{m=n\}} 1_{\{m'=n'\}} K(A_m, A_{m'}) \mathcal{L}_{A_n^*} \mathcal{L}_{A_{n'}^*}$$

$$= \sum_{1 \le m,m' \le d(d+1)/2} K(A_m, A_{m'}) \mathcal{L}_{A_m^*} \mathcal{L}_{A_{m'}^*}.$$

(2) We now fix any $g \in \mathrm{PSO}(1,d)$ and apply (1), now taking $\tilde{A}_\ell := \mathrm{Ad}(g)(A_\ell)$. By
eqn (1.5), we have $K(\mathrm{Ad}(g)(A_\ell^*), \tilde{A}_\ell) = K(A_\ell^*, A_\ell) = 1_{\{\ell=\ell'\}}$, which shows that $\tilde{A}_\ell^* =$
$\mathrm{Ad}(g)(A_\ell^*)$. Note also that for any C^3 function F on $\mathrm{PSO}(1,d)$ and any $A, A' \in \mathrm{so}(1,d)$,
we have for any $h \in \mathrm{PSO}(1,d)$

$$\mathcal{L}_A(F \circ \mathrm{Ad}(g))(h) = \frac{d_0}{d\varepsilon} F(gh \, \exp[\varepsilon A]g^{-1}) = \frac{d_0}{d\varepsilon} F(ghg^{-1} \, \exp[\varepsilon g A g^{-1}])$$

$$= [\mathcal{L}_{\mathrm{Ad}(g)(A)} F] \circ \mathrm{Ad}(g)(h),$$

whence

$$\mathcal{L}_A \mathcal{L}_{A'}(F \circ \mathrm{Ad}(g)) = \mathcal{L}_A\left([\mathcal{L}_{\mathrm{Ad}(g)(A')} F] \circ \mathrm{Ad}(g) \right) = [\mathcal{L}_{\mathrm{Ad}(g)(A)} \mathcal{L}_{\mathrm{Ad}(g)(A')} F] \circ \mathrm{Ad}(g).$$

Using eqn (1.5) again, we therefore obtain

$$\Xi\big(F \circ \mathrm{Ad}(g)\big) = \left[\sum_{\ell,\ell'} K(A_\ell, A_{\ell'}) \mathcal{L}_{\mathrm{Ad}(g)(A_\ell^*)} \mathcal{L}_{\mathrm{Ad}(g)(A_{\ell'}^*)} F \right] \circ \mathrm{Ad}(g)$$

$$= \left[\sum_{\ell,\ell'} K\big(\mathrm{Ad}(g)(A_\ell), \mathrm{Ad}(g)(A_{\ell'})\big) \mathcal{L}_{\tilde{A}_\ell^*} \mathcal{L}_{\tilde{A}_{\ell'}^*} F \right] \circ \mathrm{Ad}(g)$$

$$= \left[\sum_{\ell,\ell'} K(\tilde{A}_\ell, \tilde{A}_{\ell'}) \mathcal{L}_{\tilde{A}_\ell^*} \mathcal{L}_{\tilde{A}_{\ell'}^*} F \right] \circ \mathrm{Ad}(g) = [\Xi F] \circ \mathrm{Ad}(g)$$

by (1) above, since $(\tilde{A}_\ell)$ is clearly another basis of so$(1,d)$. Hence, taking $g = \exp(\varepsilon A)$, by eqn (1.4) we obtain $\Xi\big(F \circ \exp[\varepsilon \mathrm{ad}(A)]\big) = [\Xi F] \circ \exp[\varepsilon\, \mathrm{ad}(A)]$, and then, by differentiating at 0 with respect to ε, $\Xi(\mathcal{L}_A F) = \mathcal{L}_A[\Xi F]$, showing that the Casimir operator Ξ indeed commutes with any $\mathcal{L}_A, A \in$ so$(1,d)$. $\diamond$

3.2   The Laplace operator

We introduce here an important second-order operator, which induces the Laplacian on $\mathbb{H}^d$, as we shall see in Section 3.5.

Definition 3.2.1 *The **Laplace operator** $\mathcal{D}$ is the following second-order differential operator, defined on* PSO$(1,d)$:

$$\mathcal{D} := \sum_{j=2}^{d} (\mathcal{L}_{\tilde{E}_j})^2 + (\mathcal{L}_{E_1})^2 + (1-d)\mathcal{L}_{E_1}. \tag{3.3}$$

The left invariance of the right Lie derivatives (recall eqn (1.7)) entails that of their squares, and then that of the Laplace operator $\mathcal{D}$:

$$L_\gamma \circ \mathcal{D} = \mathcal{D} \circ L_\gamma, \text{ for any } \gamma \in \mathrm{PSO}(1,d).$$

Recall that we used the Iwasawa decomposition in Theorem 1.7.2 to define the canonical projection Iw from PSO$(1,d)$ onto $\mathbb{A}^d$. The following statement explains the relation between the Laplace operator $\mathcal{D}$ and the Casimir operator Ξ.

Theorem 3.2.2 *The Casimir operator Ξ and the Laplace operator $\mathcal{D}$ agree on functions on* PSO$(1,d)$ *which are invariant under the right action of* SO(d). *In other words,*

$$\Xi[f \circ \mathrm{Iw}] = \mathcal{D}[f \circ \mathrm{Iw}] \text{ for any } C^2 \text{ function } f \text{ on } \mathbb{A}^d.$$

Proof Fix any $\gamma \in \mathrm{PSO}(1,d)$. For $1 \le k, l \le d$, we have first

$$\mathcal{L}_{E_{kl}}[f \circ \mathrm{Iw}](\gamma) = \frac{d_0}{d\varepsilon} f \circ \mathrm{Iw}\big(\gamma \, \exp[\varepsilon E_{kl}]\big) = \frac{d_0}{d\varepsilon} f \circ \mathrm{Iw}(\gamma) = 0.$$

Now fix $2 \leq j \leq d$. Using eqn (1.4), we have

$$\mathcal{L}_{E_{1j}}\mathcal{L}_{\tilde{E}_j}[f \circ \mathrm{Iw}](\gamma) = \frac{d_0}{d\varepsilon}\frac{d_0}{d\eta} f \circ \mathrm{Iw}\Big(\gamma \ \exp[\varepsilon E_{1j}] \ \exp[\eta \tilde{E}_j]\Big)$$

$$= \frac{d_0}{d\varepsilon}\frac{d_0}{d\eta} f \circ \mathrm{Iw}\Big(\gamma \ \exp\Big[\eta \ \mathrm{Ad}\big(\exp[\varepsilon E_{1j}]\big)(\tilde{E}_j)\Big]\Big)$$

$$= \frac{d_0}{d\varepsilon}\frac{d_0}{d\eta} f \circ \mathrm{Iw}\Big(\gamma \ \exp\Big[\eta \ \exp\big[\varepsilon \ \mathrm{ad}(E_{1j})\big](\tilde{E}_j)\Big]\Big).$$

Now, $\mathrm{ad}(E_{1j})(\tilde{E}_j) = \mathrm{ad}(E_{1j})(E_j) = E_1$ and $\mathrm{ad}(E_{1j})(E_1) = -E_j$, so that

$$\exp[\varepsilon \ \mathrm{ad}(E_{1j})](\tilde{E}_j) = E_{1j} + (\cos\varepsilon)E_j + (\sin\varepsilon)E_1.$$

Hence,

$$\mathcal{L}_{E_{1j}}\mathcal{L}_{\tilde{E}_j}[f \circ \mathrm{Iw}](\gamma) = \frac{d_0}{d\varepsilon}\frac{d_0}{d\eta} f \circ \mathrm{Iw}\Big(\gamma \ \exp\Big[\eta\big[E_{1j} + (\cos\varepsilon)E_j + (\sin\varepsilon)E_1\big]\Big]\Big)$$

$$= \frac{d_0}{d\varepsilon}\Big(\mathcal{L}_{E_{1j}} + (\cos\varepsilon)\mathcal{L}_{E_j} + (\sin\varepsilon)\mathcal{L}_{E_1}\Big)[f \circ \mathrm{Iw}](\gamma) = \mathcal{L}_{E_1}[f \circ \mathrm{Iw}](\gamma).$$

By eqns (3.2) and (3.3), we deduce at once that

$$\Xi[f \circ \mathrm{Iw}] = \Big[(\mathcal{L}_{E_1})^2 + \sum_{j=2}^{d}(\mathcal{L}_{\tilde{E}_j})^2 - \sum_{j=2}^{d}\mathcal{L}_{E_{1j}}\mathcal{L}_{\tilde{E}_j}\Big][f \circ \mathrm{Iw}] = \mathcal{D}[f \circ \mathrm{Iw}]. \qquad \Diamond$$

Note that this property means that $\mathcal{D}$ appears as canonical among all similar operators, and allows us to restrict it to the affine subgroup $\mathbb{A}^d$. We now express the restrictions of Lie derivatives to $\mathbb{A}^d$, using the parametrization of the subgroup $\mathbb{A}^d$ by the Poincaré half-space $\mathbb{R}^{d-1} \times \mathbb{R}_+^*$.

Proposition 3.2.3 *Restricted to $\mathbb{A}^d$, Lie derivatives and the Laplace operator can be computed as follows: for any $z = (x, y) \in \mathbb{R}^{d-1} \times \mathbb{R}_+^*$ and for $2 \leq j \leq d$,*

$$\mathcal{L}_{E_1} f(T_z) = y\frac{\partial}{\partial y}f(T_z)\,, \mathcal{L}_{\tilde{E}_j} f(T_z) = y\frac{\partial}{\partial x_j}f(T_z), \qquad (3.4)$$

$$\mathcal{L}'_{E_1} f(T_z) = \Big[y\frac{\partial}{\partial y} + \sum_{k=2}^{d} x_k\frac{\partial}{\partial x_k}\Big]f(T_z), \mathcal{L}'_{\tilde{E}_j} f(T_z) = \frac{\partial}{\partial x_j}f(T_z), \qquad (3.5)$$

$$\mathcal{D} = y^2\Big(\frac{\partial^2}{\partial y^2} + \Delta_x^{d-1}\Big) + (2-d)y\frac{\partial}{\partial y}, \qquad (3.6)$$

where Δ_x^{d-1} denotes the Euclidean Laplacian of $\mathbb{R}^{d-1}$ acting on the coordinate x.

Proof For any C^2 function f on $\mathbb{A}^d$ and any $z = (x, y) \in \mathbb{R}^{d-1} \times \mathbb{R}_+^*$, by eqn (1.17) we have

$$\mathcal{L}_{E_1} f(T_z) = \frac{d_0}{d\varepsilon} f(T_z T_{0,e^\varepsilon}) = \frac{d_0}{d\varepsilon} f(T_{x,e^\varepsilon y}) = y \frac{\partial}{\partial y} f(T_z)$$

and

$$\mathcal{L}_{\tilde{E}_j} f(T_z) = \frac{d_0}{d\varepsilon} f(T_z T_{\varepsilon e_j, 1}) = \frac{d_0}{d\varepsilon} f(T_{x+y\varepsilon e_j, y}) = y \frac{\partial}{\partial x_j} f(T_z).$$

The computation with respect to left derivatives is very similar. Equation (3.6) for $\mathcal{D}$ follows at once from eqns (3.4) and (3.3). $\diamond$

3.3 Haar measure of PSO(1, d)

A **left Haar measure** λ on a locally compact group G is a positive Radon measure which is invariant under left translations:

$$\int_G L_g F \, d\lambda = \int_G F \, d\lambda \quad \text{for any } g \in G \text{ and any test function } F \text{ on } G.$$

Of course, the analogous notion with respect to the right-hand side is that of a **right Haar measure**, and a **Haar measure** is both a left and a right Haar measure.

For any measure λ on a group G, denote by $\check{\lambda}$ its image under the inverse map $(\gamma \mapsto \gamma^{-1})$, so that $\int_G F \, d\check{\lambda} := \int_G F(\gamma^{-1}) \lambda(d\gamma)$, for any test function F on G.

A Lie group is said to be **semisimple** when its Lie algebra is. By Proposition 1.3.2, this applies to PSO(1, d). A classical theorem (see Section [He], Section X.1, Proposition 1.4 or [Kn], Corollary 8.31) states that any semisimple Lie group is **unimodular**, which means that its left and right Haar measures are the same. We give a short proof of this in Theorem 3.3.4. Taking advantage of this, we present in Theorem 3.3.5 a Haar measure for PSO(1, d), and provide an explicit expression for it in terms of Iwasawa coordinates.

We first have the following well-known result.

Proposition 3.3.1 (i) *The image $\check{\lambda}$ of a left Haar measure λ by the inverse map is a right Haar measure.*

(ii) *A left Haar measure on a locally compact group is unique, up to a multiplicative constant.*

Proof Take any compactly supported continuous function F on G and any $g \in G$.

(i) Let λ be any left Haar measure on G. We then have

$$\int_G F(\gamma g) \check{\lambda}(d\gamma) = \int_G F(\gamma^{-1} g) \lambda(d\gamma) = \int_G F((g^{-1}\gamma)^{-1}) \lambda(d\gamma)$$

$$= \int_G F(\gamma^{-1}) \lambda(d\gamma) = \int_G F \, d\check{\lambda}.$$

(*ii*) Let λ_1, λ_2 be left Haar measures on a locally compact group G. Consider any compactly supported continuous function f on G such that $\lambda_1(f) := \int_G f d\lambda_1 \neq 0$, and for any $g \in G$, set $\varphi_f(g) := \lambda_1(f)^{-1} \int_G R_g f \, d\lambda_2$. We now show that φ_f is continuous on G. For that, we take any $g_0 \in G$, any $\varepsilon > 0$ and a compact neighbourhood V of the unit element. On the one hand, by the uniform continuity of f, we have $\sup_G |R_g f - R_{g_0} f| < \varepsilon$ if $g^{-1} g_0$ remains in some neighbourhood $V_\varepsilon \subset V$ of the unit element, and on the other hand, again for $g \in g_0 V_\varepsilon^{-1}$, $|R_g f - R_{g_0} f|$ is supported in the compact set $K := \text{Support}(f) V g_0^{-1}$, so that $g \in g_0 V_\varepsilon^{-1}$ implies $|\varphi_f(g) - \varphi_f(g_0)| \leq |\lambda_1(f)|^{-1} \lambda_2(K) \varepsilon$.

We then take any other compactly supported continuous function h on G, and note that $(g, \gamma) \mapsto f(g) h(\gamma g)$ defines a compactly supported continuous function on G^2. Using the left invariance properties of λ_1, λ_2 and (*i*), we have

$$
\lambda_1(f) \times \int_G h d\check{\lambda}_2 = \int_G f(g) \left[\int_G h(\gamma g) \check{\lambda}_2(d\gamma) \right] \lambda_1(dg)
$$

$$
= \int_G \left[\int_G f(g) h(\gamma g) \lambda_1(dg) \right] \check{\lambda}_2(d\gamma)
$$

$$
= \int_G \left[\int_G f(\gamma^{-1} g) h(g) \lambda_1(dg) \right] \check{\lambda}_2(d\gamma)
$$

$$
= \int_G h(g) \left[\int_G f(\gamma g) \lambda_2(d\gamma) \right] \lambda_1(dg) = \lambda_1(f) \int_G h \varphi_f d\lambda_1
$$

by the definition of φ_f. Hence, we obtain $\int_G h \, d\check{\lambda}_2 = \int_G h \varphi_f \, d\lambda_1$.

As a consequence, if f_1 is any other compactly supported continuous function on G such that $\lambda_1(f_1) \neq 0$, we must have $0 = \int_G h(\varphi_f - \varphi_{f_1}) \, d\lambda_1$ for any compactly supported continuous function h on G. Thus $\{\varphi_f \neq \varphi_{f_1}\}$ is a λ_1-negligible open set, and then must be empty, by the invariance of the non-null measure λ_1. This means that $\varphi := \varphi_f$ does not depend on f. By the definition of φ_f, we then obtain $\int_G f \, d\lambda_2 = \varphi(e) \int_G f \, d\lambda_1$, which holds by linearity for any compactly supported continuous function f on G, proving the proportionality. ◇

Let us establish the existence of a Haar measure in the simpler case of $SO(d)$, together with a disintegration of it.

We have already stressed (just before Proposition 2.6.2) that a rotation $\varrho \in SO(d)$ such that $\varrho e_1 \neq -e_1$ can be decomposed in a unique way as $\varrho = R_{\alpha,\sigma} \tilde{\varrho}$, with $\tilde{\varrho} \in SO(d-1)$, $\alpha \in [0, \pi[$, $\sigma \in \mathbb{S}^{d-2} \subset \mathbb{R}^{d-1} \equiv \{e_0, e_1\}^\perp$, and $R_{\alpha,\sigma} \in SO(d)$ denoting a planar rotation by an angle α in the plane oriented by (e_1, σ) (for $\alpha \neq 0$). Thus $R_{\alpha,\sigma} e_1 = \varrho e_1 = (\cos\alpha) e_1 + (\sin\alpha)\sigma$. Note that (α, σ) is determined by $R_{\alpha,\sigma}$, and that $R_{\alpha,\sigma}$ is determined by $\varrho e_1 = R_{\alpha,\sigma} e_1 =: \tilde{\sigma} \in \mathbb{S}^{d-1}$, so that we can write $R_{\alpha,\sigma} \equiv R_{\tilde{\sigma}}$ as well.

Proposition 3.3.2 *There exists a unique Haar probability measure $d\varrho$ on $SO(d)$. It is invariant under the inverse map. Using the decomposition $\varrho = R_{\tilde{\sigma}} \tilde{\varrho}$, which identifies $\varrho \in SO(d)$ with $(\tilde{\sigma}, \tilde{\varrho}) \in \mathbb{S}^{d-1} \times SO(d-1)$, we have $d\varrho = d\tilde{\sigma} \, d\tilde{\varrho}$, where $d\tilde{\sigma}$ denotes the uniform probability measure on the sphere $\mathbb{S}^{d-1}$ and $d\tilde{\varrho}$ denotes the Haar probability measure on $SO(d-1)$. Moreover, $R_{\tilde{\sigma}} = R_{\alpha,\sigma}$*

can be written in a unique way as $\mathcal{R}_{\tilde{\sigma}} = \exp\left[\alpha \sum_{k=2}^{d} \sigma_k E_{1k}\right]$, *with* $(\alpha, \sigma) \in$
$[0, \pi] \times \mathbb{S}^{d-2}$, *and we have the disintegration*

$$d\tilde{\sigma} = \frac{\Gamma(d/2)}{\sqrt{\pi}\Gamma((d-1)/2)}(\sin \alpha)^{d-2} d\alpha \, d\sigma.$$

Proof The property is trivial on $\mathrm{SO}(2)$. Let us proceed by induction on d, showing that if $d\tilde{\varrho}$ is a $(\tilde{\varrho} \mapsto \tilde{\varrho}^{-1})$-invariant Haar measure on $\mathrm{SO}(d-1)$, then $d\varrho := d\tilde{\sigma} \, d\tilde{\varrho}$ is a $(\varrho \mapsto \varrho^{-1})$-invariant Haar measure $d\varrho$ on $\mathrm{SO}(d)$. This will be sufficient, by Proposition 3.3.1.

Take any positive measurable function $F = F(\varrho)$ on $\mathrm{SO}(d)$, identified with $F(\tilde{\sigma}, \tilde{\varrho})$ on $\mathbb{S}^{d-1} \times \mathrm{SO}(d-1)$, and $\varrho_0 = \mathcal{R}_{\tilde{\sigma}_0}\tilde{\varrho}_0 \in \mathrm{SO}(d)$. Set $\tilde{\varrho}_1 = \tilde{\varrho}_1(\varrho_0, \tilde{\sigma})$ $:= \mathcal{R}_{\varrho_0\tilde{\sigma}}^{-1}\varrho_0\mathcal{R}_{\tilde{\sigma}} \in \mathrm{SO}(d-1)$, so that $\varrho_0\varrho = \mathcal{R}_{\tilde{\sigma}_0}\tilde{\varrho}_0\mathcal{R}_{\tilde{\sigma}}\tilde{\varrho} = \mathcal{R}_{\varrho_0\tilde{\sigma}}\tilde{\varrho}_1(\varrho_0, \tilde{\sigma})\tilde{\varrho}$. We then have

$$\int F(\varrho_0\varrho) \, d\varrho = \int F\big(\varrho_0\tilde{\sigma}, \tilde{\varrho}_1(\varrho_0, \tilde{\sigma})\tilde{\varrho}\big) \, d\tilde{\sigma} \, d\tilde{\varrho} = \int \left[\int F\big(\varrho_0\tilde{\sigma}, \tilde{\varrho}_1(\varrho_0, \tilde{\sigma})\tilde{\varrho}\big) \, d\tilde{\varrho}\right] d\tilde{\sigma}$$

$$= \int F(\varrho_0\tilde{\sigma}, \tilde{\varrho}) \, d\tilde{\sigma} \, d\tilde{\varrho} = \int F(\tilde{\sigma}, \tilde{\varrho}) \, d\tilde{\sigma} \, d\tilde{\varrho} = \int F(\varrho) \, d\varrho,$$

where we have used successively the left Haar property of $d\tilde{\varrho}$ and the invariance of $d\tilde{\sigma}$ under any $\varrho_0 \in \mathrm{SO}(d)$ (which holds by the very definition of $d\tilde{\sigma}$). We then set $\check{F}(\varrho) := F(\varrho^{-1})$, and for any $\tilde{\sigma} \in \mathbb{S}^{d-1}$, we denote by $\check{\sigma} := -2\langle\tilde{\sigma}, e_1\rangle e_1 - \tilde{\sigma}$ its symmetrical counterpart with respect to $\mathbb{R}e_1$. By the invariance of $d\tilde{\varrho}$ under $\tilde{\varrho} \mapsto \tilde{\varrho}^{-1}$ and the invariance of $d\tilde{\sigma}$ under $\varrho \in \mathrm{SO}(d)$, we have

$$\int F \, d\varrho = \int F(\mathcal{R}_{\tilde{\sigma}} \, \tilde{\varrho}^{-1}) \, d\tilde{\sigma} \, d\tilde{\varrho} = \int \check{F}(\tilde{\varrho} \, \mathcal{R}_{\tilde{\sigma}}) \, d\tilde{\sigma} \, d\tilde{\varrho} = \int \check{F}(\mathcal{R}_{\tilde{\varrho}\tilde{\sigma}} \, \tilde{\varrho}) \, d\tilde{\sigma} \, d\tilde{\varrho} = \int \check{F} \, d\varrho,$$

showing the invariance of $d\varrho$ under $\varrho \mapsto \varrho^{-1}$. By the above, we finally have

$$\int F(\varrho\varrho_0) \, d\varrho = \int F(\varrho^{-1}\varrho_0) \, d\varrho = \int \check{F}(\varrho_0^{-1}\varrho) \, d\varrho = \int \check{F}(\varrho) \, d\varrho = \int F(\varrho) \, d\varrho.$$

Hence $d\varrho$ is a Haar measure on $\mathrm{SO}(d)$, and a probability measure if $d\tilde{\varrho}$ is.

Moreover, we have

$$\exp\left[\alpha \sum_{k=2}^{d} \sigma_k E_{1k}\right] e_1 = (\cos \alpha)e_1 - \sin \alpha \sum_{k=2}^{d} \sigma_k e_k,$$

which equals $\tilde{\sigma}$ for a unique $(\alpha, \sigma) \in [0, \pi] \times \mathbb{S}^{d-2}$, implying the required formula $\mathcal{R}_{\tilde{\sigma}} = \exp\left[\alpha \sum_{k=2}^{d} \sigma_k E_{1k}\right]$. When we perform this change of variable (using for example the fact that $d\sigma = |\mathbb{S}^{d-2}|^{-1}|\sigma_1|^{-1}d\sigma_2 \ldots d\sigma_{d-1}$, and similarly for $d\tilde{\sigma}$), this implies also that $d\tilde{\sigma} = c_d(\sin\alpha)^{d-2}\, d\alpha\, d\sigma$, with a normalization constant c_d given in the statement. $\diamond$

The beginning of Proposition 3.3.2 is actually a particular case of a general statement (see [He], Section X.1). We now present a general proof for semisimple Lie groups, of both unimodularity and invariance under the inverse map, in the spirit of [He], Section X.1.

We begin with the following lemma.

Lemma 3.3.3 *Let λ be a left Haar measure on a Lie group G. Then, for any $g \in G$ and any test function F on G, we have*

$$\int_G F(\gamma g)\lambda(d\gamma) = \left|\det[\mathrm{Ad}(g)]\right| \times \int_G F(\gamma)\lambda(d\gamma).$$

Proof Changing the variable $\gamma \in G$ to $g\gamma g^{-1}$, we obtain

$$\int F(\gamma g)\lambda(d\gamma) = \int F(g^{-1}\gamma g)\lambda(d\gamma) = \int F(\gamma)\lambda\big(\mathrm{Ad}(g)(d\gamma)\big) = \int F \times |\det\ \mathrm{Ad}(g)|\ d\lambda.$$

Note, finally, that $|\det\ \mathrm{Ad}(g)|$ does not depend on $\gamma \in G$, since $\mathrm{Ad}(g)$ operates on $\mathcal{G}$ here. $\diamond$

As a consequence of Lemma 3.3.3, we see that a left Haar measure on a Lie group G is also a right Haar measure if and only if $\left|\det\mathrm{Ad}(g)\right| = 1$ for any $g \in G$. This entails the following general result.

Theorem 3.3.4 *If a Lie group G is semisimple, then it is unimodular and, moreover, its Haar measures are invariant under the inverse map $G \ni \gamma \mapsto \gamma^{-1}$.*

Proof (i) If we fix some basis of $\mathcal{G}$ and denote by M and $\mathcal{K}$ the matrices of $\mathrm{Ad}(g)$ and of the Killing form, respectively, then the Ad-invariance of the Killing form K on $\mathcal{G}$ (recall Section 1.1.1) implies ${}^t MKM = \mathcal{K}$, whence $(\det M^2 - 1)\ \det\mathcal{K} = 0$, and hence $|\det\ M| = 1$ by the hypothesis. Using Lemma 3.3.3, this proves that G is unimodular.

(ii) By (i) above and by Proposition 3.3.1, we know that any left Haar measure λ on G is proportional to its image $\check{\lambda}$ under the inverse map: $\lambda = c\check{\lambda}$. Otherwise, we choose a basis $(A_1, \ldots, A_n)$ of $\mathcal{G}$, a small $\varepsilon > 0$ such that exp is a smooth diffeomorphism on $\left\{\sum_{j=1}^n \alpha^j A_j \,\middle|\, (\alpha^1, \ldots, \alpha^n) \in B(0;\varepsilon) \subset \mathbb{R}^n\right\}$, and a continuous function F supported on this set, and we set $\tilde{F}(\alpha^1, \ldots, \alpha^n) := F\!\left(\exp\left[\sum_{j=1}^n \alpha^j A_j\right]\right)$. We then have $\lambda(d\gamma) = \kappa\, d\alpha^1 \ldots d\alpha^n$ at the unit element $\mathbf{1}$ of G, for some $\kappa > 0$, so that

$$c\int F\, d\check{\lambda} = \int F(\gamma)\lambda(d\gamma) = \kappa \int_{B(0;\varepsilon)} \tilde{F}(\alpha^1, \ldots, \alpha^n)\, d\alpha^1 \ldots d\alpha^n + \mathcal{O}(\varepsilon)$$

$$= \kappa \int_{B(0;\varepsilon)} \tilde{F}(-\alpha^1, \ldots, -\alpha^n)\, d\alpha^1 \ldots d\alpha^n + \mathcal{O}(\varepsilon)$$

$$= \int F(\gamma^{-1})\lambda(d\gamma) + \mathcal{O}(\varepsilon) = \int F\, d\check{\lambda} + \mathcal{O}(\varepsilon),$$

whence $c = 1 + \mathcal{O}(\varepsilon)$, and the conclusion follows by letting ε go to 0. $\diamond$

This theorem applies in particular to $\mathrm{PSO}(1, d)$, which is semisimple by Proposition 1.3.2, and hence unimodular.

In the following theorem, we give an explicit expression for the Haar measure of $\mathrm{PSO}(1, d)$ in terms of its Iwasawa decomposition.

Theorem 3.3.5 *Let a measure* λ *be defined on* $\mathrm{PSO}(1, d)$ *using the Iwasawa decomposition (recall Theorem 1.7.2), by*

$$\int_{\mathrm{PSO}(1,d)} F \, d\lambda := \int_{\mathbb{R}^{d-1} \times \mathbb{R}_+^* \times \mathrm{SO}(d)} F(T_{x,y}\varrho) y^{-d} \, dx \, dy \, d\varrho. \qquad (3.7)$$

The measure λ *is a Haar measure on* $\mathrm{PSO}(1, d)$ *and, moreover, we have* $\lambda = \check{\lambda}$.

We begin with the affine subgroup $\mathbb{A}^d$, which is not unimodular.

Lemma 3.3.6 *The Lie subgroup* $\mathbb{A}^d = \{T_{x,y} | (x, y) \in \mathbb{R}^{d-1} \times \mathbb{R}_+^*\}$ *(recall Proposition 1.4.3 and eqn (1.17)) admits* $y^{-d} \, dx \, dy$ *as a left Haar measure and* $y^{-1} \, dx \, dy$ *as a right Haar measure (where* $dx \, dy$ *denotes the Lebesgue measure on* $\mathbb{R}^{d-1} \times \mathbb{R}_+^* \subset \mathbb{R}^d$*).*

Since these two measures are clearly non-proportional (recall that $d \geq 2$), Proposition 3.3.1 ensures that $\mathbb{A}^d$ is not unimodular.

Proof Take any compactly supported continuous function F on $\mathbb{A}^d$, and any $(x_0, y_0) \in \mathbb{R}^{d-1} \times \mathbb{R}_+^*$. Using the fact that $T_{x,y} T_{x',y'} = T_{x+yx',yy'}$, we have at once

$$\int_{\mathbb{R}^{d-1} \times \mathbb{R}_+^*} L_{T_{x_0,y_0}} F(T_{x,y}) y^{-d} \, dx \, dy = \int_{\mathbb{R}^{d-1} \times \mathbb{R}_+^*} F(T_{x_0+y_0 x, y_0 y}) y^{-d} \, dx \, dy$$

$$= \int_{\mathbb{R}^{d-1} \times \mathbb{R}_+^*} F(T_{x,y}) y^{-d} \, dx \, dy,$$

whereas

$$\int_{\mathbb{R}^{d-1} \times \mathbb{R}_+^*} R_{T_{x_0,y_0}} F(T_{x,y}) y^{-1} dx \, dy = \int_{\mathbb{R}^{d-1} \times \mathbb{R}_+^*} F(T_{x+yx_0, yy_0}) y^{-1} \, dx \, dy$$

$$= \int_{\mathbb{R}^{d-1} \times \mathbb{R}_+^*} F(T_{x,y}) y^{-1} dx \, dy. \qquad \diamondsuit$$

Lemma 3.3.7 *The Laplace operator* $\mathcal{D}$ *(recall Definition 3.2.1) of the affine subgroup* $\mathbb{A}^d$ *is self-adjoint with respect to the left Haar measure* $y^{-d} \, dx \, dy$ *(of Lemma 3.3.6).*

Proof For any test function f, h on $\mathbb{A}^d$, by eqn (3.6) we have

$$\int_{\mathbb{R}^{d-1}\times\mathbb{R}_+^*} \mathcal{D}f(T_{x,y}) \times h(T_{x,y})y^{-d} \; dx \; dy$$

$$= \int \left[\frac{\partial^2}{\partial y^2} + \Delta_x^{d-1}\right] f(T_{x,y}) \times h(T_{x,y})y^{2-d} \; dx \; dy + (2-d)\int \frac{\partial}{\partial y} f(T_{x,y})$$

$$\times \; h(T_{x,y})y^{1-d} \; dx \; dy$$

$$= \int f(T_{x,y}) \left[\frac{\partial^2}{\partial y^2} + \Delta_x^{d-1}\right] \left(h(T_{x,y})y^{2-d}\right) dx \; dy - (2-d)\int f(T_{x,y})$$

$$\times \; \frac{\partial}{\partial y}\left(h(T_{x,y})y^{1-d}\right) dx \; dy$$

$$= \int_{\mathbb{R}^{d-1}\times\mathbb{R}_+^*} f(T_{x,y}) \times \mathcal{D}h(T_{x,y})y^{-d} \; dx \; dy. \qquad \diamondsuit$$

Lemma 3.3.8 *For any test function F on $\mathrm{PSO}(1,d)$, we have*

$$\int_{\mathrm{PSO}(1,d)} F \; d\check{\lambda} = \int_{\mathbb{R}^{d-1}\times\mathbb{R}_+^*\times\mathrm{SO}(d)} F(\varrho T_{x,y})y^{-1}dxdyd\varrho.$$

Proof Using Proposition 3.3.2, we have

$$\int_{\mathrm{PSO}(1,d)} F \; d\check{\lambda} = \int_{\mathbb{R}^{d-1}\times\mathbb{R}_+^*\times\mathrm{SO}(d)} F(\varrho^{-1}T_{x,y}^{-1})y^{-d} \; dx \; dy \; d\varrho$$

$$= \int_{\mathbb{R}^{d-1}\times\mathbb{R}_+^*\times\mathrm{SO}(d)} F(\varrho T_{-xy^{-1},y^{-1}})y^{-d} \; dx \; dy \; d\varrho$$

$$= \int_{\mathbb{R}^{d-1}\times\mathbb{R}_+^*\times\mathrm{SO}(d)} F(\varrho T_{x,y^{-1}})y^{-1} \; dx \; dy \; d\varrho = \int_{\mathbb{R}^{d-1}\times\mathbb{R}_+^*\times\mathrm{SO}(d)} F(\varrho T_{x,y})y^{-1} \; dx \; dy \; d\varrho.$$

$$\diamondsuit$$

The proof of Theorem 3.3.5 requires in addition the following infinitesimal commutation relations.

Lemma 3.3.9 *For any $(x, y) \in \mathbb{R}^{d-1} \times \mathbb{R}_+^*$ and for $2 \le j \le d$, we have*

$$\mathrm{Ad}(T_{x,y}^{-1})(E_{1j}) = \left(\frac{1}{2y} - \frac{y}{2} - \frac{|x|^2}{2y}\right)\tilde{E}_j + y \; E_{1j} + x_j \; E_1 - \sum_{k=2}^{d} x_k \; E_{jk} + \frac{x_j}{y}\sum_{k=2}^{d} x_k\tilde{E}_k,$$

and for $2 \le i < j \le d$,

$$\mathrm{Ad}\big(T_{x,y}^{-1}\big)(E_{ij}) = E_{ij} - \frac{x_i}{y}\tilde{E}_j + \frac{x_j}{y}\tilde{E}_i.$$

Proof Denote $\sum_{j=2}^{d} x_j \tilde{E}_j$ by $\tilde{E}(x)$. By eqns (1.13) and (1.14), we have

$$\mathrm{ad}\left[\tilde{E}(x)\right](E_1) = -\tilde{E}(x),\ \mathrm{ad}\left[\tilde{E}(x)\right](E_{1j}) = \sum_{k=2}^{d} x_k E_{jk} - x_j E_1, \mathrm{ad}\left[\tilde{E}(x)\right](E_{ij})$$

$$= x_i \tilde{E}_j - x_j \tilde{E}_i,$$

whence

$$\mathrm{ad}\big(\tilde{E}(x)\big)^2(E_{1j}) = 2x_j \tilde{E}(x) - |x|^2 \tilde{E}_j$$

and

$$\exp\left[\mathrm{ad}(\tilde{E}(-x))\right](E_{1j}) = E_{1j} + x_j E_1 - \sum_{k=2}^{d} x_k E_{jk} + x_j \tilde{E}(x) - \tfrac{1}{2}|x|^2 \tilde{E}_j.$$

Similarly,

$$\mathrm{ad}(E_1)(E_{1j}) = E_j, \mathrm{ad}(E_1)^2(E_{1j}) = E_{1j}$$

and then, by expanding the exponential,

$$\exp\left[\log(\tfrac{1}{y})\mathrm{ad}(E_1)\right](E_{1j}) = \tfrac{1}{2}(\tfrac{1}{y} - y)\tilde{E}_j + yE_{1j}.$$

The first claim now follows from eqn (1.4). Similarly, for $2 \le i < j \le d$,

$$\mathrm{ad}\big(\tilde{E}(x)\big)(E_{ij}) = x_i \tilde{E}_j - x_j \tilde{E}_i,\ \text{whence } \mathrm{ad}(\tilde{E}(x))^2(E_{ij}) = 0,$$

and then

$$\mathrm{Ad}(\theta_x^+)(E_{ij}) = \exp\left[\mathrm{ad}(\tilde{E}(x))\right](E_{ij}) = E_{ij} + x_i \tilde{E}_j - x_j \tilde{E}_i,$$

Hence, finally, using the fact that E_{ij} commutes with $\theta_{1/y}$,

$$\mathrm{Ad}\big(T_{x,y}^{-1}\big)(E_{ij}) = \mathrm{Ad}\big(\theta_{-x/y}^+\big)(E_{ij}) = E_{ij} - \frac{x_i}{y}\tilde{E}_j + \frac{x_j}{y}\tilde{E}_i. \qquad \diamond$$

Lemma 3.3.10 *Infinitesimal invariance implies global invariance: in particular, if $\int \mathcal{L}_{E_{ij}} F\, d\check{\lambda} = 0$ for any test function F on $\mathrm{PSO}(1, d)$ and any $1 \le i < j \le d$, then $\int R_\varrho F\, d\check{\lambda} = \int F\, d\check{\lambda}$ for any test function F and any $\varrho \in \mathrm{SO}(d)$.*

Proof Infinitesimal invariance implies that for any $s \ge 0$,

$$\frac{d}{ds}\int F\big(\gamma\ \exp[s\ E_{ij}]\big)\, d\check{\lambda}(\gamma) = \int \mathcal{L}_{E_{ij}} F_s\, d\check{\lambda} = 0,$$

where $F_s(\gamma) := F\big(\gamma\ \exp[s\ E_{ij}]\big)$. This is sufficient, since the elements $\exp[s\ E_{ij}]$ generate $\mathrm{SO}(d)$. $\qquad \diamond$

End of the proof of Theorem 3.3.5 By Lemmas 3.3.6 and 3.3.8, $\check{\lambda}$ is clearly right-invariant under $\mathbb{A}^d$. Thus, to prove that $\check{\lambda}$ is a right Haar measure, it is enough to show its right invariance under $SO(d)$ and, by Lemma 3.3.10, to show its infinitesimal right invariance under $so(d)$. Now, for any test function F on $PSO(1, d)$ and any $1 \le i < j \le d$, by eqn (1.8) we have

$$\int \mathcal{L}_{E_{ij}} F \, d\check{\lambda} = \int \mathcal{L}_{E_{ij}} F(\varrho T_{x,y}) \, \frac{dx \, dy}{y} \, d\varrho = \int R_{T_{x,y}}(\mathcal{L}_{E_{ij}} F)(\varrho) \, \frac{dx \, dy}{y} \, d\varrho$$

$$= \int \mathcal{L}_{\mathrm{Ad}(T_{x,y})(E_{ij})}(R_{T_{x,y}} F)(\varrho) \, \frac{dx \, dy}{y} \, d\varrho$$

$$= \int \mathcal{L}'_{\mathrm{Ad}(T_{x,y})(E_{ij})}(L_\varrho F)(T_{x,y}) \, \frac{dx \, dy}{y} \, d\varrho,$$

by eqn (1.10). Otherwise, by Lemma 3.3.9 (applied to $T_{-xy^{-1}, y^{-1}} = T_{x,y}^{-1}$), we have

$$\mathrm{Ad}(T_{x,y})(E_{1j}) = \left(\frac{y}{2} - \frac{1}{2y} - \frac{|x|^2}{2y} \right) \tilde{E}_j + \frac{1}{y} E_{1j} - \frac{x_j}{y} E_1 + \frac{1}{y} \sum_{k=2}^d x_k E_{jk} + \frac{x_j}{y} \sum_{k=2}^d x_k \tilde{E}_k,$$

and for $2 \le i < j \le d$,

$$\mathrm{Ad}(T_{x,y})(E_{ij}) = E_{ij} + x_i \tilde{E}_j - x_j \tilde{E}_i.$$

By the invariance of $d\varrho$, the integrals containing a rotational derivative $\mathcal{L}'_{E_{ij}}$ vanish. Hence, we have only to show that

$$0 = \int \left[\left(\frac{y}{2} - \frac{1}{2y} - \frac{|x|^2}{2y} \right) \mathcal{L}'_{\tilde{E}_j} - \frac{x_j}{y} \mathcal{L}'_{E_1} + \sum_{k=2}^d \frac{x_k x_j}{y} \mathcal{L}'_{\tilde{E}_k} \right] (L_\varrho F)(T_{x,y}) \, \frac{dx \, dy}{y} \, d\varrho$$

and

$$0 = \int \left[x_i \mathcal{L}'_{\tilde{E}_j} - x_j \mathcal{L}'_{\tilde{E}_i} \right] (L_\varrho F)(T_{x,y}) \, \frac{dx \, dy}{y} \, d\varrho.$$

Now, using eqn (3.5), we have only to show that

$$0 = \int \frac{\partial}{\partial x_j} L_\varrho F(T_{x,y}) \left(\frac{y}{2} - \frac{1}{2y} - \frac{|x|^2}{2y} \right) \frac{dx \, dy}{y} \, d\varrho - \int \frac{\partial}{\partial y} L_\varrho F(T_{x,y}) x_j \frac{dx \, dy}{y} \, d\varrho,$$

and that

$$0 = \int_{\mathbb{R}^{d-1} \times \mathbb{R}^*_+ \times SO(d)} \frac{\partial}{\partial x_j} L_\varrho F(T_{x,y}) x_i \, \frac{dx \, dy}{y} \, d\varrho$$

$$- \int_{\mathbb{R}^{d-1} \times \mathbb{R}^*_+ \times SO(d)} \frac{\partial}{\partial x_i} L_\varrho F(T_{x,y}) x_j \, \frac{dx \, dy}{y} \, d\varrho.$$

Both formulae follow from obvious integrations by parts.

We have thus shown that $\check{\lambda}$ is a right Haar measure on $\mathrm{PSO}(1,d)$. The proof is completed by applying Proposition 1.3.2 and Theorem 3.3.4. $\qquad\diamond$

Corollary 3.3.11 *The Lie derivatives are antisymmetric with respect to the Haar measure λ. The Casimir operator Ξ is self-adjoint with respect to the Haar measure λ.*

Proof For any test functions F, G on $\mathrm{PSO}(1,d)$, and $A \in \mathrm{so}(1,d)$, by right invariance we have

$$\int \mathcal{L}_A F \times G \ d\lambda = \frac{d_0}{d\varepsilon} \int F\big(\gamma \ \exp[\varepsilon A]\big) G(\gamma) \lambda(d\gamma)$$

$$= \frac{d_0}{d\varepsilon} \int F(\gamma) G\big(\gamma \ \exp[-\varepsilon A]\big) \lambda(d\gamma) = -\int F \times \mathcal{L}_A G \ d\lambda,$$

whence $\int (\mathcal{L}_A)^2 F \times G \ d\lambda = \int F \times (\mathcal{L}_A)^2 G \ d\lambda$, and then the claim follows by eqn (3.1). $\qquad\diamond$

3.4 The spherical Laplacian $\Delta^{\mathbb{S}^{d-1}}$

Recall that we encountered the Casimir operator on $\mathrm{SO}(d)$, $\Xi_0 := \sum_{1 \le k < \ell \le d} (\mathcal{L}_{E_{k\ell}})^2$, as a part of the Casimir operator Ξ on $\mathrm{PSO}(1,d)$ in eqn (3.1). We use it now to derive the spherical Laplacian of $\mathbb{S}^{d-1}$ by means of the projection $\varrho \mapsto \varrho e_1$ from $\mathrm{SO}(d)$ onto $\mathbb{S}^{d-1}$. To this end, note that by decomposing the generic rotation $\varrho \in \mathrm{SO}(d)$ as in Proposition 3.3.2,

$$\varrho = \mathcal{R}_{\tilde{\sigma}} \tilde{\varrho} = \exp\left[\sum_{\ell=2}^{d} x_\ell E_{1\ell}\right] \tilde{\varrho},$$

with $\tilde{\sigma} \in \mathbb{S}^{d-1}, x \in \mathbb{R}^{d-1}, \tilde{\varrho} \in \mathrm{SO}(d-1)$, for any $f \in C^2(\mathbb{S}^{d-1})$ we have

$$\mathcal{L}_{E_{k\ell}} e_1^* f(\varrho) = \frac{d_0}{d\varepsilon} f\Big(\mathcal{R}_{\tilde{\sigma}} \ \exp\big[\varepsilon \ \mathrm{Ad}(\tilde{\varrho})[E_{k\ell}]\big] e_1\Big) = 0 \text{ for } k, \ell \ge 2.$$

Moreover, it is clear that (as in the proof of Proposition 3.3.2)

$$\varrho e_1 = \mathcal{R}_{\tilde{\sigma}} e_1 = \tilde{\sigma} = \big(\cos|x|\big) e_1 - \frac{\sin|x|}{|x|} \sum_{\ell=2}^{d} x_\ell \ e_\ell.$$

Thus we see that $\Xi_0 e_1^* f(\varrho) = \sum_{\ell=2}^{d} (\mathcal{L}_{E_{1\ell}})^2 e_1^* f(\varrho)$ depends on ϱ only through $\varrho e_1 \equiv \tilde{\sigma} \in \mathbb{S}^{d-1} \equiv \mathrm{SO}(d)/\mathrm{SO}(d-1)$, thereby specifying an operator on $\mathbb{S}^{d-1}$. This allows the following definition.

Definition 3.4.1 *Setting* $e_1^* f(\varrho) := f(\varrho e_1)$ *for any* $\varrho \in \mathrm{SO}(d)$, *we define the spherical Laplacian* $\Delta^{\mathbb{S}^{d-1}}$ *as follows: for any* $f \in C^2(\mathbb{S}^{d-1})$,

$$e_1^*(\Delta^{\mathbb{S}^{d-1}} f) := \Xi_0 e_1^* f = \sum_{\ell=2}^{d} (\mathcal{L}_{E_{1\ell}})^2 e_1^* f.$$

This definition and the left invariance of the right Lie derivatives imply at once the $\mathrm{SO}(d)$-covariance of the spherical Laplacian: for any $f \in C^2(\mathbb{S}^{d-1})$ and $\varrho \in \mathrm{SO}(d)$, we have $\Delta^{\mathbb{S}^{d-1}}(f \circ \varrho) = [\Delta^{\mathbb{S}^{d-1}} f] \circ \varrho$. Note also that a straightforward adaptation of Corollary 3.3.11 proves that the Casimir operator Ξ_0 is self-adjoint with respect to the Haar measure of $\mathrm{SO}(d)$.

The following proposition provides an expression for the spherical Laplacian in Euclidean coordinates, and also some useful classical decompositions, which confirm in particular that it can also be derived from the ambient Euclidean structure of $\mathbb{R}^d \supset \mathbb{S}^{d-1}$.

Proposition 3.4.2 (*i*) *Writing* $(\tilde{\sigma}_1, \ldots, \tilde{\sigma}_d)$ *for the coordinates in* $\mathbb{R}^d$ *of* $\tilde{\sigma} \in \mathbb{S}^{d-1}$, *we have*

$$\Delta_{\tilde{\sigma}}^{\mathbb{S}^{d-1}} = \sum_{2 \le j,k \le d} (\delta_{jk} - \tilde{\sigma}_j\, \tilde{\sigma}_k) \frac{\partial^2}{\partial \tilde{\sigma}_j\, \partial \tilde{\sigma}_k} - (d-1) \sum_{k=2}^{d} \tilde{\sigma}_k \frac{\partial}{\partial \tilde{\sigma}_k}.$$

(*ii*) *The Euclidean Laplacian* Δ^d *of* $\mathbb{R}^d$ *can be decomposed in polar coordinates* $(R, \tilde{\sigma}) \in \mathbb{R}_+ \times \mathbb{S}^{d-1}$ *according to*

$$\Delta^d = \frac{\partial^2}{\partial R^2} + \frac{d-1}{R}\frac{\partial}{\partial R} + \frac{1}{R^2}\Delta_{\tilde{\sigma}}^{\mathbb{S}^{d-1}}.$$

(*iii*) *The spherical Laplacian* $\Delta^{\mathbb{S}^{d-1}}$ *can be decomposed in spherical polar coordinates* $(\alpha, \sigma) \in [0, \pi] \times \mathbb{S}^{d-2}$ $\big($*writing* $\tilde{\sigma} \equiv (\cos\alpha, \sigma\sin\alpha) \in \mathbb{S}^{d-1} \subset \mathbb{R}^d\big)$ *as*

$$\Delta_{\tilde{\sigma}}^{\mathbb{S}^{d-1}} = \frac{\partial^2}{\partial\alpha^2} + (d-2)\cot g\,\alpha\frac{\partial}{\partial\alpha} + \frac{1}{\sin^2\alpha}\Delta_\sigma^{\mathbb{S}^{d-2}}.$$

Proof (*i*) We have for $2 \le k, \ell \le d$

$$\mathcal{L}_{E_{1\ell}}\tilde{\sigma}_k = -\mathcal{L}_{E_{1\ell}}\langle\varrho e_1, e_k\rangle = -\frac{d_0}{d\varepsilon}\langle\varrho \exp[\varepsilon\, E_{1\ell}]e_1, e_k\rangle = \langle\varrho e_\ell, e_k\rangle.$$

Hence, using the spherical coordinates $(\tilde{\sigma}_2, \ldots, \tilde{\sigma}_d)$, we have

$$\mathcal{L}_{E_{1\ell}}e_1^* f(\varrho) = \mathcal{L}_{E_{1\ell}}f\left(\sum_{k=1}^{d}\tilde{\sigma}_k e_k\right) = \sum_{k=2}^{d}(\mathcal{L}_{E_{1\ell}}\tilde{\sigma}_k)\frac{\partial}{\partial\tilde{\sigma}_k}f(\tilde{\sigma}) = \sum_{k=2}^{d}\langle\varrho e_\ell, e_k\rangle\frac{\partial}{\partial\tilde{\sigma}_k}f(\tilde{\sigma}).$$

On the other hand,

$$(\mathcal{L}_{E_{1\ell}})^2 \tilde{\sigma}_k = \frac{d_0}{d\varepsilon}\big\langle \varrho \, \exp[\varepsilon E_{1\ell}]e_\ell, e_k \big\rangle = \langle \varrho e_1, e_k \rangle = \langle \tilde{\sigma}, e_k \rangle = -\tilde{\sigma}_k,$$

and for $2 \le j, k \le d$,

$$\sum_{\ell=2}^{d}(\mathcal{L}_{E_{1\ell}}\tilde{\sigma}_j)(\mathcal{L}_{E_{1\ell}}\tilde{\sigma}_k) = \sum_{\ell=2}^{d}\langle \varrho e_\ell, e_j \rangle \langle \varrho e_\ell, e_k \rangle$$

$$= -\langle e_j, e_k \rangle - \langle \varrho e_1, e_j \rangle \langle \varrho e_1, e_k \rangle = \delta_{jk} - \tilde{\sigma}_j \tilde{\sigma}_k.$$

Thus we finally obtain the expression given in part (i) in the statement, since

$$\sum_{\ell=2}^{d}(\mathcal{L}_{E_{1\ell}})^2 = \sum_{\ell=2}^{d}\left[\sum_{k=2}^{d}(\mathcal{L}_{E_{1\ell}}\tilde{\sigma}_k)\frac{\partial}{\partial\tilde{\sigma}_k}\right]^2 = \sum_{2\le j,k\le d}(\delta_{jk}-\tilde{\sigma}_j\tilde{\sigma}_k)\frac{\partial^2}{\partial\tilde{\sigma}_j\,\partial\tilde{\sigma}_k} + \sum_{k,\ell=2}^{d}(\mathcal{L}_{E_{1\ell}})^2\tilde{\sigma}_k\frac{\partial}{\partial\tilde{\sigma}_k}.$$

(ii) Starting from the Euclidean Laplacian Δ^d, we change from the variables $(X_1, \ldots, X_d)$ to the variables $(R, \tilde{\sigma}_2, \ldots, \tilde{\sigma}_d)$, with $\tilde{\sigma}_\ell = X_\ell/R$ and $R = \sqrt{X_1^2 + \cdots + X_d^2}$; we have

$$\frac{\partial}{\partial X_\ell} = \tilde{\sigma}_\ell\frac{\partial}{\partial R} + \sum_{j=2}^{d}\frac{\delta_{j\ell} - \tilde{\sigma}_j\tilde{\sigma}_\ell}{R}\frac{\partial}{\partial\tilde{\sigma}_j},$$

and then

$$\frac{\partial^2}{\partial X_\ell^2} = \tilde{\sigma}_\ell^2\frac{\partial^2}{\partial R^2} + \frac{2}{R}\sum_{j=2}^{d}(\delta_{j\ell} - \tilde{\sigma}_\ell^2)\tilde{\sigma}_j\frac{\partial^2}{\partial\tilde{\sigma}_j\,\partial R}$$

$$+ \frac{1}{R^2}\sum_{2\le j,k\le d}\left[\delta_{j\ell}\delta_{jk} + [\tilde{\sigma}_\ell^2 - \delta_{j\ell} - \delta_{k\ell}]\tilde{\sigma}_j\tilde{\sigma}_k\right]\frac{\partial^2}{\partial\tilde{\sigma}_j\,\partial\tilde{\sigma}_k} + \frac{1-\tilde{\sigma}_\ell^2}{R}\frac{\partial}{\partial R}$$

$$- \frac{1}{R^2}\sum_{j=2}^{d}\left[\delta_{j\ell} - \tilde{\sigma}_\ell^2 + \sum_{k=2}^{d}\Big((1+\delta_{j\ell})\delta_{k\ell} - \delta_{jk}\tilde{\sigma}_\ell^2 + (\delta_{1\ell} - \delta_{k\ell})\tilde{\sigma}_k^2\Big)\right]\tilde{\sigma}_j\frac{\partial}{\partial\tilde{\sigma}_j}.$$

By summing over ℓ, the required expression for (ii) now follows straightforwardly from (i).

(iii) We consider the additional variable $r := R\sin\alpha$, and change from the variables (X_1, r) to the variables (R, α). We have $R = \sqrt{X_1^2 + r^2}$ and $\alpha = \arccos\left(X_1\big/\sqrt{X_1^2 + r^2}\right)$, so that

$$\frac{\partial}{\partial X_1} = \cos\alpha\frac{\partial}{\partial R} - \frac{\sin\alpha}{R}\frac{\partial}{\partial\alpha} \quad \text{and} \quad \frac{\partial}{\partial X_1} = \sin\alpha\frac{\partial}{\partial R} + \frac{\cos\alpha}{R}\frac{\partial}{\partial\alpha},$$

$$\frac{\partial^2}{\partial X_1^2} = \cos^2\alpha\frac{\partial^2}{\partial R^2} + \frac{\sin^2\alpha}{R^2}\frac{\partial^2}{\partial\alpha^2} - \frac{\sin(2\alpha)}{R}\frac{\partial^2}{\partial R\,\partial\alpha} + \frac{\sin(2\alpha)}{R^2}\frac{\partial}{\partial\alpha} + \frac{\sin^2\alpha}{R}\frac{\partial}{\partial R},$$

$$\frac{\partial^2}{\partial r^2} = \sin^2\alpha\frac{\partial^2}{\partial R^2} + \frac{\cos^2\alpha}{R^2}\frac{\partial^2}{\partial\alpha^2} + \frac{\sin(2\alpha)}{R}\frac{\partial^2}{\partial R\,\partial\alpha} - \frac{\sin(2\alpha)}{R^2}\frac{\partial}{\partial\alpha} + \frac{\cos^2\alpha}{R}\frac{\partial}{\partial R}.$$

Hence, by (ii) (applied to $\mathbb{R}^{d-1}$),

$$\Delta^d = \frac{\partial^2}{\partial X_1^2} + \Delta^{d-1} = \frac{\partial^2}{\partial X_1^2} + \frac{\partial^2}{\partial r^2} + \frac{d-2}{r}\frac{\partial}{\partial r} + \frac{1}{r^2}\Delta_\sigma^{\mathbb{S}^{d-2}}$$

$$= \frac{\partial^2}{\partial R^2} + \frac{d-1}{R}\frac{\partial}{\partial R} + \frac{1}{R^2}\left[\frac{\partial^2}{\partial \alpha^2} + (d-2)\,\mathrm{cotg}\,\alpha\frac{\partial}{\partial \alpha} + \frac{1}{\sin^2\alpha}\Delta_\sigma^{\mathbb{S}^{d-2}}\right],$$

which proves (iii), by using (ii) again. $\diamond$

3.5 The hyperbolic Laplacian Δ

We extend the identification of Remark 1.3.3 between $\mathrm{PSO}(1,d)$ and $\mathbb{F}^d$ to functions, defining for any function h on $\mathrm{PSO}(1,d)$ a function ιh on $\mathbb{F}^d$ by

$$\iota h(\beta) := h(\tilde{\beta}), \text{ for any } \beta \in \mathbb{F}^d. \tag{3.8}$$

Remark 3.5.1 The following statements about a function h on $\mathrm{PSO}(1,d)$ are equivalent:

(i) h is invariant under the right action of $\mathrm{SO}(d)$;

(ii) the restriction $g := h\big|_{\mathbb{A}^d}$ of h to $\mathbb{A}^d$ is such that $h = g \circ \mathrm{Iw}$;

(iii) there exists a function f on $\mathbb{H}^d$ such that $\iota h = f \circ \pi_0$.

The equivalence between (i) and (ii) is obvious from the very definition of the Iwasawa projection Iw, and the equivalence between (ii) and (iii) is easily seen as follows: the function h is invariant under the right action of $\mathrm{SO}(d)$ if and only if ιh on $\mathbb{F}^d$ is, and the latter is clearly equivalent to (iii).

Theorem 3.5.2 *There exists a second-order differential operator Δ on $\mathbb{H}^d$, called the **hyperbolic Laplacian** , such that (i) Δ is covariant with respect to hyperbolic isometries: we have $\Delta(f \circ \gamma) = (\Delta f) \circ \gamma$ for any $\gamma \in \mathrm{PSO}(1,d)$, for any C^2 function f on $\mathbb{H}^d$, together with (ii)*

$$(\Delta f) \circ \pi_0 = \iota \circ \Xi \circ \iota^{-1}(f \circ \pi_0) = \iota \circ \mathcal{D} \circ \iota^{-1}(f \circ \pi_0) \text{ on } \mathbb{F}^d. \tag{3.9}$$

Proof Write $\iota^{-1}(f \circ \pi_0) =: h$, so that by Remark 3.5.1, h is right $\mathrm{SO}(d)$-invariant, and $\iota h = f \circ \pi_0$. Moreover, by Theorem 3.2.2, we have $\mathcal{D}h = \Xi h$. Then, using the right invariance of Ξ, we obtain

$$\Xi h = \Xi R_\varrho h = R_\varrho \Xi h, \text{ for any } \varrho \in \mathrm{SO}(d).$$

This proves the existence of a Δf satisfying (ii), again by Remark 3.5.1. Then, for any $\gamma \in \mathrm{PSO}(1,d)$, we set $f_\gamma(p) := f(\gamma p) = f \circ \gamma(p)$, and recall Remark 1.3.3: $\widetilde{\gamma(\beta)} = \gamma\tilde{\beta}$. Thus

$$[f_\gamma \circ \pi_0](\beta) = f(\gamma\beta_0) = [f \circ \pi_0](\gamma(\beta)) = \iota h(\gamma(\beta)) = h(\widetilde{\gamma(\beta)}) = h(\gamma\tilde{\beta}) = L_\gamma h(\tilde{\beta}),$$

whence by eqn (3.8) $\iota^{-1}(f_\gamma \circ \pi_0) = L_\gamma h$. The covariance property (i) of Δ then follows from the left invariance of Ξ:

$$\Delta(f_\gamma)(\beta_0) = [\Delta(f_\gamma) \circ \pi_0](\beta) = \Xi(\iota^{-1}(f_\gamma \circ \pi_0))(\tilde{\beta}) = \Xi(L_\gamma h)(\tilde{\beta})$$
$$= L_\gamma \Xi h(\tilde{\beta}) = \Xi h(\gamma \tilde{\beta}) = \Xi h(\widetilde{\gamma(\beta)}) = [(\Delta f) \circ \pi_0](\gamma(\beta)) = (\Delta f)_\gamma(\beta_0). \quad \Diamond$$

Let us now express the hyperbolic Laplacian in Poincaré coordinates and also give an alternative definition of it, which underlines its analogy to the spherical Laplacian (recall Definition 3.4.1, where we used the projection $\varrho \mapsto \varrho e_1$ from $SO(d)$ onto $\mathbb{S}^{d-1}$).

Corollary 3.5.3 (i) *In Poincaré coordinates (recall Propositions 1.6.2 and 3.2.3), the hyperbolic Laplacian is expressed by*

$$\Delta = y^2 \left(\frac{\partial^2}{\partial y^2} + \Delta_x^{d-1} \right) + (2 - d) y \frac{\partial}{\partial y}. \tag{3.10}$$

(ii) *Using the projection $\gamma \mapsto \gamma\, e_0$ from $PSO(1, d)$ onto $\mathbb{H}^d$, we set $e_0^* f(\gamma) :=$ $f(\gamma e_0)$ for any $f \in C^2(\mathbb{H}^d)$ and any $\gamma \in PSO(1, d)$ or, equivalently (using the notation of Remark 1.3.3),*

$$e_0^* f(\tilde{\beta}) := f(\tilde{\beta} e_0) = f \circ \pi_0(\beta) \text{ for any } \beta \in \mathbb{F}^d.$$

Then we have the following, which is equivalent to the definition in eqn (3.9):

$$e_0^*(\Delta f) = \Xi e_0^* f = \mathcal{D} e_0^* f \text{ on } PSO(1, d).$$

Proof (i) To obtain eqn (3.10), note that by the above proof of Theorem 3.5.2 (with $\gamma = T_z$ and the canonical basis as β), we have

$$(\Delta f)(T_z e_0) = (\Delta f)_{T_z}(e_0) = \Delta(f_{T_z})(e_0) = \Xi(\iota^{-1}(f_{T_z} \circ \pi_0))(\mathbf{1})$$
$$= \Xi(L_{T_z} h)(\mathbf{1}) = L_{T_z} \Xi h(\mathbf{1}) = \Xi h(T_z) = \mathcal{D} h(T_z).$$

Hence, noting that by the above we also have $f(T_z e_0) = h(T_z)$, we can apply eqn (3.6).

(ii) The equivalence of eqn (3.10) with (3.9) is clear from eqn (3.8), according to which we have $e_0^* f = \iota^{-1} f \circ \pi_0$ on $PSO(1, d)$. $\quad \Diamond$

We observe now that the Cartan decomposition of $PSO(1, d)$ (recall Theorem 1.7.1) naturally induces **polar coordinates** in $\mathbb{H}^d$: any $p \in \mathbb{H}^d$ can be written uniquely as $p = (\operatorname{ch} r) e_0 + (\operatorname{sh} r)\phi$, with coordinates $(r, \phi) \in \mathbb{R}_+ \times \mathbb{S}^{d-1}$ (and, of course, a singularity at the origin e_0).

In fact, the Cartan decomposition and Proposition 1.6.2 yield directly $p = \varrho\theta_r \varrho' e_0 = \varrho\theta_r e_0$, whence $\phi = \varrho e_1$. We then have the following result.

Proposition 3.5.4 *Using the polar coordinates $(r, \phi) \in \mathbb{R}_+ \times \mathbb{S}^{d-1}$, and the spherical Laplacian $\Delta_\phi^{\mathbb{S}^{d-1}}$ of Section 3.4 acting on the coordinate ϕ, we have*

$$\Delta = \frac{\partial^2}{\partial r^2} + (d-1) \coth r \, \frac{\partial}{\partial r} + (\operatorname{sh} r)^{-2} \Delta_\phi^{\mathbb{S}^{d-1}}. \tag{3.11}$$

Proof We use eqn (3.10) and Proposition 1.6.1, and perform a change of variable.

Writing $\phi = \sum_{i=1}^d \phi_i e_i$, this change between Poincaré coordinates $(y,x) = (y, x_2, \dots, x_d)$ and polar coordinates $(r, \phi_2, \dots, \phi_d)$ is defined by $\pi_0(T_{x,y}) = \pi_0(R_\phi \theta_r)$, where $R_\phi \in SO(d)$ denotes the Euclidean rotation in the Euclidean plane $\{e_1, \phi\} \subset \mathbb{R}^d$ mapping e_1 to ϕ, that is to say,

$$\left[\frac{y^2 + |x|^2 + 1}{2y}\right] e_0 + \left[\frac{y^2 + |x|^2 - 1}{2y}\right] e_1 + \sum_{j=2}^d \frac{x_j}{y} e_j = \operatorname{ch} r \, e_0 + \sum_{j=1}^d \phi_j \operatorname{sh} r \, e_j,$$

meaning that

$$\operatorname{ch} r = \frac{y^2 + |x|^2 + 1}{2y}, \ \frac{x_j}{y} = \phi_j \operatorname{sh} r \ \text{ for } 2 \le j \le d, \text{ and } \ \frac{y^2 + |x|^2 - 1}{2y} = \phi_1 \operatorname{sh} r.$$

In particular, we have $y^{-1} = \operatorname{ch} r - \phi_1 \operatorname{sh} r$ and $|x|^2 = y^2(1 - \phi_1^2) \operatorname{sh}^2 r$.

Differentiating the above formulae for $\operatorname{ch} r$ and x_j/y, we find

$$\frac{\partial r}{\partial x_j} = \phi_j \text{ for } 2 \le j \le d, \ \frac{\partial r}{\partial y} = \phi_1 \operatorname{ch} r - \operatorname{sh} r, \ \frac{\partial \phi_i}{\partial y} = \frac{-\phi_i \phi_1}{\operatorname{sh} r}$$

and

$$\frac{\partial \phi_i}{\partial x_j} = (\delta_{ij} - \phi_i \phi_j) \coth r - \delta_{ij} \phi_1 \ \text{ for } 2 \le i, j \le d.$$

Hence

$$\frac{\partial}{\partial y} = [\phi_1 \operatorname{ch} r - \operatorname{sh} r]\frac{\partial}{\partial r} - \frac{\phi_1}{\operatorname{sh} r} \sum_{i=2}^d \phi_i \frac{\partial}{\partial \phi_i}$$

and for $2 \le j \le d$,

$$\frac{\partial}{\partial x_j} = \phi_j \frac{\partial}{\partial r} + \sum_{i=2}^d \left[(\delta_{ij} - \phi_i \phi_j) \coth r - \delta_{ij} \phi_1\right] \frac{\partial}{\partial \phi_i}.$$

This implies, in turn,

$$\frac{\partial^2}{\partial y^2} = [\phi_1 \operatorname{ch} r - \operatorname{sh} r]^2 \frac{\partial^2}{\partial r^2} + \frac{\phi_1^2}{\operatorname{sh}^2 r} \sum_{2 \le i,k \le d} \phi_i \phi_k \frac{\partial^2}{\partial \phi_i \, \partial \phi_k} - 2\phi_1(\phi_1 \coth r - 1) \sum_{i=2}^d \phi_i \frac{\partial^2}{\partial \phi_i \, \partial r}$$

$$+ \left[(1 + \phi_1^2)\frac{\operatorname{sh}(2r)}{2} - \phi_1 \operatorname{ch}(2r) + (1 - \phi_1^2) \coth r\right]\frac{\partial}{\partial r}$$

$$+ \left[\phi_1^2 \coth^2 r - \phi_1 \coth r + \frac{2\phi_1^2 - 1}{\operatorname{sh}^2 r}\right] \sum_{i=2}^d \phi_i \frac{\partial}{\partial \phi_i},$$

$$\frac{\partial^2}{\partial x_j^2} = \phi_j^2 \frac{\partial^2}{\partial r^2} + 2\sum_{i=2}^{d}\left[(\delta_{ij}-\phi_j^2)\coth r - \delta_{ij}\phi_1\right]\phi_i \frac{\partial^2}{\partial \phi_i\, \partial r} + \left[(1-\phi_j^2)\coth r - \phi_1\right]\frac{\partial}{\partial r}$$

$$+ \sum_{2\leq i,k\leq d}\left[(\delta_{ij}\delta_{ik} + [\phi_j^2-\delta_{ij}-\delta_{kj}]\phi_i\phi_k)\coth^2 r\right.$$

$$\left. - (\delta_{ij}+\delta_{kj})(\delta_{ik}-\phi_i\phi_k)\phi_1\coth r + \delta_{ij}\delta_{ik}\phi_1^2\right]\frac{\partial^2}{\partial \phi_i\, \partial \phi_k}$$

$$- \sum_{k=2}^{d}\left[\frac{\delta_{jk}-\phi_j^2}{\mathrm{sh}^2 r} - \sum_{i=2}^{d}\left((2\phi_j^2-\delta_{jk}-1)\coth^2 r + (2\delta_{jk}+1)\phi_1\coth r - \delta_{jk}\right)\right]\phi_k\frac{\partial}{\partial \phi_k},$$

$$\sum_{j=2}^{d}\frac{\partial^2}{\partial x_j^2} = (1-\phi_1^2)\frac{\partial^2}{\partial r^2} + 2\phi_1\,[\phi_1\coth r - 1]\sum_{i=2}^{d}\phi_i\frac{\partial^2}{\partial \phi_i\, \partial r} + \left[(d-2+\phi_1^2)\coth r - (d-1)\phi_1\right]\frac{\partial}{\partial r}$$

$$+ \sum_{2\leq i,k\leq d}\left[\left(\delta_{ik}-[1+\phi_1^2]\phi_i\phi_k\right)\coth^2 r - 2(\delta_{ik}-\phi_i\phi_k)\phi_1\coth r + \delta_{ik}\phi_1^2\right]\frac{\partial^2}{\partial \phi_i\, \partial \phi_k}$$

$$- \sum_{k=2}^{d}\left[\frac{\phi_1^2}{\mathrm{sh}^2 r} + 2\phi_1^2\coth^2 r - (d+1)\phi_1\coth r + (d-2)\coth^2 r + 1\right]\phi_k\frac{\partial}{\partial \phi_k}.$$

Substituting the above expressions in eqn (3.10), we find

$$\Delta = \frac{\partial^2}{\partial r^2} + (d-1)\coth r\frac{\partial}{\partial r} + \frac{1}{\mathrm{sh}^2 r}\left[\sum_{2\leq i,k\leq d}(\delta_{ik}-\phi_i\phi_k)\frac{\partial^2}{\partial \phi_i\, \partial \phi_k} - (d-1)\sum_{k=2}^{d}\phi_k\frac{\partial}{\partial \phi_k}\right],$$

which, by Proposition 3.4.2(i), yields eqn (3.11), as wanted. $\diamond$

3.6   Harmonic, Liouville and volume measures

The measures that we introduce in this section play a fundamental role in the theory, from the geometrical as well as the probabilistic or dynamical point of view.

3.6.1   Harmonic measures and the Poisson kernel   Let us begin with visual measures of the hyperbolic boundary: the harmonic measure μ_p must be understood as the measure of the boundary viewed from the point p. Recall from Section 2.2.2 that given any $p \in \mathbb{H}^d$, J_p maps $\partial\mathbb{H}^d$ onto the unit sphere of $p^\perp$. Let $\sigma_{p^\perp}$ denote the normalized Lebesgue measure on this sphere.

Definition 3.6.1.1   *The **harmonic measure** μ_p on $\partial\mathbb{H}^d$ is the image of $\sigma_{p^\perp}$ under J_p^{-1}.*

Using Proposition 2.2.2.1, this means that we can say for short $\mu_p(d\eta) = dp_\eta$. We first have the following lemma.

Lemma 3.6.1.2 *In Poincaré coordinates, the harmonic measure μ_{e_0} is expressed by*

$$\mu_{e_0}(d\eta) = \frac{2^{d-2}\Gamma(d/2)}{\pi^{d/2}}\left(1+|u|^2\right)^{1-d} du,$$

where $u \in \mathbb{R}^{d-1}$ is the Poincaré coordinate of $\eta \in \partial\mathbb{H}^d$ (recall Proposition 2.2.2.6).

Proof By eqn (2.5), using Proposition 2.2.2.1, we have

$$u^j = \frac{-\langle \eta_{e_0}, e_j \rangle}{\langle \eta_{e_0}, e_0 + e_1 \rangle},$$

whence

$$|u|^2 = \frac{\langle \eta_{e_0}, e_0 \rangle^2 - \langle \eta_{e_0}, e_1 \rangle^2}{\langle \eta_{e_0}, e_0 + e_1 \rangle^2} = \frac{\langle \eta_{e_0}, e_0 - e_1 \rangle}{\langle \eta_{e_0}, e_0 + e_1 \rangle},$$

and then

$$1 + |u|^2 = \frac{2}{\langle \eta_{e_0}, e_0 + e_1 \rangle} = \frac{2}{1 + \langle \sigma, e_1 \rangle},$$

setting $\sigma := (e_0)_\eta \in \mathbb{S}^{d-1}$. We perform the change of variable $\sigma \mapsto u$. Since $\eta_{e_0} = \sigma + e_0$, we have

$$u^j = \frac{-\langle \sigma, e_j \rangle}{1 + \langle \sigma, e_1 \rangle}, \text{ or } u = \frac{\sigma + \langle \sigma, e_1 \rangle e_1}{1 + \langle \sigma, e_1 \rangle} = e_1 + \frac{\sigma - e_1}{1 + \langle \sigma, e_1 \rangle} =: \varphi(\sigma).$$

Then the differential $d\varphi_\sigma$ acts from the Euclidean $(d-1)$-space $\{e_0, \sigma\}^\perp$ onto the Euclidean space $\mathbb{R}^{d-1}$ by

$$d\varphi_\sigma = [1 + \langle \sigma, e_1 \rangle]^{-1} \times \left[1 - \frac{\sigma - e_1}{1 + \langle \sigma, e_1 \rangle} \langle \cdot, e_1 \rangle \right] = [1 + \langle \sigma, e_1 \rangle]^{-1} \times \ell,$$

with

$$\ell := 1 - \frac{\sigma - e_1}{1 + \langle \sigma, e_1 \rangle} \langle \cdot, e_1 \rangle.$$

Now, for any $v \in \{e_0, \sigma\}^\perp$, we have

$$\langle \ell(v), \ell(v) \rangle = \langle v, v \rangle + 2\frac{\langle v, e_1 \rangle}{1 + \langle \sigma, e_1 \rangle} \langle v, e_1 - \sigma \rangle + \frac{\langle v, e_1 \rangle^2}{[1 + \langle \sigma, e_1 \rangle]^2} \langle e_1 - \sigma, e_1 - \sigma \rangle$$

$$= \langle v, v \rangle.$$

Hence, ℓ is an isometry from $\{e_0, \sigma\}^\perp$ onto $\{e_0, e_1\}^\perp \equiv \mathbb{R}^{d-1}$. This proves that the Jacobian of φ at σ is $[1 + \langle \sigma, e_1 \rangle]^{1-d}$, meaning by Definition 3.6.1.1 and the above, that

$$\frac{2\pi^{d/2}}{\Gamma(d/2)}\mu_{e_0}(d\eta) = |\mathbb{S}^{d-1}|\mu_{e_0}(d\eta) = d\sigma = [1 + \langle \sigma, e_1\rangle]^{d-1}du$$

$$= \left(\frac{2}{1 + |u|^2}\right)^{d-1}du.$$

$\diamond$

Remark 3.6.1.3 It is clear from Definition 3.6.1.1 that for any hyperbolic isometry g, since $J_{g(p)} \circ g = g \circ J_p$ on $\partial\mathbb{H}^d$, $\mu_{g(p)}$ is the image of μ_p under g: $\mu_p \circ g^{-1} = \mu_{g(p)}$. This is called the **geometric property** of harmonic measures. Moreover, $\{\mu_p | p \in \mathbb{H}^d\}$ is the unique family of probability measures on $\partial\mathbb{H}^d$ which possesses this property.

In fact, the geometric property determines the whole family from μ_{e_0}, and prescribes that μ_{e_0} is $SO(d)$-invariant. Thus, $\mu_{e_0} \circ J_{e_0}$ must be an $SO(d)$-invariant probability measure on $\mathbb{S}^{d-1}$, and hence the normalized uniform measure. $\diamond$

We now obtain an expression for harmonic measures in Poincaré coordinates, which leads to the Poisson kernel in a natural way, together with a classical expression for it.

Proposition 3.6.1.4 *In Poincaré coordinates, the harmonic measure μ_p (based at $p \in \mathbb{H}^d$ which has Poincaré coordinates $(x, y) \in \mathbb{R}^{d-1} \times \mathbb{R}_+^*$, recall Proposition 1.6.1) can be expressed as*

$$\mu_p(d\eta) = \frac{2^{d-2}\Gamma(d/2)}{\pi^{d/2}}\left(\frac{y}{y^2 + |x - u|^2}\right)^{d-1}du, \qquad (3.12)$$

*where $u \in \mathbb{R}^{d-1}$ is the Poincaré coordinate of $\eta \in \partial\mathbb{H}^d$ (recall Proposition 2.2.2.6). The density which appears in eqn (3.12) is the **Poisson kernel** of $\mathbb{R}^{d-1} \times \mathbb{R}_+^*$.*

Proof Lemma 3.6.1.2 corresponds of course to $p = e_0$. To proceed to the case of an arbitrary $p \in \mathbb{H}^d$, we use the geometric property stated above, which tells us that μ_p is the image of μ_{e_0} under the isometry $T_{x,y}$. Hence, using Proposition 2.2.2.6 and Lemma 3.6.1.2, we obtain for any test function f on $\partial\mathbb{H}^d$

$$|\mathbb{S}^{d-1}|\int_{\partial\mathbb{H}^d} f d\mu_p = |\mathbb{S}^{d-1}|\int f(T_{x,y}(\eta))\mu_{e_0}(d\eta)$$

$$= \int_{\mathbb{R}^{d-1}} f(T_{x,y}\mathbb{R}_+\theta_u^+(e_0 - e_1))\left[\frac{2}{1 + |u|^2}\right]^{d-1}du$$

$$= \int f[T_{x+yu,y}\mathbb{R}_+(e_0 - e_1)]\left[\frac{2}{1 + |u|^2}\right]^{d-1}du$$

$$= \int f[T_{v,y}\mathbb{R}_+(e_0 - e_1)]\left[\frac{2}{1 + |v - x/y|^2}\right]^{d-1}d\left(\frac{v}{y}\right)$$

$$= \int f\left[T_{u,y}\mathbb{R}_+(e_0 - e_1)\right]\left[\frac{2y}{y^2 + |u - x|^2}\right]^{d-1} du$$

$$= \int f\left[\theta_u^+\mathbb{R}_+(e_0 - e_1)\right]\left[\frac{2y}{y^2 + |u - x|^2}\right]^{d-1} du. \qquad \diamond$$

It follows that all harmonic measures are equivalent. More precisely, we have the following result.

Proposition 3.6.1.5 *The harmonic measures μ_p are absolutely continuous with respect to each other. More precisely, (using η_q of Proposition 2.2.2.1), we have the following relations: for any $p, q \in \mathbb{H}^d$,*

$$\mu_q(d\eta) = \langle p, \eta_q\rangle^{d-1}\mu_p(d\eta) = e^{(d-1)B_\eta(p,q)}\mu_p(d\eta). \qquad (3.13)$$

Proof By Proposition 3.6.1.4, we have

$$\frac{\mu_{e_0}(d\eta)}{\mu_p(d\eta)} = \left[\frac{y^2 + |u - x|^2}{y(1 + |u|^2)}\right]^{d-1},$$

in Poincaré coordinates. Now, using eqns (1.18) and (2.4), we have

$$(1 + |u|^2) \times \langle p, \eta_{e_0}\rangle = \left[(y^2 + |x|^2 + 1)(|u|^2 + 1) - (y^2 + |x|^2 - 1)(|u|^2 - 1)\right.$$

$$\left. - 4\sum_{j=2}^d x^j u^j\right]\Big/(2y)$$

$$= \left[y^2 + |x|^2 + |u|^2 + 2\langle x, u\rangle\right]\Big/y = \frac{y^2 + |x - u|^2}{y}.$$

This shows eqn (3.13) for $q = e_0$. The general case follows directly, since

$$\frac{\mu_q(d\eta)}{\mu_p(d\eta)} = \frac{\mu_{e_0}(d\eta)}{\mu_p(d\eta)}\Big/\frac{\mu_{e_0}(d\eta)}{\mu_q(d\eta)} = \left[\frac{\langle p, \eta_{e_0}\rangle}{\langle q, \eta_{e_0}\rangle}\right]^{d-1} = \langle p, \eta_q\rangle^{d-1};$$

in fact, since η_q and η_{e_0} must be collinear, we immediately have $\eta_q = \langle q, \eta_{e_0}\rangle^{-1}\eta_{e_0}$, and then $\langle p, \eta_q\rangle = \langle p, \eta_{e_0}\rangle/\langle q, \eta_{e_0}\rangle$. The alternative expression using the Busemann function is obtained using merely Proposition 2.7.4. $\qquad \diamond$

3.6.2   Liouville and volume measures

The Liouville and volume measures on $\mathbb{F}^d$ and $\mathbb{H}^d$, respectively, are deduced from the Haar measure of $\mathrm{PSO}(1, d)$.

Definition 3.6.2.1 *The **Liouville measure** $\tilde{\lambda}$ of $\mathbb{F}^d$ is the measure on $\mathbb{F}^d$ induced by the Haar measure λ of $\mathrm{PSO}(1, d)$ using the bijection $\beta \leftrightarrow \tilde{\beta}$ of Remark 1.3.3.*

*The **volume measure** of $\mathbb{H}^d$ is the image measure of the Liouville measure under the canonical projection π_0 (from $\mathbb{F}^d$ onto $\mathbb{H}^d$). It will be denoted by dp (so that $dp = \tilde{\lambda} \circ \pi_0^{-1}$).*

Note that the Liouville measure $\tilde{\lambda}$ on $\mathbb{F}^d$ is clearly invariant under the left and right actions of $\mathrm{PSO}(1, d)$, and, in particular, under the geodesic and horocyclic flows. This fundamental fact follows immediately from Definitions 3.6.2.1 and 2.3.1 and Theorem 3.3.5. With Proposition 2.1.2, this entails the invariance of the volume measure of $\mathbb{H}^d$ under hyperbolic isometries.

Note also that we have the commutative diagram shown in Figure 3.1, in which $\beta_0 = T_{x,y} e_0$ has Poincaré coordinates (x, y).

Proposition 3.6.2.2 (i) *The volume measure of $\mathbb{H}^d$ can be expressed in terms of the Poincaré coordinates $(x, y) \in \mathbb{R}^{d-1} \times \mathbb{R}_+^*$ as*

$$dp = y^{-d} \, dx \, dy.$$

(ii) *The Liouville measure $\tilde{\lambda}$ of $\mathbb{F}^d$ disintegrates into $\tilde{\lambda} = dp \, d\varrho$, where $\varrho \in \mathrm{SO}(d)$ denotes the second Iwasawa coordinate and $d\varrho$ denotes the normalized Haar measure of $\mathrm{SO}(d)$ (recall Proposition 3.3.2).*

(iii) *Using the polar coordinates (r, ϕ) of Proposition 3.5.4, the volume measure of $\mathbb{H}^d$ can be expressed as $dp = (\mathrm{sh}\, r)^{d-1} \, dr \, d\phi$.*

Proof (i) The volume measure is induced by the left Haar measure $y^{-d} dx dy$ on $\mathbb{A}^d$ by means of Poincaré coordinates, owing to Proposition 1.6.2, Lemma 3.3.6, and eqn (3.7). As a matter of fact, for any test function f on $\mathbb{H}^d$ (extending π_0 by setting $\pi_0(\tilde{\beta}) := \pi_0(\beta) = \beta_0 = \tilde{\beta}(e_0)$), we have

$$\int_{\mathbb{H}^d} f \, dp = \int_{\mathbb{F}^d} f \circ \pi_0 \, d\tilde{\lambda} = \int_{\mathrm{PSO}(1,d)} f \circ \pi_0(\gamma) \lambda(d\gamma)$$

$$= \int_{\mathbb{R}^{d-1} \times \mathbb{R}_+^* \times \mathrm{SO}(d)} f \circ \pi_0(T_{x,y}\varrho) y^{-d} \, dx \, dy \, d\varrho = \int_{\mathbb{R}^{d-1} \times \mathbb{R}_+^*} f(T_{x,y}e_0) y^{-d} \, dx \, dy.$$

(ii) follows immediately, using the definitions of the measures λ and $\tilde{\lambda}$ in eqn (3.7) and Definition 3.6.2.1.

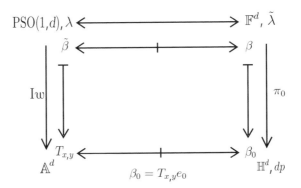

Fig. 3.1 Commutative diagram relating $\mathrm{PSO}(1, d)$, $\mathbb{F}^d$, $\mathbb{H}^d$, $\mathbb{A}^d$, Iw and π_0

(iii) follows directly from the same change of variable as that in Proposition 3.5.4. Recall that we have $y^{-1} = \operatorname{ch} r - \phi_1 \operatorname{sh} r$,

$$\frac{\partial r}{\partial x_j} = \phi_j \text{ for } 2 \le j \le d, \quad \frac{\partial r}{\partial y} = \phi_1 \operatorname{ch} r - \operatorname{sh} r, \quad \frac{\partial \phi_i}{\partial y} = \frac{-\phi_i \phi_1}{\operatorname{sh} r}$$

and

$$\frac{\partial \phi_i}{\partial x_j} = (\delta_{ij} - \phi_i \phi_j) \coth r - \delta_{ij}\phi_1 \text{ for } 2 \le i, j \le d.$$

Hence the Jacobian

$$J := \frac{dr \, d\phi_2 \ldots d\phi_d}{dy \, dx_2 \ldots dx_d}$$

is such that

$$J^{-1} = (\operatorname{sh} r)^{1-d} \det \begin{pmatrix} \phi_1 \operatorname{ch} r - \operatorname{sh} r & -\phi_1\phi_2 & \ldots & -\phi_1\phi_d \\ \phi_2 & (1-\phi_2^2)\operatorname{ch} r - \phi_1 \operatorname{sh} r \ldots & & -\phi_2\phi_d \operatorname{ch} r \\ \ldots & \ldots & \ldots & \ldots \\ \phi_d & -\phi_2\phi_d \operatorname{ch} r & \ldots (1-\phi_d^2)\operatorname{ch} r - \phi_1 \operatorname{sh} r \end{pmatrix}$$

$$= (\operatorname{sh} r)^{1-d} \det \begin{pmatrix} \phi_1 \operatorname{ch} r - \operatorname{sh} r & -\phi_1\phi_2 & \ldots & -\phi_1\phi_d \\ \phi_2 - \phi_2/\phi_1 \operatorname{ch} r(\phi_1\operatorname{ch} r - \operatorname{sh} r) & \operatorname{ch} r - \phi_1\operatorname{sh} r & \ldots & 0 \\ \ldots & & \ldots & \ldots \\ \phi_d - \phi_d/\phi_1 \operatorname{ch} r(\phi_1\operatorname{ch} r - \operatorname{sh} r) & 0 & \ldots & \operatorname{ch} r - \phi_1\operatorname{sh} r \end{pmatrix}$$

$$= (\operatorname{sh} r)^{1-d} y^{1-d} \det \begin{pmatrix} \phi_1\operatorname{ch} r - \operatorname{sh} r & -\phi_1\phi_2 & \ldots & -\phi_1\phi_d \\ \phi_2/\phi_1\operatorname{sh} r & 1 & \ldots & 0 \\ \ldots & \ldots & \ldots & \ldots \\ \phi_d/\phi_1\operatorname{sh} r & 0 & \ldots & 1 \end{pmatrix}$$

$$= (y \operatorname{sh} r)^{1-d}\left(\phi_1 \operatorname{ch} r - \operatorname{sh} r + \phi_1\phi_2 \frac{\phi_2}{\phi_1}\operatorname{sh} r + \ldots + \phi_1\phi_d \frac{\phi_d}{\phi_1}\operatorname{sh} r \right) = (\operatorname{sh} r)^{1-d} y^{-d}\phi_1.$$

This shows that

$$dp = y^{-d} \, dx \, dy = y^{-d} J \, dr \, d\phi_2 \ldots d\phi_d = (\operatorname{sh} r)^{d-1}\phi_1^{-1} dr \, d\phi_2 \ldots d\phi_d,$$

whence the result, since $\phi_1^{-1} d\phi_2 \ldots d\phi_d = d\phi$, as is easily computed by changing the Euclidean variables $(X_1, \ldots, X_d)$ to the variables $(r, \phi_2, \ldots, \phi_d)$ as in the last part of the proof of Proposition 3.5.4. Recall that we have

$$\frac{\partial}{\partial X_i} = \phi_i \frac{\partial}{\partial r} + \sum_{j=2}^{d} \frac{\delta_{ij} - \phi_i\phi_j}{r} \frac{\partial}{\partial \phi_j},$$

whence the Jacobian

$$\frac{dr\, d\phi_2 \dots d\phi_d}{dX_1 \dots dX_d} = r^{1-d} \det \begin{pmatrix} \phi_1 & -\phi_1\phi_2 & \dots & -\phi_1\phi_d \\ \phi_2 & 1 - \phi_2^2 & \dots & -\phi_2\phi_d \\ \dots & \dots & \dots & \dots \\ \phi_d & -\phi_2\phi_d & \dots & 1 - \phi_d^2 \end{pmatrix} = r^{1-d}\phi_1,$$

and then $dX_1 \dots dX_d = r^{d-1}\phi_1^{-1}\, dr\, d\phi_2 \dots d\phi_d = r^{d-1}\, dr\, d\phi$.   ◇

Proposition 3.6.2.3 *The hyperbolic Laplacian Δ (recall Section 3.5) is self-adjoint with respect to the volume measure dp. Similarly, the spherical Laplacian $\Delta^{\mathbb{S}^{d-1}}$ (recall Section 3.4) is self-adjoint with respect to the volume measure of $\mathbb{S}^{d-1}$.*

Proof The first statement is an immediate consequence of eqns (3.6) and (3.10), Lemma 3.3.7, and Proposition 3.6.2.2(i). The second statement is easily proved, either by induction on the dimension $(d-1)$ using Propositions 3.3.2 and 3.4.2(iii) or from the first statement using eqn (3.11) and Proposition 3.6.2.2(iii) (the details are left as an exercise). It can also be proved by using the self-adjointness of Ξ_0 (as observed just after Definition 3.4.1, as an adaptation of Corollary 3.3.11), Definition 3.4.1 and Proposition 3.3.2, as follows:

$$\int_{\mathbb{S}^{d-1}} \Delta^{\mathbb{S}^{d-1}} f(\tilde{\sigma}) \times h(\tilde{\sigma})\, d\tilde{\sigma} = \int_{SO(d)} \Xi_0(e_1^* f)(\varrho) \times (e_1^* h)(\varrho)\, d\varrho$$

$$= \int_{SO(d)} (e_1^* f)(\varrho) \times \Xi_0(e_1^* h)(\varrho)\, d\varrho = \int_{\mathbb{S}^{d-1}} f(\tilde{\sigma}) \times \Delta^{\mathbb{S}^{d-1}} h(\tilde{\sigma}) d\tilde{\sigma}.$$
   ◇

We now specify the relation between the harmonic measures μ_p of Definition 3.6.1.1 and the Liouville measure of Definition 3.6.2.1.

Proposition 3.6.2.4 *(disintegration of the Liouville measure) The projection on the unit tangent bundle $\mathbb{H}^d \times \partial\mathbb{H}^d$ (recall Section 2.2.2) of the Liouville measure $\tilde{\lambda}$ can be written as*

$$\tilde{\lambda} \circ \pi_1^{-1}(dp, d\eta) = \mu_p(d\eta)\, dp. \tag{3.14}$$

In Poincaré coordinates, this becomes

$$\tilde{\lambda} \circ \pi_1^{-1}(dp, d\eta) = 2^{d-2}\Gamma(\tfrac{d}{2})\pi^{-d/2} \left(y^2 + |x - u|^2\right)^{1-d} y^{-1}\, dx\, dy\, du. \tag{3.15}$$

Proof We start from Proposition 3.6.2.2(ii): $\tilde{\lambda} = dp\, d\varrho$.

A Lorentz frame $\beta \in \mathbb{F}^d$ is naturally decomposed into $\beta_0 \in \mathbb{H}^d$, β_1 in the hyperplane $\beta_0^\perp$ and the frame $(\beta_2, \dots, \beta_d) \in \{\beta_0, \beta_1\}^\perp$. The right action of $SO(d)$ preserves β_0, and rotates β_1 within the unit sphere of $\beta_0^\perp$, while the right action of $SO(d-1)$ rotates $(\beta_2, \dots, \beta_d)$. Accordingly, $d\varrho$ is the product of the normalized volume measure $d\sigma_{\beta_0^\perp}$ on the unit sphere of $\beta_0^\perp$ and the normalized Haar measure of SO$(d-1)$ (see Proposition 3.3.2). Equation (3.14) then follows directly from the very definition of the harmonic measure in Definition 3.6.1.1. The expression in Poincaré coordinates in eqn (3.15) then results immediately from eqn (3.14) and Propositions 3.6.1.4 and 3.6.2.2(i).   ◇

Let us now introduce an alternative coordinate system on the unit tangent bundle $\mathbb{H}^d \times \partial \mathbb{H}^d$ (recall Section 2.2.2).

Definition 3.6.2.5 *For any fixed $q \in \mathbb{H}^d$ and any $(p, \eta) \in \mathbb{H}^d \times \partial \mathbb{H}^d$, denote by η' the other end of the geodesic determined by (p, η), and denote by $s = s_q(p, \eta)$ the algebraic hyperbolic distance to p from the orthogonal projection of q on the oriented geodesic (η', η). This provides global coordinates $(\eta', \eta, s) \in \partial \mathbb{H}^d \bar{\times} \partial \mathbb{H}^d \times \mathbb{R}$ on the unit tangent bundle $\mathbb{H}^d \times \partial \mathbb{H}^d$, where $\partial \mathbb{H}^d \bar{\times} \partial \mathbb{H}^d$ denotes the Cartesian square $\partial \mathbb{H}^d \times \partial \mathbb{H}^d$ without its diagonal.*

Note that in this new coordinate system, the geodesic flow acts particularly simply (recall Section 2.3): $(\eta', \eta, s)\theta_t = (\eta', \eta, s+t)$, for any $s, t \in \mathbb{R}$.

We shall compute an expression for the Liouville measure in this new coordinate system. We begin with the following sequel to Proposition 3.6.1.4.

Proposition 3.6.2.6 *Using the Poincaré coordinates (u', u) of Proposition 2.2.2.6 for (η', η), the Liouville measure on $\mathbb{H}^d \times \partial \mathbb{H}^d$ can be written as*

$$\tilde{\lambda} \circ \pi_1^{-1}(du', du, ds) = 2^{d-2}\Gamma(d/2)\pi^{-d/2}|u' - u|^{2(1-d)} \, du' \, du \, ds. \qquad (3.16)$$

Proof We start from the expression in eqn (3.15) in the coordinates (y, x, u). We perform a change of variables $(y, x, u) \mapsto (s, u', u)$. We take $q = e_0$ as the reference point of Definition 3.6.2.5.

We observe in Figure 3.2 that the Pythagorean theorem provides the equation

$$u' = u + \frac{y^2 + |x - u|^2}{|x - u|^2}(x - u).$$

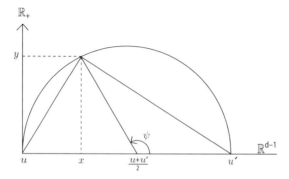

Fig. 3.2 Change of variables in the proof of Proposition 3.6.2.6

We use the polar angle ψ as an intermediate coordinate, replacing the arc-length coordinate s for the time being. This satisfies

$$x = \tfrac{1}{2}(u' - u)\,\cos\psi + \tfrac{1}{2}(u + u')\,;\, y = \tfrac{1}{2}\,|u' - u|\sin\psi.$$

We then have

$$y^2 + |x - u|^2 = \tfrac{1}{4}\,|u' - u|^2[\sin^2\psi + (1 + \cos\psi)^2] = \tfrac{1}{2}|u' - u|^2(1 + \cos\psi),$$

whence

$$\left(y^2 + |x - u|^2\right)^{1-d} y^{-1} = 2^d|u' - u|^{1-2d}(1 + \cos\psi)^{1-d}(\sin\psi)^{-1}.$$

On the other hand, by Remark 2.1.4, we have

$$ds = \frac{|u' - u|d\,\psi}{2y} = \frac{d\psi}{\sin\psi}.$$

Let us now compute the Jacobian J_d of the change of variables $(x, y) \mapsto (u', \psi)$, for fixed u. We have

$$J_d = \det\begin{bmatrix} \frac{\partial x}{\partial u'} & \frac{\partial y}{\partial \psi} \end{bmatrix} = \det\left[\frac{1}{2}\begin{pmatrix}(1 + \cos\psi)\mathbf{1_{d-1}} & \frac{\sin\psi}{|u'-u|}(u' - u) \\ {}^t(u - u')\sin\psi & |u' - u|\cos\psi\end{pmatrix}\right] =: 2^{-d}\delta_d,$$

where $\mathbf{1_{d-1}}$ denotes the unit of $\mathrm{GL}(d-1)$, and the vector $(u' - u)$ is seen as a column. Expanding the determinant δ_d with respect to its first column yields the following recursion formula:

$$\delta_d = (1 + \cos\psi)\delta_{d-1} + (1 + \cos\psi)^{d-2}\frac{\sin^2\psi}{|u' - u|}(u' - u)_1^2,$$

with $\delta_1 = |u' - u|\cos\psi$. Hence, by induction

$$\delta_d = (1 + \cos\psi)^{d-1}\delta_1 + (1 + \cos\psi)^{d-2}|u' - u|\sin^2\psi = (1 + \cos\psi)^{d-1}|u' - u|.$$

Hence we obtain from the above

$$\tilde\lambda \circ \pi_1^{-1}(dp, d\eta) = \frac{2^{d-2}\Gamma(d/2)}{\pi^{-d/2}}(y^2 + |x - u|^2)^{1-d}y^{-1}dx\,dy\,du$$

$$= \frac{2^{d-2}\Gamma(d/2)}{\pi^{-d/2}}|u' - u|^{2-2d}\frac{d\psi}{\sin\psi}du'du = \frac{2^{d-2}\Gamma(d/2)}{\pi^{-d/2}}|u' - u|^{2-2d}\,ds\,du'\,du.$$

$$\diamond$$

We end with an intrinsic expression for eqn (3.16).

Proposition 3.6.2.7 *In the coordinate system of Definition 3.6.2.5, depending on some reference point $q \in \mathbb{H}^d$, the Liouville measure on $\mathbb{H}^d \times \partial\mathbb{H}^d$ can be written as*

$$\tilde{\lambda} \circ \pi_1^{-1}(d\eta', d\eta, ds) = 2\pi^{d/2}\Gamma(d/2)^{-1}\langle\eta_q', \eta_q\rangle^{1-d}\mu_q(d\eta')\mu_q(d\eta) \ ds.$$

Proof Using the absolute continuity relation of eqn (3.13), $\mu_q(d\eta) = \langle e_0, \eta_q\rangle^{d-1}\mu_{e_0}(d\eta)$, and $\eta_q = \langle e_0, \eta_q\rangle\eta_{e_0}$ (as already noted at the end of the proof of Proposition 3.6.1.5), we first obtain

$$\langle\eta_q', \eta_q\rangle^{1-d}\mu_q(d\eta')\mu_q(d\eta) = \langle\eta_{e_0}', \eta_{e_0}\rangle^{1-d}\mu_{e_0}(d\eta')\mu_{e_0}(d\eta).$$

Hence we have only to consider the case $q = e_0$. We derive the intrinsic expression from Proposition 3.6.2.6 and Lemma 3.6.1.2. We have, in Poincaré coordinates (recall eqn (2.4)),

$$\frac{\langle\eta_{e_0}', \eta_{e_0}\rangle^{1-d}\mu_{e_0}(d\eta')\mu_{e_0}(d\eta)}{du'du} = \left\langle\frac{\theta_u^+(e_0 - e_1)}{1 + |u|^2}, \frac{\theta_{u'}^+(e_0 - e_1)}{1 + |u'|^2}\right\rangle^{1-d} \times \frac{\left[2^{d-2}\Gamma(d/2)\pi^{-d/2}\right]^2}{[(1 + |u|^2)(1 + |u'|^2)]^{d-1}}$$

$$= 4^{d-2}\Gamma(d/2)^2\pi^{-d}\left\langle\theta_{u-u'}^+(e_0 - e_1), e_0 - e_1\right\rangle^{1-d}$$

$$= 2^{d-3}\Gamma(d/2)^2\pi^{-d}|u - u'|^{2(1-d)}. \qquad \diamondsuit$$

3.7   Notes and comments

Casimir operators are usually presented in the more general context of a semisimple Lie group and the enveloping algebra formalism, in the form of the so-called Casimir element; see [Bi], [Ho] and [Kn].

The invariance of the Haar measure is considerably simpler in two dimensions ($d = 2$), where we can use the PSL(2) model. See, for example, [Ni], Section 10.2.

The construction of the Laplacian via the Casimir operator transported to a frame bundle has a certain analogy with the construction of the Laplacian on a Riemannian manifold as the projection of the Bochner horizontal Laplacian on the orthonormal frame bundle (see [KN]). Note, however, that the foliated Laplace operator $\mathcal{D}$ constructed in Section 3.2 is *not* the horizontal Laplacian.

The disintegration of the Liouville measure (Proposition 3.6.2.4) is classical; see for example [Ni], Section 8.1. For a general Riemannian manifold $\mathcal{M}$, the Liouville measure is defined on the unit frame bundle $T^1\mathcal{M}$. It is invariant under the geodesic flow, which describes the evolution of the position and velocity of a free particle. It can be extended to the orthonormal frame bundle $O\mathcal{M}$ by tensorization with the Haar measure on SO($d - 1$). The geodesic flow can also be extended (via parallel transport) to the orthonormal frame bundle $O\mathcal{M}$, where it describes the motion of a solid. It preserves the extended Liouville measure. See, for example, [KN].

Remark 3.7.1 On a general Riemannian (or Lorentzian) manifold $\mathcal{M}$, the metric (or pseudo-metric) is given infinitesimally, on the tangent bundle $T\mathcal{M}$. In a given chart, this corresponds to a symmetric matrix $((g_{jk})) = ((g_{jk}(m)))_{m\in\mathcal{M}}$ (i.e., a quadratic form on $T_m\mathcal{M}$). For example, in the particular case of $\mathbb{H}^d$, this amounts to the expression $ds = \sqrt{|dx|^2 + dy^2}/y$ of Remark 2.1.4 for the hyperbolic norm of the line element given by the Poincaré coordinates $\big((x,y),(dx,dy)\big)$. For $\mathbb{R}^{1,d}$ and the polar coordinates (r,ϱ,ϕ) (recall Section 3.5), the expression for the Lorentzian pseudo-metric is easily seen to be $\langle d\xi, d\xi\rangle = dr^2 - r^2\big(d\rho^2 + \mathrm{sh}^2\rho\,|d\phi|^2\big)$.

Then eqns (3.10) and (3.11) for the hyperbolic Laplacian Δ $\big($and similarly for the formulae of Proposition 3.4.2 relating to the spherical Laplacian $\Delta^{\mathbb{S}^{d-1}}$, and eqn (A.6) for the d'Alembertian $\square$; see Section A.3$\big)$ can be directly deduced from the general expression

$$\big|\det\!\big((g_{jk}(m))\big)\big|^{-1/2} \times \frac{\partial}{\partial m_j}\left[\big|\det\!\big((g_{jk}(m))\big)\big|^{1/2} \times g^{jk}(m)\frac{\partial}{\partial m_k}\right],$$

where $\big((g^{jk}(m))\big)$ denotes the inverse matrix $\big((g_{jk}(m))\big)^{-1}$.

Accordingly, the general expression for the volume measure (from which, for example, the formulae of Proposition 3.6.2.2 can be directly deduced) is $\big|\det\!\big((g_{jk}(m))\big)\big|^{1/2}\,dm$.

Chapter Four
Kleinian groups

Hyperbolic space has an infinite volume. To observe interesting dynamical properties such as ergodicity and mixing, which will be studied in Chapter 5, we will consider periodic functions, i.e., functions on spaces obtained by taking a quotient. Such quotient spaces, associated with a discrete subgroup of isometries, are locally identical to hyperbolic space but can have a finite volume. Geodesic flows on Riemann surfaces of this kind (when $d = 2$) are the simplest examples of mixing unstable flows defined by differential equations.

In this chapter, we deal with the geometric theory of Kleinian groups and their fundamental domains. We begin with the example of the parabolic tessellation of the hyperbolic plane by means of $2n$-gons bounded by full geodesic lines, namely ideal $2n$-gons. We discuss Dirichlet polyhedra and modular groups, with $\Gamma(2)$, $\Gamma(1)$, $D\Gamma(1)$ and $\Gamma(3)$ as the main examples.

4.1 Terminology

Definition 4.1.1 *A **Kleinian group** is a discrete subgroup of* $\mathrm{PSO}(1, d)$. *In other words, when considered as a subset of* $\mathrm{PSO}(1, d)$, *a Kleinian group is a subgroup that has no accumulation point.*

Obviously, a Kleinian group is countable. Note also that a subgroup Γ is discrete if and only if the unit element $\mathbf{1}$ is not the limit of any injective sequence $(\gamma_n) \subset \Gamma$.

Indeed, if an injective sequence $(\gamma_n) \subset \Gamma$ converges to some $g \in \mathrm{PSO}(1, d)$, then we can find inductively a sequence $(\varphi_n) \subset \mathbb{N}^*$ such that $(\gamma_n^{-1}\gamma_{n+\varphi_n})$ is injective, and it is clear that it converges to $\mathbf{1}$ as $n \to \infty$.

The main object of study associated with a Kleinian group Γ is its action on $\mathbb{H}^d$, and its **orbit space**, that is, the space $\Gamma \backslash \mathbb{H}^d := \{\Gamma p | p \in \mathbb{H}^d\}$ of the orbits of $\mathbb{H}^d$ under the (left) action of Γ. It is natural, in order to visualize this orbit space, to consider subsets of $\mathbb{H}^d$ which essentially contain one and only one representative of each orbit. More precisely, a **fundamental domain** for a Kleinian group Γ is a connected open subset $D \subset \mathbb{H}^d$ with zero boundary volume, such that $\gamma D \cap \gamma' D = \emptyset$ for any $\gamma \neq \gamma' \in \Gamma$ and $\bigcup_{\gamma \in \Gamma} \gamma \overline{D} = \mathbb{H}^d$. From the definition of a fundamental domain, the sequence of isometric images of D under Γ covers $\mathbb{H}^d$ (up to a negligible subset) without overlapping, thereby drawing a paving, or **tessellation**, of $\mathbb{H}^d$.

Of course, there are fundamental domains D which are simpler than others and are more convenient for most purposes. Of particular interest are those which are **polyhedral**, i.e., bounded by a finite set of geodesic hyperplanes.

Naturally, we say that a subset of $\mathbb{H}^d$ is **convex** if it contains the geodesic segment linking any pair of its points. By the definition of geodesics in Definition 2.2.1.1 and the identification of $\mathbb{H}^d$ with a part of the projective Minkowski space, this amounts to convexity within the projective Minkowski space.

We use the term **convex polyhedron** to denote any convex closed subset $\mathcal{P}$ of $\mathbb{H}^d$, possibly unbounded, that has a boundary made from a finite number of (convex subsets of) geodesic hyperplanes of $\mathbb{H}^d$, namely its **sides**. Within the projective Minkowski space, this can be identified with the intersection of $\mathbb{H}^d$ with a finite number of half-spaces (containing a common timelike ray).

A convex polyhedron is said to be **ideal** when all its vertices are **ideal**, meaning that they belong to $\partial\mathbb{H}^d$ (i.e., are light rays).

A **fundamental polyhedron** for Γ is a convex polyhedron, the interior of which is a fundamental domain for Γ, satisfying the following additional property: for each side S of $\mathcal{P}$, there exists some $\gamma \in \Gamma$ such that $S = \mathcal{P} \cap \gamma\mathcal{P}$. A Kleinian group Γ that admits a fundamental polyhedron is said to be **geometrically finite**.

Note that there are several other definitions of geometrical finiteness, which are essentially (but not exactly) equivalent (see in particular [B1]), and that the present definition is more restrictive for $d \geq 4$ than the ones considered in [B1] and [Rac]. For a more precise account, see [B1], Section 5 and [Rac], Section 12.3. We shall consider only geometrically finite Kleinian groups.

4.2 Dirichlet polyhedra

We describe here an important way of obtaining fundamental polyhedra for a Kleinian group Γ. We begin with a characterization of Kleinian groups. A subgroup Γ of $\mathrm{PSO}(1, d)$ is said to **act discontinuously** (or **act properly discontinuously**) on the hyperbolic space $\mathbb{H}^d$ if for any compact subset $K \subset \mathbb{H}^d$, the set $\{g \in \Gamma \,|\, K \cap gK \neq \emptyset\}$ is finite.

Lemma 4.2.1 *A subgroup Γ of $\mathrm{PSO}(1, d)$ is Kleinian if and only if it acts discontinuously on the hyperbolic space $\mathbb{H}^d$. In particular, any orbit Γp of an infinite Kleinian group Γ goes to infinity in $\mathbb{H}^d$. The torsion elements (i.e., the non-unit elements of finite order) of a Kleinian group are precisely its rotations.*

Proof If the action of Γ were not discontinuous, we would have an injective sequence $(g_n) \subset \Gamma$ and a convergent sequence in $\mathbb{H}^d$, $x_n \to x \in \mathbb{H}^d$, such that $g_n x_n \to y \in \mathbb{H}^d$. Thus we would have $(g_n x)$ bounded, and (g_n) equicontinuous and closed (since it would be discrete) in $C_c(\mathbb{H}^d, \mathbb{H}^d)$, and therefore compact by Ascoli's theorem and hence finite (since discrete), a contradiction. In particular, if an orbit Γp does not go to infinity, then there exists a compact set K_0 such that $K_0 \cap \Gamma p$ is infinite, and then applying the above criterion with $K = \{p\} \cup K_0$ would yield the finiteness of Γ. This criterion implies, moreover, that any rotation of Γ must generate a finite subgroup, i.e., it must be a torsion element, and conversely, by Theorem 1.5.1, any torsion element can only be a rotation.

Conversely, if Γ acts discontinuously on $\mathbb{H}^d$, then any injective sequence $(g_n) \subset \Gamma$ converging to the identity map, applied to any compact neighbourhood of any $x \in \mathbb{H}^d$, would yield a contradiction, showing that Γ has to be discrete. $\diamond$

Note, however, that a Kleinian group generally does not act discontinuously on the boundary $\partial \mathbb{H}^d$.

For any $p \in \mathbb{H}^d$ which is not fixed by any element of a Kleinian group Γ, the so-called **Dirichlet polyhedron** centred at p is

$$\mathcal{P}(\Gamma, p) := \bigcap_{\gamma \in \Gamma} \left\{ q \in \mathbb{H}^d \,\middle|\, \mathrm{dist}(p, q) \leq \mathrm{dist}(\gamma p, q) \right\}. \tag{4.1}$$

Proposition 4.2.2 *A Dirichlet polyhedron with respect to Γ is a fundamental polyhedron for Γ.*

Proof For any $p, p' \in \mathbb{H}^d$, the half-space $\{ q \in \mathbb{H}^d \,|\, \mathrm{dist}(p, q) \leq \mathrm{dist}(p', q) \}$ is convex and bounded by the geodesic hyperplane $\{ q \in \mathbb{H}^d \,|\, \mathrm{dist}(p, q) = \mathrm{dist}(p', q) \}$ (the geodesic joining two points of this hyperplane is contained in it), as is easily seen by choosing a half-space Poincaré model (i.e., Poincaré coordinates in some frame; see p. 34) in which p, p' have the same vertical coordinate. Hence $\mathcal{P}(\Gamma, p)$ is clearly convex and closed, as it is an intersection of geodesic hyperplanes in $\mathbb{H}^d$.

Then, for any $R > 0$, by Lemma 4.2.1 the orbit Γp has finitely many points $\gamma_0 p, \ldots, \gamma_n p$ in the compact set $\overline{B}(p, 2R)$, so that $\left[\bigcap_{j=0}^{n} \left\{ q \in \mathbb{H}^d \,\middle|\, \mathrm{dist}(p, q) < \mathrm{dist}(\gamma_j p, q) \right\} \right] \cap B(p, R)$ is the interior of $\mathcal{P}(\Gamma, p) \cap \overline{B}(p, R) = \left[\bigcap_{j=0}^{n} \left\{ q \in \mathbb{H}^d \,\middle|\, \mathrm{dist}(p, q) \leq \mathrm{dist}(\gamma_j p, q) \right\} \right] \cap \overline{B}(p, R)$. Hence the interior of $\mathcal{P}(\Gamma, p)$ is $\bigcap_{\gamma \in \Gamma} \left\{ q \in \mathbb{H}^d \,|\, \mathrm{dist}(p, q) < \mathrm{dist}(\gamma p, q) \right\}$, and $\mathcal{P}(\Gamma, p)$ is the closure of its interior. In particular, $\mathcal{P}(\Gamma, p)$ has a locally finite number of sides, each of which is convex and included in a hyperplane $\{ q \in \mathbb{H}^d \,|\, \mathrm{dist}(p, q) = \mathrm{dist}(\gamma p, q) \}$. And, since $\gamma \mathcal{P}(\Gamma, p) = \mathcal{P}(\Gamma, \gamma p)$ for any $\gamma \in \Gamma$, we have $\mathcal{P}(\Gamma, p) \cap \{ q \in \mathbb{H}^d \,|\, \mathrm{dist}(p, q) = \mathrm{dist}(\gamma p, q) \} = \mathcal{P}(\Gamma, p) \cap \gamma \mathcal{P}(\Gamma, p)$.

It is clear that for any $\gamma \in \Gamma \setminus \mathbf{1}$ and any $q \in \mathcal{P}(\Gamma, p)$, γq cannot be in the interior of $\mathcal{P}(\Gamma, p)$: this would imply that $\mathrm{dist}(p, \gamma q) < \mathrm{dist}(\gamma p, \gamma q) = \mathrm{dist}(p, q) \leq \mathrm{dist}(\gamma^{-1} p, q) = \mathrm{dist}(p, \gamma q)$. Finally, for any $q \in \mathbb{H}^d$, by discreteness, the minimal distance from q to the orbit Γp is attained for some γ_0: $\mathrm{dist}(\Gamma p, q) = \mathrm{dist}(\gamma_0 p, q)$, so that $q \in \gamma_0 \mathcal{P}(\Gamma, p)$. $\diamond$

4.3   Parabolic tessellation by an ideal $2n$-gon

When $d = 2$, Kleinian groups are often called **Fuchsian**. In this section we describe in detail a planar example, which exhibits a particular type of Fuchsian group. All the figures in the present chapter use the Poincaré models $\mathbb{B}^2$ (disc) and $\mathbb{R} \times \mathbb{R}_+^*$ (upper half-plane) of Section 2.5.

Theorem 4.3.1 *Let $n \geq 2$, and consider $2n$ light rays, say $\eta_1, \eta_2, \ldots, \eta_{2n}$, arranged clockwise along $\partial \mathbb{H}^2$ (seen as a circle). Let P_0 denote the ideal $2n$-gon $[\eta_1, \eta_2] \cup \ldots [\eta_{2n-1}, \eta_{2n}] \cup [\eta_{2n}, \eta_1]$, and for $j \in \{1, \ldots, n\}$, let φ_j denote the unique parabolic isometry that fixes η_{2j} and maps η_{2j+1} onto η_{2j-1} (letting*

$\eta_{2n+j} \equiv \eta_j$). Then the group Ξ_n generated by $\{\varphi_j | 1 \leq j \leq n\}$ is freely generated by the φ_j's, is Fuchsian, and admits the convex hull of P_0 as a fundamental polygon. It has $(n+1)$ inequivalent ideal vertices, and area $2(n-1)\pi$.

Note that $P_0 \equiv [\eta_1, \eta_2] \cup \ldots [\eta_{2n-1}, \eta_{2n}] \cup [\eta_{2n}, \eta_1]$ (where each $[\eta_j, \eta_{j+1}]$ denotes the geodesic from η_j to η_{j+1}) has no double points. We identify P_0 with the ordered set of its vertices $P_0 \equiv \{\eta_1, \eta_2, \ldots, \eta_{2n}\}$, and similarly for any (clockwise-arranged) finite subset of the circle $\partial \mathbb{H}^2$. Note also that for each $j \in \{1, \ldots, n\}$, there exists a unique parabolic isometry φ_j (recall the third case in Proposition 2.2.1.3; this is also clear in the half-plane $\mathbb{R} \times \mathbb{R}_+^*$, if we choose Poincaré coordinates (i.e., a frame; see p. 34 and Proposition 2.2.2.6) such that $\eta_{2j} \equiv \infty$).

Proof Observe that $\varphi_j(P_0) = \{\eta_{2j-1}, \varphi_j(\eta_{2j+2}), \ldots, \varphi_j(\eta_{2j-1}), \eta_{2j}\}$ is arranged clockwise, and is contained in the arc of $\partial \mathbb{H}^2$ which joins η_{2j-1} to η_{2j} and does not contain the points $\eta_{2j+1}, \ldots, \eta_{2j-2}$. Similarly, $\varphi_j^{-1}(P_0) = \{\eta_{2j}, \varphi_j^{-1}(\eta_{2j+1}), \ldots, \varphi_j^{-1}(\eta_{2j-2}), \eta_{2j+1}\}$ is arranged clockwise, and is contained in the arc of $\partial \mathbb{H}^2$ which joins η_{2j} to η_{2j+1} and does not contain the points $\eta_{2j+1}, \ldots, \eta_{2j-2}$. (See Figure 4.1 for the case $n = 3$, depicted in the disc model $\mathbb{B}^2$.) Thus, the convex hulls of $P_0, \varphi_1(P_0), \varphi_1^{-1}(P_0), \ldots, \varphi_n(P_0), \varphi_n^{-1}(P_0)$ have pairwise disjoint interiors, and their union constitutes the convex hull of the ideal $2n(2n-1)$-gon

$$P_1 := P_0 \cup \{\varphi_j(\eta_{2j+k+1}), \varphi_j^{-1}(\eta_{2j+k}) \mid 1 \leq j \leq n, 1 \leq k \leq 2n-2\}.$$

Then, the same procedure is applied to each such $2n$-gon $\varphi_j^{\pm 1}(P_0)$: for $1 \leq \ell < n$, this is done by means of the parabolic isometry $\varphi_j \varphi_{j+\ell} \varphi_j^{-1}$ that fixes $\varphi_j(\eta_{2j+2\ell})$ and maps $\varphi_j(\eta_{2j+2\ell+1})$ onto $\varphi_j(\eta_{2j+2\ell-1})$, and the parabolic isometry $\varphi_j^{-1} \varphi_{j+\ell} \varphi_j$ that fixes $\varphi_j^{-1}(\eta_{2j+2\ell})$ and maps $\varphi_j^{-1}(\eta_{2j+2\ell+1})$ onto $\varphi_j^{-1}(\eta_{2j+2\ell-1})$.

Hence, taking the images of $\varphi_j(P_0)$ by $\varphi_j \varphi_{j+\ell}^{\pm 1} \varphi_j^{-1}$ and then the images of $\varphi_j^{-1}(P_0)$ by $\varphi_j^{-1} \varphi_{j+\ell}^{\pm 1} \varphi_j$ for $1 \leq \ell < n$ and $1 \leq j \leq n$ yields, in clockwise order, the following isometric $2n$-gons, which have the interiors of their convex hulls pairwise disjoint and disjoint from the convex hulls of $P_0, \varphi_1(P_0), \varphi_1^{-1}(P_0), \ldots, \varphi_n(P_0), \varphi_n^{-1}(P_0)$:

$$\varphi_j \varphi_{j+1}(P_0), \varphi_j \varphi_{j+1}^{-1}(P_0), \ldots, \varphi_j \varphi_{j-1}^{-1}(P_0), \varphi_j^2(P_0), \varphi_j^{-2}(P_0),$$

$$\varphi_j^{-1} \varphi_{j+1}(P_0), \ldots, \varphi_j^{-1} \varphi_{j-1}^{-1}(P_0).$$

Continuing by induction, at the m-th step we obtain isometric $2n$-gons

$$\varphi_{j_1}^{\varepsilon_1} \varphi_{j_1+j_2}^{\varepsilon_2} \cdots \varphi_{j_1+j_2+\cdots+j_m}^{\varepsilon_m}(P_0), \quad 1 \leq j_k \leq n, \varepsilon_k = \pm 1,$$

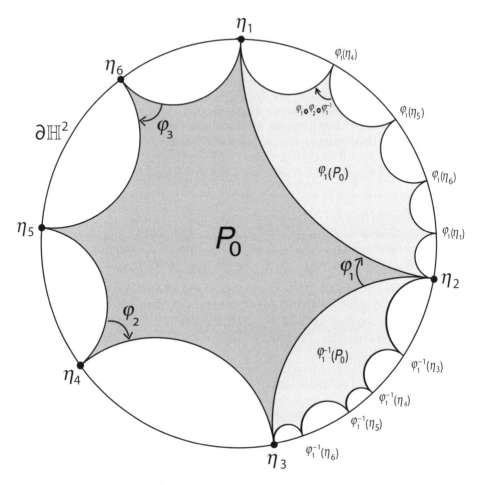

Fig. 4.1 Tessellation of $\mathbb{H}^2$ by an ideal hexagon P_0, with parabolic pairings $\varphi_1, \varphi_2, \varphi_3$

indexed by words of length m on the alphabet $\{\varphi_1^{\pm 1}, \ldots, \varphi_n^{\pm 1}\}$, such that the convex hulls of all isometric $2n$-gons associated with the words of length at most m have pairwise disjoint interiors. The union of these convex hulls, corresponding to all words of length at most m, is the (necessarily closed) convex hull of the ideal polygon P_m made up of all vertices of all polygons $\varphi_{j_1}^{\varepsilon_1} \varphi_{j_2}^{\varepsilon_2} \ldots \varphi_{j_k}^{\varepsilon_k}(P_0)$, $0 \le k \le m, j_\ell \in \mathbb{Z}/(2n)\mathbb{Z}, \varepsilon_k = \pm 1$, arranged in clockwise order along $\partial \mathbb{H}^2$.

As a consequence of this construction, we can observe that for p in the convex hull of P_0 and for any non-trivial reduced word $\varphi_{j_1}^{\varepsilon_1} \varphi_{j_2}^{\varepsilon_2} \ldots \varphi_{j_k}^{\varepsilon_k}$, the point $\varphi_{j_1}^{\varepsilon_1} \varphi_{j_2}^{\varepsilon_2} \ldots \varphi_{j_k}^{\varepsilon_k}(p)$ never belongs to the (closed) convex hull of P_0. This implies immediately the following lemma. $\diamondsuit$

Lemma 4.3.2 *The group Ξ_n is free, with rank n, and discrete.*

Note that the even and the odd (ideal) vertices of P_0 are intrinsically distinguished by the free group Ξ_n: each even vertex η_{2j} is fixed by the parabolic isometry φ_j. On the other hand, all odd vertices are equivalent under the group Ξ_n, and inequivalent to any even vertex, whereas two distinct even vertices of P_0 are inequivalent. In particular, under the action of Ξ_n, there are exactly $(n+1)$ classes of vertices of P_0, or of any P_m, or of $\bigcup_{m \in \mathbb{N}} P_m$.

Lemma 4.3.3 *The hyperbolic area of the ideal polygon P_0 equals $2(n-1)\pi$.*

Proof The ideal polygon P_0 can obviously be divided into $2(n-1)$ ideal triangles, so that it is enough to verify that the area of any ideal triangle equals 2π. This is easily seen by using a Poincaré half-plane model in which one edge of the triangle is ∞. By using a dilatation (recall this is a hyperbolic isometry), we have only to compute the area S of the geodesic triangle $\{0, 1, \infty\}$ (half of the quadrangle P_0 in Figure 4.3). Finally, by Proposition 3.6.2.2(*i*);

$$
S = \int_0^1 \int_{\sqrt{x-x^2}}^\infty \frac{dy}{y^2} \, dx = 2 \int_0^{1/2} \frac{dx}{\sqrt{x-x^2}} = \pi.
$$

$\diamond$

The proof of Theorem 4.3.1 ends with the following argument.

Lemma 4.3.4 *The increasing limit (as $m \to \infty$) of the convex hull $\overline{P_m}$ of the ideal polygon P_m, inductively defined as above, covers the entire hyperbolic plane $\mathbb{H}^2$.*

Proof Suppose that $\mathbb{H}^2 \setminus \bigcup_{m \in \mathbb{N}} \overline{P_m}$ contains a point p_∞. Choose a $p_0 \in \mathbb{H}^2 \cap \partial P_0$, and consider the geodesic arc $\gamma := [p_0, p_\infty]$, which has a finite length $\ell := \mathrm{dist}\,(p_0, p_\infty)$. This arc cuts a sequence of fundamental domains $P'_1 = \psi_1(P_0), \ldots, P'_k = \psi_k(P_0), \ldots$ (we denote the interior of the convex hull of P_0 by P_0 also here), which is necessarily infinite since $p_\infty \notin \bigcup_{m \in \mathbb{N}} \overline{P_m}$, and which (by the convexity of P_0) is such that $j \neq k \Rightarrow P'_j \cap P'_k = \emptyset$. For any $k \geq 1$, choose $p_k \in \gamma \cap \overline{P'_k} \cap \overline{P'_{k+1}}$, and for $k \geq 2$, set $p'_k := \psi_k^{-1}(p_{k-1})$, $p''_k := \psi_k^{-1}(p_k)$. We thus have $\sum_{k \geq 1} \mathrm{dist}\,(p'_k, p''_k) \leq \ell < \infty$, with $p'_k, p''_k \in \partial P_0$, not on the same edge of ∂P_0.

With each ideal vertex η_j, we associate a horoball $\mathcal{H}^+_j$ based at η_j (recall Definition 2.3.5), in such a way that these horoballs are pairwise disjoint. We need the following observation.

Lemma 4.3.5 *Let $Q := \partial P_0 \setminus \bigcup_{j=1}^{2n} \mathcal{H}^+_j$. There exists an $\varepsilon > 0$ such that the hyperbolic distance from any $p \in Q$ to $\partial P_0 \setminus E_p$ is greater than or equal to ε, where E_p denotes the edge of ∂P_0 containing p.*

Proof It is enough to consider a given light ray η and a Lorentz frame such that the Poincaré coordinate of η is ∞, the equation of the horoball $\mathcal{H}^+_\eta$ is $\{y \geq 1\}$

and the equations of the edges of P_0 starting at η are $\{x = 0\}$ and $\{x = a\}$, for some positive a. We then have to minimize dist $\big((0, y); (a, y')\big)$, for $y \leq 1$ and $y' > 0$. Now, by eqn (2.1) this equals $\operatorname{argch}\big[(a^2 + y^2 + |y'|^2)/2yy'\big]$, the minimum of which is $\min\limits_{y \leq 1} \operatorname{argch}\sqrt{1 + (a/y)^2} = \operatorname{argch}\sqrt{1 + a^2} > 0.$ $\diamond$

End of the proof of Lemma 4.3.4   The beginning of this proof and Lemma 4.3.5 entail the existence of some $k_0 \geq 1$ such that for any $k \geq k_0$, the geodesic arc $[p'_k, p''_k]$ is included in some horoball $\mathcal{H}^+_{j_k}$. Equivalently, the geodesic arc $[p_{k-1}, p_k]$ must be included in the horoball $\psi_k(\mathcal{H}^+_{j_k})$. Since, for any k, the horoballs $\psi_{k+1}(\mathcal{H}^+_j)$ are pairwise disjoint, and since $\psi_k(\mathcal{H}^+_{j_k})$ (containing p_k) must be one of them, we must have $\psi_k(\mathcal{H}^+_{j_k}) = \psi_{k+1}(\mathcal{H}^+_{j_{k+1}})$. This implies that the geodesic arc $[p_{k_0}, p_\infty[$ is included in the horoball $\psi_{k_0}(\mathcal{H}^+_{j_{k_0}})$. Now, this same geodesic arc must intersect an infinite sequence of images of P_0, which are necessarily of type $\psi^i \psi_{k_0}(P_0)$, for some parabolic isometry ψ that fixes the base of $\psi_{k_0}(\mathcal{H}^+_{j_{k_0}})$. In fact, the subgroup of Ξ_n that fixes a given ideal vertex is conjugate to the subgroup that fixes η_2, which reduces to the subgroup $\varphi_1^{\mathbb{Z}}$ generated by φ_1.

The desired contradiction is that no geodesic $\tilde{\gamma}$ can intersect an infinite sequence of geodesics $\psi^i(\gamma)$, provided ψ is parabolic and fixes η, and γ is a geodesic ending at η (where η is not an end of $\tilde{\gamma}$). In fact, by the discreteness of the parabolic subgroup $\{\psi^i | i \in \mathbb{Z}\}$ and by the compactness of $\partial \mathbb{H}^2$, the half-space of $\partial \mathbb{H}^2$ bounded by $\tilde{\gamma}$ and not containing η can contain only a finite number of ends of the geodesics $\psi^i(\gamma)$ (as can clearly be seen in a picture). $\diamond$

Remark 4.3.6   The Fuchsian group Ξ_n contains no rotations, but contains boosts.

Proof   For the absence of rotations, consider a point $z \in \mathbb{H}^2$, fixed by some $\varphi \in \Xi_n \setminus \{1\}$. By Lemma 4.3.4, $z = \psi(p)$ for some $p \in \overline{P_0}$ and $\psi \in \Xi_n$, so that $\psi^{-1}\varphi\psi$ fixes p and then must be trivial by the observation made just before Lemma 4.3.2, a contradiction.

As to the existence of boosts, consider any non-rotation isometry ψ. Taking the Poincaré half-plane model $\mathbb{R} \times \mathbb{R}^*_+$ with ∞ fixed by ψ, it is easily seen that ψ becomes an element of PSL(2) (recall Proposition 1.1.5.1) which can be identified with a translation or a dilatation, so that in both cases, any sequence $\psi^k(z_0)$ goes to some boundary point $z_\infty \in \partial \mathbb{H}^2$ as k goes to infinity. We can apply this in particular to $\varphi_2^{-1}\varphi_1$ and to $z_0 = \eta_4$. Noticing (see Figure 4.1) that $\varphi_1(\overline{\eta_4\eta_5}) \subset \overline{\eta_1\eta_2}$ and $\varphi_2^{-1}(\overline{\eta_1\eta_2}) \subset \overline{\eta_4\eta_5}$, we obtain $z_\infty \in \overline{\eta_4\eta_5}$, which is fixed by $\varphi_2^{-1}\varphi_1$. Similarly, noticing that $\varphi_2(\overline{\eta_2\eta_3}) \subset \overline{\eta_3\eta_4}$ and $\varphi_1^{-1}(\overline{\eta_3\eta_4}) \subset \overline{\eta_2\eta_3}$, we obtain $z'_\infty = \lim\limits_{k \to \infty} (\varphi_1^{-1}\varphi_2)^k(\eta_2) \in \overline{\eta_2\eta_3}$, which is fixed by $\varphi_1^{-1}\varphi_2$, and hence by $\varphi_2^{-1}\varphi_1$ too. This proves that $\varphi_2^{-1}\varphi_1$ is a boost (recall Theorem 1.5.1). $\diamond$

Note that whereas there are only finitely many non-isometric tessellations (by isometric regular convex polyhedra) of the Euclidean space $\mathbb{R}^d$, there are infinitely many non-isometric tesselations (by isometric regular convex polyhedra) of the hyperbolic space $\mathbb{H}^d$, as shown by Theorem 4.3.1 for $d = 2$. See also the simple example of triangle reflection groups ([Rac], Section 7.2).

4.4   Examples of modular groups

We restrict ourselves here to $d = 2$. By Proposition 1.1.5.1, Fuchsian groups are isomorphic to discrete subgroups of $\mathrm{PSL}(2)$. Thus, in the proofs in this section, we can and shall consider discrete subgroups of $\mathrm{PSL}(2)$. $\mathrm{PSL}(2)$ is, in turn, classically identified with the group $\mathcal{H}_2$ of **homographies** of the Poincaré upper plane $\mathbb{R} \times \mathbb{R}_+^* \equiv \{z = x + \sqrt{-1}\, y \in \mathbb{C} \,|\, x \in \mathbb{R}, y \in \mathbb{R}_+^*\}$, which is the most convenient model here for the hyperbolic plane $\mathbb{H}^2$, under the isomorphism

$$\mathrm{PSL}(2) \ni \pm \begin{pmatrix} a & b \\ c & d \end{pmatrix} \longmapsto \left(z \mapsto \frac{az + b}{cz + d} \right) \in \mathcal{H}_2.$$

The first example of a Fuchsian group is the **full modular group** $\Gamma(1) := \mathrm{PSL}(2, \mathbb{Z})$, made up of all classes of matrices having integer entries. The **modular groups** are the subgroups Γ of $\Gamma(1)$ having finite index $[\Gamma(1) : \Gamma]$. Most classical examples of modular groups belong to the **principal congruence groups** $\Gamma(N)$, of classes of matrices congruent to the unit matrix modulo $N \in \mathbb{N}^*$. They are normal in $\mathrm{PSL}(2, \mathbb{Z})$.

We shall take advantage below of Theorem 1.5.1, which yields a classification of hyperbolic isometries $\big($belonging to $\mathrm{PSO}(1, 2)$ or, equivalently, to $\mathrm{PSL}(2)\big)$ into rotations, boosts and parabolic elements.

4.4.1   Plane tessellation by means of two parabolic isometries   Consider two parabolic isometries φ_1, φ_2 of the hyperbolic plane $\mathbb{H}^2$, with distinct fixed points p_1, p_2 respectively. We shall distinguish two cases. The first one, which we describe in the following proposition, proves to be a particular case of Theorem 4.3.1 (with $n = 2$). In contrast, the other case (considered in Proposition 4.4.1.3) is not of this type.

Recall Definition 2.2.3.3, restricted to $d = 2$: a harmonic quadrangle is an ideal quadrangle in $\mathbb{H}^2$, such that its four ideal vertices are determined by light rays which are harmonically conjugate or, equivalently, such that the two geodesics joining opposite vertices intersect orthogonally. Recall that the adjective 'ideal' stresses the fact that the vertices are boundary points, i.e., located on the boundary $\partial \mathbb{H}^2$.

Proposition 4.4.1.1   *The three following conditions are equivalent.*

   (i)   The isometry $\varphi_1 \varphi_2$ is parabolic (with a fixed ideal point p).
   (ii)   The isometry $\varphi_2 \varphi_1$ is parabolic (with a fixed ideal point p').
   (iii)   There exist two ideal points $p, p' \in \partial \mathbb{H}^2 \smallsetminus \{p_1, p_2\}$ such that $\varphi_2(p) = p'$, $\varphi_1(p') = p$ and the quadrangle $P_0 := \{p_1, p, p_2, p'\}$ is harmonic.

If these conditions hold, then the parabolic isometries φ_1, φ_2 are conjugate by means of some unique involutive isometry σ, i.e., $(\exists! \, \sigma \in \mathrm{PSO}(1, 2) \smallsetminus \{\mathbf{1}\})$ $\varphi_1 = \sigma \varphi_2 \sigma$ and $\sigma^2 = \mathbf{1}$, and also by means of two unique parabolic isometries g_1, g_2 $\big(\varphi_1 = g_j \varphi_2 g_j^{-1}$ for $j = 1, 2\big)$.

Proof   It is enough to consider the half-plane $\mathbb{R} \times \mathbb{R}_+^*$, with $p_1 \equiv \infty$ and $p_2 \equiv 0$. Then φ_1 is a horizontal translation, by $u \in \mathbb{R}$ say, and up to conjugating by a dilatation, we can suppose that $u = 2$. Then we have necessarily $\varphi_2(z) = z/(cz + 1)$, for some $c \in \mathbb{R}^*$. Thus,

$$\varphi_1 \varphi_2(z) = \frac{(2c + 1)z + 2}{cz + 1} \quad \text{and} \quad \varphi_2 \varphi_1(z) = \frac{z + 2}{cz + 2c + 1},$$

whence $\varphi_1\varphi_2$ is parabolic if and only if $c = -2$, and similarly for $\varphi_2\varphi_1 = \varphi_1^{-1}(\varphi_1\varphi_2)\varphi_1$. In this case, the fixed points are $p = 1$ and $p' = -1$, respectively, and we have on the one hand $\varphi_1 = (z \mapsto -1/z)\varphi_2(z \mapsto -1/z)$, and on the other hand $\varphi_1(-1) = 1$, $\varphi_2(1) = -1$, and clearly the quadrangle $\{p_1, p, p_2, p'\} = \{\infty, 1, 0, -1\}$ is harmonic.

Conversely, if $\varphi_2(p) = p'$, $\varphi_1(p') = p$ and the quadrangle $\{\infty, p, 0, p'\}$ is harmonic, then $p = p' + 2$, $p - 2 = \varphi_2(p) = p/(cp + 1)$ and $p + p' = 0$, whence $p = 1$ and $c = -2$. Then, if $\sigma^2 = 1 \neq \sigma$, it is easily seen that

$$\sigma = \pm \begin{pmatrix} a & -b \\ \frac{a^2+1}{b} & -a \end{pmatrix},$$

for real $b \neq 0$ and a, and if, moreover, $\varphi_1\sigma = \sigma\varphi_2$, then

$$\begin{pmatrix} a & 2a-b \\ \frac{a^2+1}{b} & \frac{2(a^2+1)}{b} - a \end{pmatrix} = \varepsilon \begin{pmatrix} a & -b \\ \frac{a^2+1}{b} - 2a & 2b - a \end{pmatrix}$$

for some $\varepsilon = \pm 1$. This forces $a = 0$ and then $\varepsilon = 1$, whence $b^2 = 1$; that is, $\sigma = (z \mapsto -1/z)$ is indeed the unique solution.

Finally, if $\varphi_1 g = g\varphi_2$ for some parabolic isometry g, then

$$g = \pm \begin{pmatrix} u & v \\ w & 2 - u \end{pmatrix}$$

for real u, v, w such that $(u - 1)^2 = -vw$, and

$$\begin{pmatrix} u & 2u + v \\ w & 2w - u + 2 \end{pmatrix} = \varepsilon \begin{pmatrix} u & v \\ w - 2u & 2 - u - 2v \end{pmatrix}$$

for some $\varepsilon = \pm 1$. This forces $u = 0$ and then $\varepsilon = 1$, whence $w = -v$ and then $v^2 = 1$; that is, only the following two are solutions: $g(z) = 1/(2 - z)$ and $g(z) = -1/(z + 2)$.
$\diamond$

Remark 4.4.1.2 Any involutive (non-trivial) isometry of $\mathbb{H}^2$ has a unique fixed point (belonging to $\mathbb{H}^2$), and all involutive isometries of $\mathbb{H}^2$ (the identity map excepted) constitute a conjugacy class in the Lorentz–Möbius isometry group $PSO(1, 2)$.

Proof By Theorem 1.5.1, an involutive isometry of $\mathbb{H}^d$ cannot be parabolic, nor can it be a boost, and therefore it must be conjugate to a rotation $\gamma \in SO(d)$. For $d = 2$, the only involutive (non-unit) such γ is the orthogonal symmetry with respect to the line directed by e_0 (associated with $(z \mapsto -1/z) \in PSL(2)$). The statement is now trivial.
$\diamond$

With the parabolic isometries φ_1, φ_2 remaining as above, we now consider another case.

Proposition 4.4.1.3 *The following three conditions are equivalent.*

   (i) *The isometry $\varphi_1\varphi_2$ is a boost (with fixed ideal points p_1', p_2').*
   (ii) *The isometry $\varphi_2\varphi_1$ is a boost (with fixed ideal points p_1'', p_2'').*
   (iii) *There exist four pairwise distinct ideal points $p_1', p_2', p_1'', p_2'' \in \partial\mathbb{H}^2 \setminus \{p_1, p_2\}$ such that $p_j' = \varphi_1(p_j'')$, $\varphi_2(p_j') = p_j''$.*

If these conditions hold, the quadrangles $P_0' := \{p_1, p_1', p_2, p_2''\}$ and $P_0'' := \{p_1, p_2', p_2, p_1''\}$ are harmonic. If, moreover, the geodesics $[p_1', p_2']$ and $[p_1'', p_2'']$ are disjoint, then

- φ_1, φ_2 *are conjugate by means of some involutive isometry σ, where $\varphi_1 = \sigma\varphi_2\sigma$ and $\sigma^2 = \mathbf{1}$;*
- *there exist unique boundary points $q_1, q_2 \in \partial\mathbb{H}^2 \setminus \{p_1, p_2\}$ such that the quadrangles $Q_0' := \{p_1, q_1, p_2, \varphi_1(q_1)\}$ and $Q_0'' := \{p_2, q_2, p_1, \varphi_2(q_2)\}$ are harmonic;*
- *the geodesics $[p_1, q_1]$ and $[p_1, \varphi_1(q_1)]$ are disjoint from the geodesics $[p_2, q_2]$ and $[p_2, \varphi_2(q_2)]$;*
- *the ideal points $p_1, p_1', \varphi_1(q_1), q_2, p_2', p_2, p_1'', \varphi_2(q_2), q_1, p_2''$ are all distinct and are arranged in that order along the boundary $\partial\mathbb{H}^2$ (seen as a circle); see Figure 4.2.*

Proof Again it is sufficient, as for Proposition 4.4.1.1, to consider the half-plane $\mathbb{R} \times \mathbb{R}_+^*$, with $p_1 \equiv \infty$ and $p_2 \equiv 0$, $\varphi_1(z) = z + 2$,

$$\varphi_2(z) = \frac{z}{cz+1}, \qquad \varphi_1\varphi_2(z) = \frac{(2c+1)z+2}{cz+1}$$

and

$$\varphi_2\varphi_1(z) = \frac{z+2}{cz+2c+1}.$$

Therefore $\varphi_1\varphi_2$ is a boost if and only if $c < -2$ or $c > 0$, and similarly for $\varphi_2\varphi_1$.

In this case, the fixed points are $p_1', p_2' = 1 \pm \sqrt{1 + 2/c}$ and $p_1'', p_2'' = -1 \pm \sqrt{1 + 2/c}$. Clearly, $p_j' = \varphi_1(p_j'')$, whence $\varphi_2(p_j') = p_j''$. Moreover $P_0' = \left\{\infty, 1 + \sqrt{1 + 2/c}, 0, -1 - \sqrt{1 + 2/c}\right\}$ and $P_0'' = \left\{\infty, 1 - \sqrt{1 + 2/c}, 0, -1 + \sqrt{1 + 2/c}\right\}$ are harmonic.

Conversely, if (iii) holds, then clearly $\varphi_1\varphi_2$ fixes p_1', p_2' and is conjugate to some dilatation $(z \mapsto a^2 z)$, so that it can fix a point in $\mathbb{H}^2$ if and only if it is the identity map. As φ_1, φ_2 were supposed not to have the same fixed point, this cannot happen. Hence $\varphi_1\varphi_2$ is indeed a boost.

Furthermore, the geodesics $[p_1', p_2']$ and $[p_1'', p_2'']$ are disjoint if and only if $c < -2$. Suppose that this is the case. The claim about σ follows from the identity

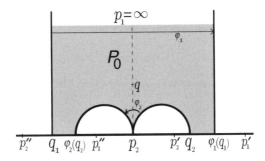

Fig. 4.2 The Dirichlet polyhedron $P_0 = \mathcal{P}(\Gamma_2, q)$ in the case of Proposition 4.4.1.3

$$\begin{pmatrix} 0 & -\sqrt{2/|c|} \\ \sqrt{|c|/2} & 0 \end{pmatrix} \times \begin{pmatrix} 1 & 0 \\ c & 1 \end{pmatrix} \times \begin{pmatrix} 0 & -\sqrt{2/|c|} \\ \sqrt{|c|/2} & 0 \end{pmatrix}^{-1} = \begin{pmatrix} 1 & 2\,\mathrm{sign}(-c) \\ 0 & 1 \end{pmatrix}.$$

Then $Q_0' = \{\infty, q_1, 0, q_1 + 2\}$ and $Q_0'' = \{0, q_2, \infty, q_2/(cq_2 + 1)\}$ are harmonic if and only if $q_1 = -1$ and $q_2 + q_2/(cq_2 + 1) = 0 \Leftrightarrow q_2 = -2/c$. As $c < -2$, the last claims follow directly.   ◇

Remark 4.4.1.4 If the parabolic isometries φ_1, φ_2 satisfy $\varphi_1 = \sigma\varphi_2\sigma$ for some involutive isometry σ, then the isometry $\varphi_1\varphi_2$ can be parabolic, or a boost with disjoint geodesics $[p_1', p_2']$ and $[p_1'', p_2'']$ (in the notation of Proposition 4.4.1.3) or a rotation.

Proof Once again we can suppose that $p_1 = \infty$ and $p_2 = 0$, $\varphi_1(z) = z + 2$,

$$\varphi_2(z) = \frac{z}{cz + 1} \quad \text{and} \quad \varphi_1\varphi_2(z) = \frac{(2c+1)z + 2}{cz + 1}.$$

As in the proof of Remark 4.4.1.2, we can take

$$\sigma = \pm \begin{pmatrix} a & -b \\ \frac{a^2+1}{b} & -a \end{pmatrix},$$

so that we have

$$\begin{pmatrix} a & -b \\ (a^2 + 1/b) & -a \end{pmatrix} \begin{pmatrix} 1 & 0 \\ c & 1 \end{pmatrix} = \pm \begin{pmatrix} 1 & 2 \\ 0 & 1 \end{pmatrix} \begin{pmatrix} a & -b \\ a^2 + 1/b & -a \end{pmatrix},$$

which is equivalent to $a = 0$ and $c = -2b^{-2}$. Hence c can be any negative real number.   ◇

Proposition 4.4.1.5 *Consider either the case of Proposition 4.4.1.1 or the case of Proposition 4.4.1.3, with disjoint geodesics $[p_1', p_2']$ and $[p_1'', p_2'']$. Then the group generated by the parabolic isometries $\{\varphi_1, \varphi_2\}$, say Γ_2, is free (with rank 2) and Fuchsian.*

Proof In the case of Proposition 4.4.1.1, we considered the harmonic quadrangle $P_0 = \{p_1, p, p_2, p'\}$. In the case of Proposition 4.4.1.3, with disjoint geodesics $[p_1', p_2']$ and $[p_1'', p_2'']$, let us consider instead the convex polygon P_0 bounded by the four sides $[p_1, q_1]$, $[p_1, \varphi_1(q_1)]$, $[p_2, q_2]$ and $[p_2, \varphi_2(q_2)]$ (in the notation of Proposition 4.4.1.4), which are pairwise disjoint (disregarding the common vertices p_1, p_2). Note that this P_0 is also bounded (at infinity) by two intervals (or arcs) of $\partial\mathbb{H}^2$ (seen as a circle), $\overline{\varphi_1(q_1)q_2}$ and $\overline{\varphi_2(q_2)q_1}$ which join q_1 to $\varphi_2(q_2)$ and q_2 to $\varphi_1(q_1)$, respectively, and do not contain the previous ideal vertices. P_0 is in fact the convex hull of

$$\check{P}_0 := \{q_1, p_1, \varphi_1(q_1), q_2, p_2, \varphi_2(q_2)\} \cup \overline{\varphi_1(q_1)q_2} \cup \overline{\varphi_2(q_2)q_1}.$$

Thus the 'harmonic' former case (which is a particular case of Theorem 4.3.1 with $n = 2$) appears as a limit of the 'non-harmonic' latter case, when $q_1, \varphi_2(q_2)$ merge into p' and $q_2, \varphi_1(q_1)$ merge into p. The proofs for the two cases can be unified by understanding that in the harmonic case, both q_1 and $\varphi_2(q_2)$ denote p' and both q_2 and $\varphi_1(q_1)$ denote p. The remainder of this proof, written for the non-harmonic case, is similar to the proof of Lemma 4.3.2.

Observe that any parabolic isometry preserves the clockwise ordering of points on $\partial\mathbb{H}^2$ (seen as a circle). Hence, the four isometries $\varphi_1, \varphi_1^{-1}, \varphi_2, \varphi_2^{-1}$ map P_0 into the closed half-planes of $\mathbb{H}^2$ not containing P_0 and delimited by $[p_1, \varphi_1(q_1)]$, $[p_1, q_1]$, $[p_2, \varphi_2(q_2)]$, $[p_2, q_2]$, respectively. Therefore, the five convex polygons P_0, $\varphi_1(P_0)$, $\varphi_1^{-1}(P_0)$, $\varphi_2(P_0)$, $\varphi_2^{-1}(P_0)$ have pairwise disjoint interiors, and their union is the convex hull of $\check{P}_0 \cup \varphi_1(\check{P}_0) \cup \varphi_1^{-1}(\check{P}_0) \cup \varphi_2(\check{P}_0) \cup \varphi_2^{-1}(\check{P}_0)$.

We then apply the same procedure to each such polygon $\varphi_j^{\varepsilon_j}(P_0)$ by means of the parabolic isometries $\varphi_j^{\varepsilon_j} \varphi_k^{\varepsilon_k} \varphi_j^{-\varepsilon_j}$ fixing $\varphi_j^{\varepsilon_j}(p_k)$, for $1 \leq j, k \leq 2$ and $\varepsilon_j, \varepsilon_k = \pm 1$ (but $\varepsilon_k \neq -\varepsilon_j$ if $k = j$, So as not to return to P_0). We thus obtain the 17 isometric convex polygons P_0, $\varphi_j^{\varepsilon_j}(P_0)$, $\varphi_j^{\varepsilon_j} \varphi_k^{\varepsilon_k}(P_0)$, which have pairwise disjoint interiors.

Continuing by induction, at the m-th step we obtain $(2 \times 3^m - 1)$ isometric polygons

$$\varphi_{j_1}^{\varepsilon_1} \varphi_{j_2}^{\varepsilon_2} \cdots \varphi_{j_k}^{\varepsilon_k}(P_0), \quad 0 \leq k \leq m, 1 \leq j_i \leq 2, \varepsilon_k = \pm 1,$$

indexed by words of length $\leq m$ on the alphabet $\{\varphi_1^{\pm 1}, \varphi_2^{\pm 1}\}$, which have pairwise disjoint interiors.

We observe that, as a consequence, for z in the interior of P_0 and for any non-trivial reduced word $\varphi_{j_1}^{\varepsilon_1} \varphi_{j_2}^{\varepsilon_2} \cdots \varphi_{j_k}^{\varepsilon_k}$, the point $\varphi_{j_1}^{\varepsilon_1} \varphi_{j_2}^{\varepsilon_2} \cdots \varphi_{j_k}^{\varepsilon_k}(z)$ never belongs to P_0. This immediately implies the statement. $\diamond$

As a simple consequence of the preceding proof, we obtain the following corollary.

Corollary 4.4.1.6 *In the case of Proposition 4.4.1.1, all ideal vertices are* **parabolic** *(i.e., fixed by some parabolic element of Γ_2), and there are exactly three classes of ideal vertices modulo Γ_2 (a set of representatives being $\{p_1, p_2, p\}$). Hence, the orbit space $\Gamma_2 \backslash \mathbb{H}^2$ has three* **cusps** *(a cusp being the image of a parabolic vertex in the orbit space).*

In the case of Proposition 4.4.1.3, there are exactly two classes of parabolic vertices modulo Γ_2 (a set of representatives being $\{p_1, p_2\}$), and the points of $\Gamma_2 q_j$ (in the notation of Proposition 4.4.1.4) are not parabolic. Hence, the orbit space $\Gamma_2 \backslash \mathbb{H}^2$ has two cusps.

Proof Only the last assertion needs an explanation. Clearly, the points of the same orbit are either all parabolic or all not. Recall from Proposition 4.4.1.3 that q_j is fixed by a boost of Γ_2. Now, no point q can be fixed by both a boost φ and a parabolic isometry ψ of the same discrete group. Indeed, taking the half-plane $\mathbb{R} \times \mathbb{R}_+^*$, up to a conjugation, we can suppose that $q = \infty$, $\varphi(z) = az$ and $\psi(z) = z + u$, with $u \in \mathbb{R}^*$ and $a > 1$ or $0 < a < 1$. But then $\{\varphi^k \psi \varphi^{-k} | k \in \mathbb{Z}\}$ clearly does not constitute a discrete subset, yielding the wanted contradiction. $\diamond$

Remark 4.4.1.7 The Fuchsian group Γ_2 contains boosts, but no rotations.

Proof By Lemma 4.2.1, a free Kleinian group cannot contain any rotation (which would be a torsion element). The existence of boosts in Γ_2 is obvious in the case of Proposition 4.4.1.3. In the case of Proposition 4.4.1.1, Remark 2.2.3.5 allows us to consider the half-plane $\mathbb{R} \times \mathbb{R}_+^*$, $P_0 = \{\infty, -1, 0, 1\}$, $\varphi_1(z) = z + 2$ and $\varphi_2(z) = z/1 - 2z$. Then

$$\varphi_1^{-1} \varphi_2 = \left(z \mapsto \frac{5z - 2}{1 - 2z}\right)$$

is a boost. This is again a particular case of Remark 4.3.6. $\diamond$

Theorem 4.4.1.8 *Consider either the case of Proposition 4.4.1.1, or the case of Proposition 4.4.1.3 with disjoint geodesics $[p_1', p_2']$ and $[p_1'', p_2'']$. Then the convex polygon P_0 defined in the proof of Proposition 4.4.1.5 is the Dirichlet polyhedron $\mathcal{P}(\Gamma_2, q)$, for* *any point $q \in \mathbb{H}^2$ of the geodesic joining the fixed points p_1, p_2 of the two parabolic isometries φ_1, φ_2 that generate Γ_2 freely.*

Proof As in Proposition 4.4.1.5, we shall unify the proof for the two cases by including the limiting former case in the latter. By Remark 2.2.3.5 and Propositions 4.4.1.1 and 4.4.1.3, it is enough to consider the half-plane $\mathbb{R} \times \mathbb{R}_+^*$, $\varphi_1(z) = z + 2$, $\varphi_2(z) = z/cz + 1$ with $c \leq -2$, and P_0 bounded by the geodesics $[\infty, -1]$, $[2/c, 0]$, $[0, -2/c]$, $[1, \infty]$ and by the intervals $\left[-1, 2/c\right]$, $\left[-2/c, 1\right]$ (which vanish in the former case) of the horizontal boundary line $\mathbb{R}$.

Consider $\lambda > 0$ and the Dirichlet polygon $\mathcal{P} := \mathcal{P}(\Gamma_2, \lambda\sqrt{-1})$, which by definition is included in $\mathcal{P}' := \mathcal{P}'_\infty \cap \mathcal{P}'_0$, with

$$\mathcal{P}'_\infty := \left\{ z \in \mathbb{R} \times \mathbb{R}_+^* \middle| \mathrm{dist}(\lambda\sqrt{-1}, z) \leq \mathrm{dist}\left(\varphi_1^{\pm 1}(\lambda\sqrt{-1}), z\right) \right\}$$

and

$$\mathcal{P}'_0 := \left\{ z \in \mathbb{R} \times \mathbb{R}_+^* \middle| \mathrm{dist}(\lambda\sqrt{-1}, z) \leq \mathrm{dist}\left(\varphi_2^{\pm 1}(\lambda\sqrt{-1}), z\right) \right\}.$$

Now, it is immediately apparent that $\mathcal{P}'_\infty = \{ z \in \mathbb{R} \times \mathbb{R}_+^* | |\Re(z)| \leq 1 \}$. And, recalling eqn (2.1) for the hyperbolic distance

$$\mathrm{ch}\left[\mathrm{dist}\left(x + y\sqrt{-1}, n + \nu\sqrt{-1}\right)\right] = \frac{(x - n)^2 + y^2 + \nu^2}{2\nu y},$$

since $\varphi_2^\varepsilon(\lambda\sqrt{-1}) = (\varepsilon c\lambda^2 + \lambda\sqrt{-1})/(1 + c^2\lambda^2)$, we obtain (for $\varepsilon = \pm 1$ and $z = x + y\sqrt{-1}$)

$$\mathrm{dist}(\lambda\sqrt{-1}, z) \leq \mathrm{dist}\left(\varphi_2^\varepsilon(\lambda\sqrt{-1}), z\right) \iff \frac{x^2 + y^2 + \lambda^2}{1 + c^2\lambda^2} \leq \left(x - \frac{\varepsilon c\lambda^2}{1 + c^2\lambda^2}\right)^2 + y^2$$

$$+ \frac{\lambda^2}{(1 + c^2\lambda^2)^2}$$

$$\iff \left(x - \frac{\varepsilon}{c}\right)^2 + y^2 \geq c^{-2}.$$

Hence we obtain $\mathcal{P} \subset \mathcal{P}'_\infty \cap \mathcal{P}'_0 = P_0$.

Conversely, for any q in the interior of P_0, by Proposition 4.2.2 there exists $\gamma \in \Gamma_2$ such that $\gamma(q) \in \mathcal{P}$, which implies in turn, by the above, that $\gamma(q) \in P_0$. Now, by the proof of Proposition 4.4.1.5, this forces γ to be the unit map, and hence $q \in \mathcal{P}$, concluding the proof that $P_0 = \mathcal{P}$ is a Dirichlet polygon for Γ_2.   ◇

Remark 4.4.1.9 Theorem 4.4.1.8 allows us to complete, in a sense, the description in Corollary 4.4.1.6 of the orbit space in the case of Proposition 4.4.1.3: $\Gamma_2 \backslash \mathbb{H}^2$ has two cusps and a so-called **funnel**, which is an unbounded part of this Riemann surface, parametrized by $\mathbb{S}^1 \times \mathbb{R}_+$, where the perimeter of $\mathbb{S}^1 \times \{y\}$ is equivalent to Ce^y as $y \to \infty$. The circle at infinity, $\Gamma_2 \backslash ([q_1, \varphi_2(q_2)] \cup [q_2, \varphi_1(q_1)])$, bounds the funnel. Note that the surface $\Gamma_2 \backslash \mathbb{H}^2$ has an infinite volume owing to its funnel, in contrast to what happens in the case of Proposition 4.4.1.1 (considered again in the following theorem); see Figure 4.2.

Theorem 4.4.1.10 *In the case of Proposition 4.4.1.1, if we choose the half-plane* $\mathbb{R} \times \mathbb{R}_+^*$ *and the harmonic quadrangle* $\{\infty, -1, 0, 1\}$, *the Fuchsian group* Γ_2 *of Proposition 4.4.1.5 is the principal congruence group* $\Gamma(2)$. *Hence,* $\Gamma(2)$ *is free and Fuchsian, has generator* $\{z \mapsto z + 2, z \mapsto z/1 - 2z\}$, *and admits the Dirichlet harmonic convex quadrangle* P_0, *which is the convex hull of* $\{\infty, -1, 0, 1\}$, *as a fundamental polygon; see Figure 4.3.*

Proof By Lemma 4.4.1.1, we know that in the case of the statement, Γ_2 is generated by $z \mapsto z + 2$ and $z \mapsto z/(1 - 2z)$, which belong to $\Gamma(2)$. It remains to show that $\Gamma(2) \subset \Gamma_2$. Suppose that we have some $\gamma_0 \in \Gamma(2) \setminus \Gamma_2$. Then, by Theorem 4.4.1.8 and Proposition 4.2.2, we would have some $\gamma_1 \in \Gamma_2$ such that $z_0 := \gamma_1 \gamma_0(\sqrt{-1}\,) \in P_0$. Hence

$$z_0 = \frac{(2a + 1)\sqrt{-1} + 2b}{2c\sqrt{-1} + 2d + 1},$$

for all non-vanishing integers a, b, c, d such that $2(ad - bc) + a + d = 0$, by the definition of $\Gamma(2)$. Thus,

$$z_0 = \frac{4(ac + bd) + 2(b + c) + \sqrt{-1}}{4(c^2 + d^2 + d) + 1} =: \frac{2m + \sqrt{-1}}{4q + 1}$$

must belong to P_0. This means that

$$|2m| \le 4q + 1 \quad \text{and} \quad \tfrac{1}{4} \le |z_0 \pm \tfrac{1}{2}|^2$$

or, equivalently,

$$|2m| \le 4q \quad \text{and} \quad 4m^2 \pm 2m(4q + 1) \ge -1,$$

and hence

$$|m| \le 2q \quad \text{and} \quad [2m \pm (4q + 1)]m \ge 0,$$

whence $m = 0$. So, we would have

$$\gamma := \gamma_1 \gamma_0 = \pm \begin{pmatrix} A & B \\ C & D \end{pmatrix} \in \Gamma(2)/\Gamma_2$$

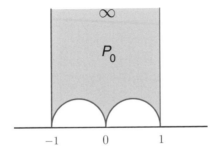

Fig. 4.3 The harmonic quadrangle P_0, the fundamental domain for $\Gamma(2)$

such that $AD - BC = 1$ and $AC + BD = 0$. Hence $BD^2 = -(1 + BC)C$ and then $C = -(C^2 + D^2)B$, and $0 = [D - (C^2 + D^2)A]B$. Now, $B = 0$ would imply $AC = 0$ and $AD = 1$, whence $C = 0$ and $\gamma = \pm 1$, which is excluded. Hence $D = (C^2 + D^2)A$, and then $(C^2 + D^2)$ divides 1, that is, $(C^2 + D^2) = 1$, $D = A$, $C = -B$. Hence $\gamma = \pm 1$ or

$$\gamma = \pm \begin{pmatrix} 0 & -1 \\ 1 & 0 \end{pmatrix} \notin \Gamma(2),$$

which is excluded. This concludes the proof. ◇

It is clear from Theorem 4.4.1.10 that taking $n = 2$ and taking the quadrangle P_0 to be harmonic in Theorem 4.3.1 yields (a conjugate of) the principal modular group $\Gamma(2)$.

4.4.2   From $\Gamma(2)$ to $\Gamma(1)$ We take a harmonic quadrangle $\{A, B, C, D\}$ (and hence a fundamental domain for $\Gamma(2)$, by Theorem 4.4.1.10) in $\mathbb{H}^2$, and denote its centre by Ω; we denote by $\sigma \equiv [A \leftrightarrow C, B \leftrightarrow D]$ the symmetry with respect to Ω. We consider also the parabolic isometry $\tau \equiv [A \mapsto A, B \mapsto C, C \mapsto D]$, and define A' to be the orthogonal projection of A on CD, A'' to be the orthogonal projection of A on BC, C' to be the orthogonal projection of C on AD, C'' to be the orthogonal projection of C on AB, $\beta := AA'' \cap CC''$ and $\delta := AA' \cap CC'$ (See Figure 4.4). The following lemma is almost obvious, by considering the actions of σ, τ on $\{A, B, C, D\}$ and knowing that any isometry maps two orthogonal geodesics to two orthogonal geodesics.

Lemma 4.4.2.1   *We have* $\beta, \delta \in BD$, $\tau(\beta) = \delta$, $\tau(A'') = A'$, $\tau(\Omega) = C'$, $\tau^{-1}(\Omega) = C''$, $\sigma(C'') = A'$ *and* $\sigma(A'') = C'$.

Now consider the following partition (up to the geodesic boundaries) of the fundamental domain delimited by $\{A, B, C, D\}$ into six subdomains delimited by the quadrangles $\{A, \Omega, \beta, C''\}$, $\{A, \Omega, \delta, C'\}$, $\{C, \Omega, \delta, A'\}$, $\{C, \Omega, \beta, A''\}$, $\{B, A'', \beta, C''\}$, $\{D, C', \delta, A'\}$. By Lemma 4.4.2.1, we have $\tau(B, A'', \beta, C'') = (C, A', \delta, \Omega)$, $\tau(A, \Omega, \beta, C'') = (A, C', \delta, \Omega)$ and, under σ,

$$(A, \Omega, \beta, C'') \leftrightarrow (C, \Omega, \delta, A'), (A, \Omega, \delta, C') \leftrightarrow (C, \Omega, \beta, A''), (B, A'', \beta, C'') \leftrightarrow (D, C', \delta, A').$$

This proves the following lemma.

Lemma 4.4.2.2   *The six subdomains listed above are isometric (see Figure 4.4).*

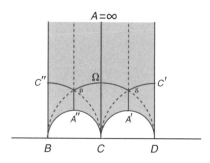

Fig. 4.4 The six isometric subdomains in Lemma 4.4.2.2

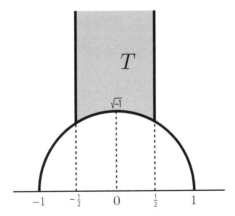

Fig. 4.5 The Dirichlet triangle T for $\Gamma(1)$

Specializing now to the harmonic quadrangle $\{A = \infty, B = -1, C = 0, D = 1\}$ in the half-plane $\mathbb{R} \times \mathbb{R}_+^*$, we have $(A, \Omega, \delta, C') = \left(\infty, \sqrt{-1}, e^{\sqrt{-1}\,\pi/3}, \sqrt{-1} + 1\right)$, and the convex triangle (δ, β, A) becomes (see Figure 4.5)

$$T := \left\{ z \in \mathbb{R} \times \mathbb{R}_+^* \,\middle|\, |\Re(z)| \leq \tfrac{1}{2}, |z| \geq 1 \right\}, \quad \text{delimited by} \quad \left(e^{\sqrt{-1}\,\pi/3}, e^{\sqrt{-1}\,2\pi/3}, \infty \right).$$

Moreover, σ can now be identified with $[z \mapsto -1/z] \in \Gamma(1)$, and τ with $[z \mapsto z + 1] \in \Gamma(1)$.

We want to prove that T is a fundamental polygon for $\Gamma(1)$, as will be stated in Theorem 4.4.1.10. To this ends let us set

$$v := \tau\sigma = \left[z \mapsto \tfrac{z-1}{z}\right]$$

and

$$\Gamma_1 := \Gamma(2) \bigcup \sigma\Gamma(2) \bigcup v\Gamma(2) \bigcup \sigma v\Gamma(2) \bigcup v^2\Gamma(2) \bigcup \sigma v^2\Gamma(2).$$

We then have $v^2(z) = -1/(z-1)$, $\sigma^2 = v^3 = 1$, $\sigma v(z) = -z/(z-1)$, $\sigma v^2(z) = z - 1$, $v\sigma(z) = z + 1$. It is very easily verified that $\sigma\Gamma(2)\sigma = \Gamma(2)$ and $v\Gamma(2)v^{-1} = \Gamma(2)$, and then that Γ_1 is a subgroup of $\Gamma(1)$, in which $\Gamma(2)$ is normal. Moreover, by Theorem 4.4.1.10, $\Gamma(2)$ is generated by $\{(\sigma v^2)^{-2}, (\sigma v)^2\}$, so that Γ_1 is the subgroup of $\Gamma(1)$ generated by $\{\sigma, v\}$.

Note that the six cosets defining Γ_1 above are distinct (all elements listed above clearly do not belong to $\Gamma(2)$), so that the quotient group $\Gamma_1/\Gamma(2)$ has order 6. Moreover, as $\sigma v\Gamma(2) = v^2\sigma\Gamma(2)$, $\Gamma_1/\Gamma(2)$ is isomorphic to the (non-Abelian) dihedral group of order 6.

By Lemma 4.4.2.2, this shows that the (convex hull of the) quadrangle $\{A, \Omega, \delta, C'\}$ of Figure 4.4 is a fundamental domain for Γ_1. Now T can plainly be deduced from (A, Ω, δ, C') by translating its right half by -1, i.e., by applying $\sigma^2 v$ to its right half. Hence the images of T under Γ_1, as well as the images of (A, Ω, δ, C') under Γ_1, cover the whole hyperbolic plane $\mathbb{R} \times \mathbb{R}_+^*$. A fortiori, the images of T under $\Gamma(1)$ cover the whole hyperbolic plane $\mathbb{R} \times \mathbb{R}_+^*$.

This and the following lemma show that the interior $\overset{o}{T}$ of T is a fundamental domain for the full modular group $\Gamma(1)$.

Lemma 4.4.2.3 *We have* $\gamma\big(\overset{o}{T}\big) \cap \overset{o}{T} = \emptyset$, *for any* $\gamma \in \Gamma(1) \smallsetminus \{\mathbf{1}\}$.

Proof Consider $z = \Re(z) + \sqrt{-1}\, \Im(z)$ belonging to the interior $\overset{o}{T}$ of T, and

$$\gamma \equiv \pm \begin{pmatrix} a & b \\ c & d \end{pmatrix} \in \Gamma(1) \smallsetminus \{\mathbf{1}\}$$

such that $\gamma(z) \in \overset{o}{T}$. This implies in particular that γ cannot be a translation ($z \mapsto z + b$), and hence that $c \neq 0$. Then, since $|z| > 1 > 2 |\Re(z)|$, we have

$$|cz + d|^2 > c^2 - |cd| + d^2 = (|c| - |d|)^2 + |cd| \geq 1,$$

whence

$$\Im\big(\gamma(z)\big) = \frac{\Im(z)}{|cz + d|^2} < \Im(z).$$

Now the same holds for $\big(\gamma(z), \gamma^{-1}\big)$, yielding also $\Im(z) < \Im\big(\gamma(z)\big)$, a contradiction. $\diamond$

Theorem 4.4.2.4 *The convex triangle* T *is a fundamental polygon for the full modular group* $\Gamma(1)$, *which is generated by* $\{\sigma, v\}$ *or, alternatively, by* $\{\sigma, \tau\}$ (*recall that* $\sigma(z) = -1/z$, $v(z) = (z - 1)/z$, $\tau(z) = z + 1$), *and admits the presentation* $\{\sigma, v | \sigma^2 = v^3 = 1\}$.
The quotient group $\Gamma(1)/\Gamma(2)$ *is isomorphic to the (non-Abelian) dihedral group of order 6. Moreover, the convex triangle* T *is the Dirichlet polygon* $\mathcal{P}\big(\Gamma(1), 2\sqrt{-1}\,\big)$.

Proof The above (from the beginning of this section, i.e., Section 4.4.2) shows that $\overset{o}{T}$ is a fundamental domain for both $\Gamma(1)$ and $\Gamma_1 \subset \Gamma(1)$, which must therefore be equal. The three sides of T are immediately obtained as $T \cap \tau(T), T \cap \tau^{-1}(T), T \cap \sigma(T)$, thereby completing the proof that T is a fundamental polygon for $\Gamma(1)$.

As to the claim about the presentation, observe that any reduced word w written with $\{\sigma, v\}$ can be written as $w = \sigma^i v^j \gamma$, with $i \in \{0, 1\}, j \in \{0, 1, 2\}$ and $\gamma \in \Gamma(2)$. Note also that we have $w = \mathbf{1}$ if and only if $w = \gamma = \mathbf{1}$, and hence, by Theorem 4.4.1.10, if and only if the word w is trivial. This shows that $\Gamma(1)$ admits the presentation $\{\sigma, v | \sigma^2 = v^3 = \mathbf{1}\}$.

Finally, let us prove the last sentence of the statement. We denote by $\mathcal{T}$ the subgroup (isomorphic to $\mathbb{Z}$) of $\Gamma(1)$ generated by the translation $\tau = (z \mapsto z + 1)$, and by $\bar{\sigma}$ the subgroup (isomorphic to $\mathbb{Z}/2\mathbb{Z}$) of $\Gamma(1)$ generated by σ. By the definition of Dirichlet polygons, we have at once

$$\mathcal{P}\big(\Gamma(1), 2\sqrt{-1}\,\big) \subset \mathcal{P}\big(\mathcal{T}, 2\sqrt{-1}\,\big) \cap \mathcal{P}\big(\bar{\sigma}, 2\sqrt{-1}\,\big).$$

Recalling eqn (2.1) for the hyperbolic distance

$$\text{ch}\Big[\text{dist}\,\big(x + \sqrt{-1}\,y, n + \nu\sqrt{-1}\,\big)\Big] = \big[(x - n)^2 + y^2 + \nu^2\big]/(2\nu y),$$

we immediately obtain $\mathcal{P}\big(\mathcal{T}, 2\sqrt{-1}\,\big) = \big\{|\Re(z)| \le \tfrac{1}{2}\big\}$ and $\mathcal{P}\big(\overline{\sigma}, 2\sqrt{-1}\,\big) = \{|z| \ge 1\}$, whence $\mathcal{P}\big(\Gamma(1), 2\sqrt{-1}\,\big) \subset T$. Now, by Proposition 4.2.2, $\mathcal{P}\big(\Gamma(1), 2\sqrt{-1}\,\big)$ is a fundamental polygon of $\Gamma(1)$, as is T by Theorem 4.4.2.4. They must therefore be equal. $\diamond$

Note that the vertices $e^{\sqrt{-1}\,\pi/3}, e^{\sqrt{-1}\,2\pi/3}$ of T are identified (in the orbit space $\Gamma(1)\backslash \mathbb{R} \times \mathbb{R}_+^*$) by the translation $\tau = v\sigma \in \Gamma(1)$, and are called **elliptic points**: they have non-trivial finite stabilizers (in $\Gamma(1)$), namely $\{1, v, v^2\}$ and $\{1, \sigma v\sigma, \sigma v^2 \sigma\}$, respectively. $\sqrt{-1}$ is another elliptic point, having stabilizer $\{1, \sigma\}$. The ideal vertex ∞ is parabolic, having as a stabilizer the infinite cyclic subgroup generated by the horizontal translation τ. Any other point of T has a trivial stabilizer in $\Gamma(1)$.

Recall from Corollary 4.4.1.6 that the image of the parabolic vertex ∞ in the orbit space $\Gamma(1)\backslash \mathbb{R} \times \mathbb{R}_+^*$ is called a cusp. This name expresses the visual idea of a boundary point at infinity, in the vicinity of which $\Gamma(1)\backslash \mathbb{R} \times \mathbb{R}_+^*$ becomes infinitely thin. Thus, $\Gamma(1)\backslash \mathbb{R} \times \mathbb{R}_+^*$ is a simply connected non-compact surface that has a unique cusp and finite volume, equal to $\pi/3$, which is the area of the triangle T (calculated with respect to the hyperbolic volume measure $y^{-2}\,dx\,dy$ of $\mathbb{R} \times \mathbb{R}_+^*$, recall Proposition 3.6.2.2).

The two elliptic points $e^{\sqrt{-1}\,\pi/3}$ and $\sqrt{-1}$ of T correspond to two singularities of the surface $\Gamma(1)\backslash \mathbb{R} \times \mathbb{R}_+^*$, which is not a Riemann surface but an **orbifold**. In contrast, $\Gamma(2)\backslash \mathbb{R} \times \mathbb{R}_+^*$ is a Riemann surface, having three cusps and surface area 2π, as is clear from Corollary 4.4.1.6 and Theorem 4.4.1.10.

4.4.3 Plane tessellation by means of two boosts, and $D\Gamma(1)$ We now consider two boosts ψ_1, ψ_2 of the hyperbolic plane $\mathbb{H}^2$, with four distinct fixed points p_1, p_1' and p_2, p_2', respectively, p_j denoting the unstable (i.e., repulsive) fixed point and p_j' denoting the stable (i.e., attractive) fixed point. We shall obtain an interesting example of a Fuchsian group which belongs neither to those which arise in Theorem 4.3.1 nor to those obtained in Section 4.4.1.

Lemma 4.4.3.1 *Suppose that the commutator $[\psi_1, \psi_2] := \psi_1 \psi_2 \psi_1^{-1} \psi_2^{-1}$ is a parabolic isometry with a fixed ideal point q, and consider the ideal quadrangle*

$$P_0' := \big\{q, \psi_2^{-1}(q), \psi_1^{-1}\psi_2^{-1}(q), \psi_1^{-1}(q)\big\}.$$

Then P_0' is harmonic if and only if ψ_1, ψ_2 are conjugate by means of some parabolic isometry. If this holds, then there exists an involutive isometry σ such that $\psi_j^{-1} = \sigma \psi_j \sigma$ and $\sigma(p_j) = p_j'$ for $j = 1, 2$, and the two quadrangles $\big\{\psi_1^{-1}(q), p_1, \psi_2^{-1}(q), p_2'\big\}$ and $\big\{\psi_1^{-1}(q), p_2, \psi_2^{-1}(q), p_1'\big\}$ are harmonic.

Proof Suppose that P_0' is a harmonic quadrangle (and, in particular, that its four vertices are pairwise distinct). Take the Poincaré half-plane model (recall Section 2.3, page 34) $\mathbb{R} \times \mathbb{R}_+^*$, with $\psi_1^{-1}(q) \equiv 0$, $\psi_2^{-1}(q) \equiv \infty$ and $q \equiv 1$. Hence $\psi_1(0) \equiv 1 \equiv \psi_2(\infty)$. Then the assumption ($P_0'$ harmonic and $[\psi_1, \psi_2](q) = q$) forces $-1 \equiv \psi_1^{-1}\psi_2^{-1}(q) = \psi_2^{-1}\psi_1^{-1}(q)$, and then $\psi_1(-1) \equiv \infty$, $\psi_2(-1) \equiv 0$. The isometry of Proposition 2.2.3.1(v), exchanging the vertices of P_0' is the involution $\sigma = [z \mapsto -1/z]$. The above constraints on ψ_1, ψ_2 force

$$\psi_1(z) = \frac{2z+1}{z+1}, \quad \psi_2(z) = \frac{z+1}{z+2},$$

whence

$$p_1 = \frac{1 - \sqrt{5}}{2}, \; p_1' = \frac{1 + \sqrt{5}}{2}, \; p_2 = \frac{-1 - \sqrt{5}}{2}, \; p_2' = \frac{-1 + \sqrt{5}}{2}.$$

Then the parabolic isometry $h := [z \mapsto z - 1]$ satisfies $\psi_2 = h\psi_1 h^{-1}$, and we have $\psi_j^{-1} = \sigma\psi_j\sigma$. The quadrangles $\{0, p_1, \infty, p_2'\}$ and $\{0, p_2, \infty, p_1'\}$ are clearly harmonic. Moreover, we have $\sigma(p_j) = p_j'$.

Conversely, suppose that there exists a parabolic isometry h satisfying $\psi_2 h = h\psi_1$. Take the Poincaré half-plane model $\mathbb{R} \times \mathbb{R}_+^*$ such that the fixed point of h is ∞, so that $h(z) = z - c$ for some $c \in \mathbb{R}^*$, and $p_1 = (1 - \sqrt{5})/2$, $p_1' = (1 + \sqrt{5})/2$. Note that this is clearly possible, since h can fix neither p_1 nor p_1' (this would imply $\{p_1, p_1'\} \cap \{p_2, p_2'\} \neq \emptyset$).

We then have

$$\psi_1(z) = \frac{(1 + \alpha)z + 1}{z + \alpha},$$

with $\alpha \in \{1, -2\}$. Now $\alpha = -2$ is excluded, since p_1 would be attractive, for example since

$$-\frac{\frac{(1+\sqrt{-5})}{2} + 1}{\frac{(1+\sqrt{-5})}{2} - 2} = \frac{-4 + \sqrt{-5}}{7}.$$

Hence

$$\psi_1(z) = \frac{2z + 1}{z + 1}.$$

Then

$$\psi_2(z) = \psi_1(z + c) - c = \frac{(2 - c)z + c - c^2 + 1}{z + c + 1}$$

and

$$[\psi_1, \psi_2](z) = \frac{(2c^3 + c^2 - 11c + 1)z + 2c^4 - 3c^3 - 11c^2 + 18c}{(c^3 - 7c)z + c^4 - 2c^3 - 6c^2 + 11c + 1}.$$

Since $[\psi_1, \psi_2]$ is parabolic, with a fixed point q, say, we have

$$\pm 2 = (2c^3 + c^2 - 11c + 1) + (c^4 - 2c^3 - 6c^2 + 11c + 1)$$
$$= c^4 - 5c^2 + 2, \; \text{i.e., } c \in \{0, \pm\sqrt{5}, \pm 2, \pm 1\},$$

and

$$q = q(c) \equiv \frac{(2c^3 + c^2 - 11c + 1) - (c^4 - 2c^3 - 6c^2 + 11c + 1)}{2(c^3 - 7c)} = \frac{c^3 - 4c^2 - 7c + 22}{2(7 - c^2)}$$
$$= \frac{2c^2 - 11}{c^2 - 7} - \frac{c}{2}.$$

Now $c \neq 0$, and (for $\varepsilon = \pm 1$)

$$q(\varepsilon\sqrt{5}) = \frac{1 - \varepsilon\sqrt{5}}{2} \in \{p_1, p_1'\}, \ q(2\varepsilon) = 1 - \varepsilon, \ q(\varepsilon) = \frac{3}{2} - \frac{\varepsilon}{2}.$$

The first case is excluded, since then the pair of fixed points of ψ_2 would be

$$\left\{ \frac{1 - 2\varepsilon\sqrt{5} \pm \sqrt{5}}{2} \right\},$$

which intersects $\{p_1, p_1'\}$. In the second case $(c = 2\varepsilon)$, we obtain

$$\psi_2(z) = \frac{2(1 - \varepsilon)z + 2\varepsilon - 3}{z + 2\varepsilon + 1}$$

and then

$$P_0' \equiv \left\{ 1 - \varepsilon, \frac{2 - \varepsilon}{1 - \varepsilon}, -\varepsilon, \frac{-\varepsilon}{\varepsilon + 1} \right\},$$

which is indeed harmonic. In the third case $(c = \varepsilon)$, we obtain

$$\psi_2(z) = \frac{2(1 - \varepsilon)z + 2\varepsilon - 3}{z + 2\varepsilon + 1}$$

and then

$$P_0' \equiv \left\{ \frac{3 - \varepsilon}{2}, \frac{2}{1 - \varepsilon}, \frac{1 + \varepsilon}{-2}, \frac{1 - \varepsilon}{1 + \varepsilon} \right\},$$

which is also harmonic. $\diamond$

Now we have the analogue of Proposition 4.4.1.5.

Proposition 4.4.3.2 *Consider, as in Lemma 4.4.3.1, two boosts ψ_1, ψ_2 having a parabolic commutator $[\psi_1, \psi_2]$. Then the group Γ_2' generated by $\{\psi_1, \psi_2\}$ is free (with rank 2) and Fuchsian.*

Proof Consider, as in Lemma 4.4.3.1, the fixed point q of $[\psi_1, \psi_2]$, and the ideal quadrangle

$$P_0' := \{q, \psi_2^{-1}(q), \psi_1^{-1}\psi_2^{-1}(q), \psi_2\psi_1^{-1}\psi_2^{-1}(q) = \psi_1^{-1}(q)\},$$

which we can see either as a quadruple of ideal vertices or as a convex polygon having these vertices. Observe now that any boost preserves the clockwise ordering of points on $\partial\mathbb{H}^2$ (seen as a circle), so that the four isometries $\psi_1, \psi_1^{-1}, \psi_2, \psi_2^{-1}$ map P_0' into closed half-planes of $\mathbb{H}^2$ not containing P_0 and delimited respectively by

$$\left[q, \psi_2^{-1}(q)\right], \left[\psi_1^{-1}(q), \psi_1^{-1}\psi_2^{-1}(q)\right], \left[q, \psi_1^{-1}(q)\right], \left[\psi_2^{-1}(q), \psi_1^{-1}\psi_2^{-1}(q)\right].$$

Hence, the five convex polygons $P_0', \psi_1(P_0'), \psi_1^{-1}(P_0'), \psi_2(P_0'), \psi_2^{-1}(P_0')$ have pairwise disjoint interiors, and their union is the convex hull of $P_0' \cup \psi_1(P_0') \cup \psi_1^{-1}(P_0') \cup \psi_2(P_0') \cup \psi_2^{-1}(P_0')$.

Then (as in the proof of Proposition 4.4.1.5) the same procedure is applied to each polygon $\psi_j^{\varepsilon_j}(P_0')$ by means of the boosts $\psi_j^{\varepsilon_j}\psi_k^{\varepsilon_k}\psi_j^{-\varepsilon_j}$, having parabolic commutators $[\psi_j^{\varepsilon_j}\psi_1\psi_j^{-\varepsilon_j},\psi_j^{\varepsilon_j}\psi_2\psi_j^{-\varepsilon_j}]=\psi_j^{\varepsilon_j}[\psi_1,\psi_2]\psi_j^{-\varepsilon_j}$, for $1\leq j,k\leq 2$ and $\varepsilon_j,\varepsilon_k=\pm 1$ (but $\varepsilon_k\neq -\varepsilon_j$ if $k=j$, so as not to return to P_0'). We thus obtain 17 isometric convex polygons $P_0',\psi_j^{\varepsilon_j}(P_0'),\psi_j^{\varepsilon_j}\psi_k^{\varepsilon_k}(P_0')$, which have pairwise disjoint interiors. Continuing in this way by induction, at the m-th step we obtain $(2\times 3^m-1)$ isometric polygons

$$\psi_{j_1}^{\varepsilon_1}\psi_{j_2}^{\varepsilon_2}\ldots\psi_{j_k}^{\varepsilon_k}(P_0'),\quad 0\leq k\leq m,1\leq j_i\leq 2,\varepsilon_k=\pm 1,$$

indexed by words of length $\leq m$ on the alphabet $\{\psi_1^{\pm 1},\psi_2^{\pm 1}\}$, which have pairwise disjoint interiors.

We observe that, as a consequence, for z in the interior of P_0 and for any non-trivial reduced word $\psi_{j_1}^{\varepsilon_1}\psi_{j_2}^{\varepsilon_2}\ldots\psi_{j_k}^{\varepsilon_k}$, the point $\psi_{j_1}^{\varepsilon_1}\psi_{j_2}^{\varepsilon_2}\ldots\psi_{j_k}^{\varepsilon_k}(z)$ never belongs to P_0'. This immediately implies the statement. $\diamond$

Note that since Γ_2' acts transitively on P_0', all ideal vertices of Γ_2' are equivalent and parabolic (recall from Corollary 4.4.1.6 that the **parabolic points** of a Kleinian group Γ are the vertices fixed by some parabolic element of Γ), so that the orbit space $\Gamma_2'\backslash\mathbb{H}^2$ has a unique cusp.

Proposition 4.4.3.3 below has to be compared with Theorem 4.4.1.10: two different Kleinian groups (which are the two rank-2 free Fuchsian subgroups $\Gamma(2)$ and $D\Gamma(1)$ of $PSL(2,\mathbb{Z})$ here) can admit the same fundamental polyhedron (which is a harmonic quadrangle here), even though they give rise to very different, non-isometric orbit spaces: $\Gamma(2)\backslash\mathbb{H}^2$ has three cusps (and genus 0), while $D\Gamma(1)\backslash\mathbb{H}^2$ has a unique cusp (and genus 1).

In fact, the two parabolic elements $\varphi_1=[z\mapsto z+2],\varphi_2=[z\mapsto z/(1-2z)]$ (of Proposition 4.4.1.10) that generate $\Gamma(2)$, on the one hand, and the two boosts ψ_1,ψ_2 (of Proposition 4.4.3.3) that generate $D\Gamma(1)$, on the other hand, realize two different pairings of the sides of the fundamental harmonic quadrangle P_0 (see Figure 4.6 and [Da]). Furthermore, $\Gamma(2)$ and $D\Gamma(1)$ differ algebraically by the fact that $\Gamma(1)/D\Gamma(1)$ is cyclic, whereas $\Gamma(1)/\Gamma(2)$ is not Abelian.

Proposition 4.4.3.3 *By Considering the half-plane* $\mathbb{R}\times\mathbb{R}_+^*$, *the harmonic quadrangle* $P_0'=\{0,-1,\infty,1\}=P_0$, *and*

$$\psi_1=\left[z\mapsto\frac{2z+1}{z+1}\right],\psi_2=\left[z\mapsto\frac{z+1}{z+2}\right]\in\Gamma(1),$$

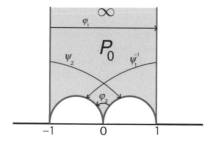

Fig. 4.6 The fundamental domain P_0 for $\Gamma(2)$ and $D\Gamma(1)$, with two distinct pairings

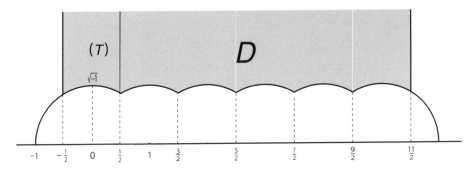

Fig. 4.7 The fundamental octagon D for $D\Gamma(1)$

we obtain a realization of the group Γ'_2 of Proposition 4.4.3.2, which the group $D\Gamma(1) = [\Gamma(1), \Gamma(1)]$, generated by the commutators of $\Gamma(1) = \mathrm{PSL}(2, \mathbb{Z})$. Moreover, $\Gamma(1)/D\Gamma(1)$ is a cyclic group of order 6.

Proof Recall from Theorem 4.4.2.4 that $\sigma = [z \mapsto -1/z]$ and $\tau = [z \mapsto z+1] = v\sigma$ generate $\Gamma(1) = \mathrm{PSL}(2, \mathbb{Z})$. Then $\psi_1 = [\tau, \sigma] = \tau\sigma\tau^{-1}\sigma^{-1} \in D\Gamma(1)$, and $\psi_2 = [\sigma, \tau^{-1}] \in D\Gamma(1)$. Hence $\Gamma'_2 \subset D\Gamma(1)$. On the other hand, since $\Gamma(1)/D\Gamma(1)$ is the Abelianized group of $\Gamma(1)$, we know from Theorem 4.4.2.4 that it admits the presentation $\{\dot\sigma, \dot v | \dot\sigma^2 = \dot v^3 = 1 = [\dot\sigma, \dot v]\}$, and hence that it is cyclic of order 6, generated by $\dot v\dot\sigma = \dot\tau$. Hence, we can obtain a fundamental polygon for $D\Gamma(1)$ by juxtaposing six isometric images of T, for example $\bigcup_{0 \le k \le 5} \tau^k T$. Finally, we find the index $[D\Gamma(1) : \Gamma'_2] = [\Gamma(1) : \Gamma'_2]/[\Gamma(1) : D\Gamma_1] = 6/6 = 1$, so that $D\Gamma(1) = \Gamma'_2$. ◇

$D\Gamma(1)$ is not a principal congruence group, but it is, however, a **congruence group**, i.e., a (modular) group containing a principal congruence group. In fact, we have $\Gamma(6) \subset D\Gamma(1)$, with $D\Gamma(1)/\Gamma(6)$ isomorphic to $\mathbb{Z}/6\mathbb{Z} \times \mathbb{Z}/2\mathbb{Z}$. More precisely, we have

$$D\Gamma(1) = \bigcup_{-3 \le k \le 2} \left[\psi_1^k \Gamma(6) \bigcup \psi_1^{k+1}\psi_2 \Gamma(6) \right]$$

(see [F3]).

Note that the subgroup $[D\Gamma(1), D\Gamma(1)]$ is not modular, having infinite index in $\Gamma(1)$, since $D\Gamma(1)/[D\Gamma(1), D\Gamma(1)]$ is isomorphic to $\mathbb{Z}^2$. Hence, the normal subgroup $[\Gamma, \Gamma]$ of a modular group Γ can be a free group, but it is generally not modular.

The proof of Proposition 4.4.3.3 demonstrates another fundamental polygon for $D\Gamma(1)$, namely the fundamental octagon $D := \bigcup_{0 \le k \le 5} \tau^k T$ shown in Figure 4.7.

4.4.4 Plane tessellation yielding $\Gamma(3)$

Here is another application of Theorem 4.3.1, which we give without its (somewhat lengthy) proof, left as an exercise.

Theorem 4.4.4.1 *For $n = 3$ and the hexagon $P_0 = \{\infty, -1, -\frac{2}{3}, 0, \frac{2}{3}, 1\}$ (in the Poincaré upper-half-plane model $\mathbb{R} \times \mathbb{R}^*_+$), Theorem 4.3.1 yields the principal congruence group $\Gamma(3)$, freely generated by the parabolic isometries*

$$\varphi_{-1} = \left[z \mapsto \frac{-2z-3}{3z+4} \right], \quad \varphi_0 = \left[z \mapsto \frac{z}{3z+1} \right], \quad \varphi_1 = \left[z \mapsto \frac{4z-3}{3z-2} \right],$$

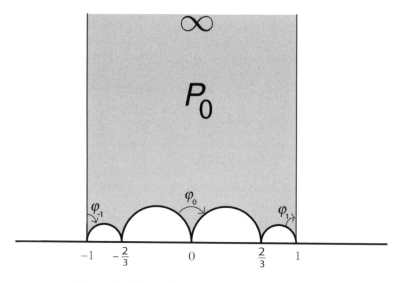

Fig. 4.8 The fundamental hexagon P_0 for $\Gamma(3)$

which admits P_0 as fundamental hexagon (see Figure 4.8). The orbit space $\Gamma(3)\backslash\mathbb{R} \times \mathbb{R}_+^*$ is a Riemann surface with four cusps, surface area 4π and genus 0. Moreover, the quotient group $\Gamma(1)/\Gamma(3)$ has order 12, and is isomorphic to the alternating group $\mathcal{A}_4$ (of even permutations on four points, also isomorphic to the group of Euclidean rotations that preserve a regular tetrahedron in $\mathbb{R}^3$).

Recall from Corollary 4.4.1.6 that a cusp is a class of equivalent (under the action of the Kleinian group considered) parabolic vertices. For example, the orbit space $\Gamma_n\backslash\mathbb{H}^2$ of Theorem 4.3.1 has $(n+1)$ cusps.

Remark 4.4.4.2 Theorem 4.3.1 cannot give rise to the the principal congruence group $\Gamma(n)$ when $n \geq 4$ (though it is still a free group), as it did in Theorems 4.4.1.10 and 4.4.4.1. For example, $\Gamma(4)$ has rank 5 (i.e., it is freely generated by five elements) and six inequivalent parabolic points, and $\Gamma(5)$ has rank 11 and 12 inequivalent parabolic points. More generally, for any $n \geq 3$, $\Gamma(n)$ is free with rank

$$\frac{n^2}{4}(\frac{n}{6}+1)\prod_{p|n}(1-p^{-2}),$$

and has

$$\frac{n^2}{2}\prod_{p|n}(1-p^{-2})$$

inequivalent parabolic points (and genus

$$1 + \frac{n^2}{4}\left(\frac{n}{6} - 1\right)\prod_{p|n}(1 - p^{-2}));$$

see [Le], Section XI.3D and [Mi], Section 4.2.

In the case of a generic modular group Γ, the **genus** (which is the maximal number of non-intersecting closed curves which can be drawn on the surface $\Gamma\backslash\mathbb{H}^2$ without disconnecting it) and the volume of $\Gamma\backslash\mathbb{H}^2$ can be expressed in terms of three parameters, namely the numbers $\nu_\infty(\Gamma)$, $\nu_2(\Gamma)$ and $\nu_3(\Gamma)$, of Γ-inequivalent cusp points of Γ and of Γ-inequivalent elliptic points of Γ of order 2 and 3, respectively. Of course, $\nu_\infty(\Gamma)$ is also the number of cusps of $\Gamma\backslash\mathbb{H}^2$.

The genus of $\Gamma\backslash\mathbb{H}^2$ is given by the formula ([Mi], Theorem 4.2.11)

$$\mathbf{g}(\Gamma) = 1 + \tfrac{1}{12}[\Gamma(1) : \Gamma] - \tfrac{1}{4}\nu_2(\Gamma) - \tfrac{1}{3}\nu_3(\Gamma) - \tfrac{1}{2}\nu_\infty(\Gamma). \tag{4.2}$$

The volume of $\Gamma\backslash\mathbb{H}^2$ is given by the formula ([Mi], Theorem 2.4.3)

$$V(\Gamma\backslash\mathbb{H}^2) = 2\pi \times \left(2\mathbf{g}(\Gamma) - 2 + \nu_\infty(\Gamma) + \tfrac{1}{2}\nu_2(\Gamma) + \tfrac{2}{3}\nu_3(\Gamma)\right). \tag{4.3}$$

In the particular case of principal congruence groups $\Gamma(N)$, very explicit formulae for $[\Gamma(1) : \Gamma(N)]$ and for $\nu_j(\Gamma(N))$ are known (see Section 4.2 of [Mi], and the above Remark 4.4.4.2).

4.5   Notes and comments

See [Rac], Section 7.1 for the example of a regular ideal dodecahedron as the fundamental domain for the Kleinian group generated by reflections in its sides. Examples in higher dimensions $d \geq 3$ are often significantly more involved. Maybe the simplest is $\mathrm{PSO}(1, d, \mathbb{Z})$ (made up of the Lorentzian matrices with integer entries), which is cofinite with a number of cusps at least of order $d^{d^2/4}$, and for which a nice algorithm to construct a fundamental domain is known by G. Collinet, using lightlike vectors that have integer coordinates and small Euclidean lengths (to be published).

In three dimensions, by means of Proposition 1.1.5.2, it is equivalent to consider subgroups of $\mathrm{PSL}_2(\mathbb{C})$. The action of

$$\gamma := \pm\begin{pmatrix} a & b \\ c & d \end{pmatrix} \in \mathrm{PSL}_2(\mathbb{C})$$

on $\mathbb{H}^3 \equiv \mathbb{C} \times \mathbb{R}_+^*$ is as follows:

$$\gamma(z, \zeta) := \left(\frac{\zeta}{|cz - d|^2 + \zeta^2\,|c|^2}, \frac{(\bar{d} - \bar{c}z)(az - b) - \zeta^2\bar{c}a}{|cz - d|^2 + \zeta^2\,|c|^2}\right).$$

Swan [Sw] studied in detail three-dimensional analogues of modular groups, associated with quadratic imaginary number fields $\mathbb{Q}[\sqrt{-m}]$, in particular the following one.

Proposition 4.5.1 *([Sw]) The group $\mathrm{PSL}_2(\mathbb{Z} + \sqrt{-2}\mathbb{Z})$ is Kleinian, is generated by*

$$\tau := \pm\begin{pmatrix} 1 & 1 \\ 0 & 1 \end{pmatrix}, \; \nu := \pm\begin{pmatrix} 1 & \sqrt{-2} \\ 0 & 1 \end{pmatrix}, \; \alpha := \pm\begin{pmatrix} 0 & -1 \\ 1 & 0 \end{pmatrix},$$

with a complete system of relations $\tau\nu = \nu\tau, \alpha^2 = (\tau\alpha)^3 = (\alpha\nu^{-1}\alpha\nu)^2 = \pm\mathbf{1}$, *and admits the fundamental polyhedron*

$$\mathcal{P} := \left\{(z,\zeta) \in \mathbb{C} \times \mathbb{R}_+^* \,\middle|\, |\Re(z)| \leq \tfrac{1}{2}, |\Im(z)| \leq \frac{1}{\sqrt{2}}, \min\left\{|z|, |z \pm 1|, |z \pm \sqrt{-2}|\right\}^2 + \zeta^2 \geq 1\right\}.$$

Note that this example (like that of the dodecahedron) is not cocompact but cofinite, since $\mathcal{P} \subset \{\zeta \geq \tfrac{1}{2}\}$.

Remark 4.5.2 Cocompact examples of Fuchsian groups and domains (including how to obtain them by the Poincaré polyhedron theorem) can be seen in [Rac], Section 11.2. Examples of Fuchsian groups with fundamental domains that have an infinite volume (beyond the example of Proposition 4.4.1.3 and Remark 4.4.1.9), namely the so-called Schottky groups, can be found in [Da].

   This chapter was included for self-containedness. It provides several examples in which the theory of Chapters 5 and 8 is applied, but it was not our aim to provide a systematic introduction to Riemann surfaces. Historical presentations are given in [Mr] and [SG].

   Classical references such as [FK] include a lot of examples of fundamental domains; see also [Kr] , [Rag], [Mas], [Rac], [MT] and [Da]. Modular groups are specifically studied in [Le], [Sc] and [Mi]. Different notions of geometrical finiteness are compared thoroughly in [B1] and [B2].

Chapter Five
Measures and flows on $\Gamma\backslash\mathbb{F}^d$

We consider here a fixed Kleinian group Γ and the associated relative measures and flows (induced by the measures of Section 3.6 and the flows of Section 2.3) on the orbit or quotient space $\Gamma\backslash\mathbb{F}^d$, and we investigate their basic properties, when Γ is geometrically finite and cofinite.

We start with general notions about left quotients, Γ-invariant sets and covolumes (Proposition 5.1.1). The first main result of this chapter is a mixing theorem for the action of the geodesic and horocyclic flows on square-integrable Γ-invariant functions, namely Theorem 5.3.1. This proceeds essentially from the commutation relations that produce the instability of the flow illustrated in Proposition 2.3.9.

The second main result of this chapter is a Poincaré inequality (i.e., the existence of a spectral gap) for the Laplacian acting on Γ-invariant functions, namely Theorem 5.4.2.5. We prove it by decomposing a fundamental domain into a compact core and the neighbourhoods of cusps (called solid cusps), which overlap. We then show that a Poincaré inequality holds in these parts, and we establish a general proposition stating that it is conserved by taking a union of overlapping domains.

5.1 Measures of Γ-invariant sets

Recall that any $\gamma \in \mathrm{PSO}(1, d)$ acts also on the frame bundle $\mathbb{F}^d$ and that plainly, if $D \subset \mathbb{H}^d$ is a fundamental domain for Γ, then the cylindrical domain $\pi_0^{-1}(D) \subset \mathbb{F}^d$ is fundamental for the (left) action of Γ on $\mathbb{F}^d$.

From now on, we shall consider mainly Γ-invariant functions. We first specify the relative (volume and Liouville) measures against which we shall integrate them. We also emphasize the invariance of these relative measures under the flows, which follows from the fact that the Kleinian group Γ acts on the left-hand side, while the flows act on the right-hand side, so that these two actions commute.

Proposition 5.1.1 *Take some Kleinian group Γ. Then the following statements hold.*

(i) *The volume of a fundamental domain D depends only on the Kleinian group Γ, it is called the **covolume** of Γ, and denoted by $\mathrm{covol}(\Gamma)$.*

(ii) *The volume measure dp of $\mathbb{H}^d$, restricted to a fundamental domain D, induces a measure $d^\Gamma p$ on left Γ-invariant Borel sets of $\mathbb{H}^d$, which depends only on Γ.*

(iii) *The Liouville measure $\tilde{\lambda}$ of $\mathbb{F}^d$, restricted to a cylindrical fundamental domain $\pi_0^{-1}(D)$, induces a relative Liouville measure $\tilde{\lambda}^\Gamma$ on left Γ-invariant Borel sets of $\mathbb{F}^d$, which depends only on Γ.*

(iv) *The Liouville measure $\tilde{\lambda}^\Gamma$ is invariant under the (right) action of the geodesic and horocyclic flows and, more generally, under the right action of any $g \in \mathrm{PSO}(1,d)$.*

Notation We denote henceforth the quotient of $\mathbb{F}^d$ under the left action of the Kleinian group Γ by $\Gamma\backslash\mathbb{F}^d$ and, similarly, denote the quotient of $\mathbb{H}^d$ by $\Gamma\backslash\mathbb{H}^d$. The Borel sets of $\Gamma\backslash\mathbb{F}^d$ and $\Gamma\backslash\mathbb{H}^d$ are then identified with the Γ-invariant Borel sets of $\mathbb{F}^d$ and $\mathbb{H}^d$, respectively, and the measures $\tilde{\lambda}^\Gamma$ and $d^\Gamma p$ are identified with measures on $\Gamma\backslash\mathbb{F}^d$ and $\Gamma\backslash\mathbb{H}^d$, respectively.

Proof Let D, D' be fundamental domains for Γ. Observe that $\mathbb{H}^d \setminus \bigcup_{g\in\Gamma} gD \subset \bigcup_{g\in\Gamma} g\,\partial D$, so that the volume of $\mathbb{H}^d \setminus \bigcup_{g\in\Gamma} gD$ must be 0. Hence, for any non-negative measurable (left) Γ-invariant function h on $\mathbb{H}^d$, we have

$$\int_D h(p)\,dp = \int_{D\cap\bigcup_{g\in\Gamma} gD'} h(p)\,dp = \sum_{g\in\Gamma} \int_{D\cap gD'} h(p)\,dp$$

$$= \sum_{g\in\Gamma} \int_{g^{-1}D\cap D'} h(p)\,dp = \int_{D'} h(p)\,dp.$$

Taking $h \equiv 1$ proves that the volumes of D and D' are equal, proving (i). And this shows more generally that

$$\int h(p)\,d^\Gamma p := \int_D h(p)\,dp$$

is indeed uniquely defined, proving (ii).

(iii) We take any (left) Γ-invariant test function f on $\mathbb{F}^d$, and two fundamental domains D, D' for Γ. Using the invariance of $\tilde{\lambda}$ (recall Section 3.6.2) and of f and proceeding as above, we have

$$\int f \times 1_D \circ \pi_0\,d\tilde{\lambda} = \sum_{g\in\Gamma} \int f\, 1_{D\cap gD'} \circ \pi_0\,d\tilde{\lambda}$$

$$= \sum_{g\in\Gamma} \int f\, 1_{g^{-1}D\cap D'} \circ \pi_0\,d\tilde{\lambda} = \int f \times 1_{D'} \circ \pi_0\,d\tilde{\lambda}.$$

(iv) We now prove the invariance under the flows and, more generally, that the right action of any $g \in \mathrm{PSO}(1,d)$ preserves the relative Liouville measure $\tilde{\lambda}^\Gamma$. We take a Γ-invariant test function f on $\mathbb{F}^d$, $g \in \mathrm{PSO}(1,d)$, and for any $\beta \in \mathbb{F}^d$, we set $f^g(\beta) := f(\beta g)$. Then, for any cylindrical fundamental domain $\tilde{D} := \pi_0^{-1}(D)$, $\tilde{D}g = \pi_0^{-1}(Dg)$ is another cylindrical fundamental domain for the action of Γ on $\mathbb{F}^d$, so that the right invariance of $\tilde{\lambda}$ implies

$$\int f \, d\tilde{\lambda}^\Gamma = \int_{\tilde{D}} f \, d\tilde{\lambda} = \int_{\tilde{D}g} f^g \, d\tilde{\lambda} = \int f^g \, d\tilde{\lambda}^\Gamma.$$

◇

A Kleinian group Γ is said to be **cocompact** if it has a relatively compact fundamental domain. A Kleinian group Γ is said to be **cofinite** if it has a finite covolume.

5.2 Ergodicity

Let us first introduce the important mixing and ergodic properties.

Definition 5.2.1 *Consider any probability space $(E, \mathcal{E}, \mu)$ and a one-parameter group (g_t) of measure-preserving maps such that for any $f \in L^2(\mathcal{E}, \mu)$, $t \mapsto \langle f \circ g_t, f \rangle_{L^2}$ is continuous. The action of (g_t) is said to be **mixing** if, for any $f, \varphi \in L^2$,*

$$\lim_{t \to \pm\infty} \langle f \circ g_t, \varphi \rangle_{L^2} = \langle f, 1 \rangle_{L^2} \langle 1, \varphi \rangle_{L^2}.$$

A function $f \in L^2$ is said to be (g_t)-invariant if $f \circ g_t = f$ almost surely for any $t \in \mathbb{R}$.
   *The action of (g_t) is said to be **ergodic** if any (g_t)-invariant $f \in L^2(\mu)$ is almost surely constant.*

We have the following very easy implication.

Proposition 5.2.2 *If (g_t) is mixing, then it is ergodic.*

We recall the fundamental ergodic theorem, due to Birkhoff. This states that if ergodicity holds, then temporal means converge to spatial means.

Theorem 5.2.3 (ergodic theorem) *If (g_t) is ergodic and if $(t, \beta) \mapsto g_t(\beta)$ is measurable on $\mathbb{R} \times \Gamma\backslash\mathbb{F}^d$, then for any $f \in L^2$ we have*

$$\lim_{t \to \infty} \frac{1}{t} \int_0^t f \circ g_s \, ds = \int f \, d\mu, \quad \mu - \text{almost surely and in } L^2.$$

Proof (due to Riesz and Yosida-Kakutani; see [Ka]) Set $F := \int_0^1 f \circ g_s \, ds$.

(1) (due to Garsia) For any $n \in \mathbb{N}^*$, set

$$F_n := \sup_{k \in \{1, \dots, n\}} \int_0^k f \circ g_s \, ds.$$

As

$$\int_0^{k+1} f \circ g_s \, ds = \int_0^1 f \circ g_s \, ds + \left[\int_0^k f \circ g_s \, ds \right] \circ g_1,$$

we have $F_n \leq F_{n+1} \leq F + F_n^+ \circ g_1$, whence

$$\int_{\{F_n > 0\}} F \, d\mu \geq \int_{\{F_n > 0\}} F_n \, d\mu - \int F_n^+ \circ g_1 \, d\mu = 0.$$

Since $\|F\|_2 \leq \|f\|_2 < \infty$, by letting $n \to \infty$ we obtain by dominated convergence the so-called 'maximal ergodic lemma',

$$\int_{\left\{\sup\limits_{n \geq 1} 1/n \sum_{k=0}^{n-1} F \circ g_k > 0\right\}} F \, d\mu = \int_{\left\{\sup\limits_{k \in \mathbb{N}^*} \int_0^k f \circ g_s \, ds > 0\right\}} F \, d\mu \geq 0.$$

(2) For any rational numbers $a < b$, consider the following (g_t)-invariant measurable set:

$$E_{ab} := \left\{ \liminf_{n \to \infty} \frac{1}{n} \sum_{k=0}^{n-1} F \circ g_k < a < b < \limsup_{n \to \infty} \frac{1}{n} \sum_{k=0}^{n-1} F \circ g_k \right\}.$$

Applying the above maximal ergodic lemma, with μ replaced by its restriction to E_{ab}, to $(F - b)$ and $(a - F)$ yields

$$\int_{E_{ab}} (F - b) \, d\mu \geq 0 \quad \text{and} \quad \int_{E_{ab}} (a - F) \, d\mu \geq 0,$$

whence

$$\mu(E_{ab}) = \frac{-1}{b - a} \int_{E_{ab}} (F - b + a - F) \, d\mu \leq 0,$$

and then $\mu(E_{ab}) = 0$. This proves the μ-almost sure convergence of the sequence

$$\frac{1}{n} \int_0^n f \circ g_s \, ds = \frac{1}{n} \sum_{k=0}^{n-1} F \circ g_k.$$

Applying this convergence to f^+ instead of f and using (with $[t] := \max\{n \in \mathbb{N} \,|\, n \leq t\}$, for any $t > 0$)

$$\int_0^{[t]} f^+ \circ g_s \, ds \leq \int_0^t f^+ \circ g_s \, ds \leq \int_0^{[t]+1} f^+ \circ g_s \, ds,$$

we deduce at once the μ-almost sure existence of

$$\lim_{t \to \infty} \frac{1}{t} \int_0^t f^+ \circ g_s \, ds,$$

and similarly that of

$$\lim_{t \to \infty} \frac{1}{t} \int_0^t f^- \circ g_s \, ds,$$

and hence of

$$h := \lim_{t \to \infty} \frac{1}{t} \int_0^t f \circ g_s \, ds.$$

Clearly, the measurable function h is (g_t)-invariant and belongs to L^2 by Fatou's lemma, and hence is μ-almost surely constant by the ergodicity assumption.

Finally, for any μ-measurable set A and any $t, R > 0$, we have

$$\int_A \left| \frac{1}{t} \int_0^t f \circ g_s \, ds \right|^2 d\mu \le \frac{1}{t} \int_0^t \int_A |f \circ g_s|^2 \, d\mu \, ds$$

$$= \frac{1}{t} \int_0^t \int_{g_s(A)} |f|^2 \, d\mu \, ds \le R^2 \mu(A) + \int_{\{|f| > R\}} |f|^2 \, d\mu,$$

which (by taking a large enough R and then a small enough $\mu(A)$) shows the uniform integrability of the family

$$\left(\frac{1}{t} \int_0^t f \circ g_s \, ds \right)^2_{t > 0}.$$

The convergence to h in $L^2(\mu)$ follows, and then

$$h = \int h \, d\mu = \lim_{t \to \infty} \int \frac{1}{t} \int_0^t f \circ g_s \, ds \, d\mu = \lim_{t \to \infty} \frac{1}{t} \int_0^t \int f \, d\mu \, ds = \int f \, d\mu.$$

$\diamondsuit$

5.3 A mixing theorem

Consider a cofinite Kleinian group Γ, the normalized induced measure $\mu := \tilde{\lambda}^\Gamma / |\tilde{\lambda}^\Gamma| = \text{covol}(\Gamma)^{-1} \tilde{\lambda}^\Gamma$, the Hilbert space $L^2 \equiv L^2(\Gamma \backslash \mathbb{F}^d, \mu)$, with inner product $\langle f, \varphi \rangle_{L^2} := \int f \varphi \, d\mu$, and a one-parameter subgroup (g_t) of $\text{PSO}(1, d)$.

Theorem 5.3.1 *The geodesic flow (θ_t) and any non-trivial one-parameter subgroup (θ_{tu}^+) of the horocyclic flow are mixing, and hence ergodic.*

The proof of this theorem needs several steps. We follow mainly an elegant, elementary proof by Guivarc'h [Gu], which shows in fact a similar but stronger result: the above theorem is true not only for $\text{PSO}(1, d)$ but also for $\text{SL}(d + 1)$. Focusing on $\text{PSO}(1, d)$ allows us to simplify the proof (when $d > 2$).

Consider the unitary representation $\gamma \mapsto \left(f \mapsto \gamma f := f \circ \gamma\right)$, from $\mathrm{PSO}(1,d)$ into the group of unitary endomorphisms of

$$L_0^2 = L_0^2(\mu) = \{f \in L^2 \,|\, \langle f, 1 \rangle_{L^2} = 0\}.$$

Note that this representation has no non-zero $\mathrm{PSO}(1,d)$-invariant vector: since the action of $\mathrm{PSO}(1,d)$ is transitive, any $\mathrm{PSO}(1,d)$-invariant function $f \in L_0^2$ has to be constant, and then equal to zero.

We say that a sequence $(\gamma_n = \varrho_n \theta_{r_n} \varrho_n')$ in $\mathrm{PSO}(1,d)$ (using the Cartan decomposition, recall Theorem 1.7.1) 'goes to infinity' if and only if $r_n \to \infty$. It is clear from the examples following Theorem 1.7.1 that the one-parameter subgroups (θ_t) and (θ_{tu}^+) (for any non-zero $u \in \mathbb{R}^{d-1}$) go to infinity as $t \to \pm\infty$, so that the proof of the theorem follows immediately from the following proposition.

Proposition 5.3.2 *If $(\gamma_n) \subset \mathrm{PSO}(1,d)$ is a sequence that goes to infinity, then for any $f, \varphi \in L_0^2$ we have $\lim\limits_{n\to\infty} \langle \gamma_n f, \varphi \rangle_{L^2} = 0$.*

To establish Proposition 5.3.2, we shall use the following lemma.

Lemma 5.3.3 *Suppose that there exists no non-zero vector $f \in L_0^2$ fixed by all elements of $\mathbb{T}_{d-1} = \{\theta_u^+ \equiv (f \mapsto f \circ \theta_u^+) \,|\, u \in \mathbb{R}^{d-1}\}$. Then, for any $f, \varphi \in L_0^2$, we have $\lim\limits_{t\to\infty} \langle \theta_t f, \varphi \rangle_{L^2} = 0$.*

Proof For any $f, \psi \in L_0^2$ and $u \in \mathbb{R}^{d-1}$, using the commutation relation in eqn (1.16), we have

$$\left|\langle (\theta_u^+ \psi - \psi), \theta_t f \rangle_{L^2}\right| = \left|\langle (\theta_u^+ - \mathbf{1})\theta_t \theta_{-t} \psi, \theta_t f \rangle_{L^2}\right|$$

$$= \left|\langle \theta_{-t}\psi, \theta_{-t}(\theta_{-u}^+ - \mathbf{1})\theta_t f \rangle_{L^2}\right| \le \|\theta_{-t}\psi\|_{L^2} \times \|\theta_{-t}(\theta_{-u}^+ - \mathbf{1})\theta_t f\|_{L^2}$$

$$= \|\psi\|_{L^2} \times \|\theta_{-u\,e^{-t}}^+ f - f\|_{L^2} \longrightarrow 0.$$

The vector space V generated by $\{(\theta_u^+ \psi - \psi) | \psi \in L_0^2, u \in \mathbb{R}^{d-1}\}$ is dense in L_0^2. In fact, if $h \in V^\perp$, then for any $u \in \mathbb{R}^{d-1}$ we have $\langle \psi, \theta_{-u}^+ h \rangle_{L^2} = \langle \theta_u^+ \psi, h \rangle_{L^2} = \langle \psi, h \rangle_{L^2}$ for any $\psi \in L_0^2$, whence $\theta_{-u}^+ h = h$. By the hypothesis, this forces $h = 0$. The statement follows at once. $\diamond$

Lemma 5.3.4 *Let $(\gamma_n = \varrho_n \theta_{r_n} \tilde{\varrho}_n) \subset \mathrm{PSO}(1,d)$ be a sequence such that $\lim\limits_{n\to\infty} \langle \theta_{r_n} f, \varphi \rangle_{L^2} = 0$ for any $f, \varphi \in L_0^2$. Then $\lim\limits_{n\to\infty} \langle \gamma_n f, \varphi \rangle_{L^2} = 0$, for any $f, \varphi \in L_0^2$.*

Proof We take $f, \varphi \in L_0^2$. Up to extracting sub-sequences, we may suppose that $\varrho_n \to \varrho \in \mathrm{SO}(d)$, $\tilde{\varrho}_n \to \tilde{\varrho} \in \mathrm{SO}(d)$, and $\langle \gamma_n f, \varphi \rangle_{L^2} \to \ell \in \overline{\mathbb{R}}$. We then obtain

$$|\langle \gamma_n f, \varphi \rangle_{L^2}| \le |\langle \theta_{r_n} \tilde{\varrho}_n f, (\varrho_n^{-1} - \varrho^{-1})\varphi \rangle_{L^2}|$$

$$+ |\langle (\tilde{\varrho}_n - \tilde{\varrho})f, \theta_{-r_n} \varrho^{-1}\varphi \rangle_{L^2}| + |\langle \theta_{r_n} \tilde{\varrho} f, \varrho^{-1}\varphi \rangle_{L^2}|$$

$$\leq \|f\|_{L^2} \times \|(\varrho_n^{-1} - \varrho^{-1})\varphi\|_{L^2} + \|(\tilde{\varrho}_n - \tilde{\varrho})f\|_{L^2} \times \|\varphi\|_{L^2} + |\langle \theta_{r_n} \, \tilde{\varrho} f, \varrho^{-1}\varphi\rangle_{L^2}|$$

which goes to 0. This shows that 0 is the only limit of the sequence $\langle \gamma_n f, \varphi\rangle_{L^2}$, whence the result. $\diamond$

Now, applying Lemmas 5.3.3 and 5.3.4, we see that the proof reduces to proving that there exists no non-zero vector fixed by all elements of $\mathbb{T}_{d-1}$. As we have already noticed that there exists no non-zero $\mathrm{PSO}(1,d)$-invariant vector, this reduces to the following lemma.

Lemma 5.3.5 *Any function $f \in L_0^2$ such that $\theta_u^+ f \equiv f \circ \theta_u^+ = f$ (for any $u \in \mathbb{R}^{d-1}$) is invariant under the right action of $\mathrm{PSO}(1,d)$.*

Proof (1) We choose $f \in L_0^2$ such that $f \circ \theta_u^+ = f$, for any $u \in \mathbb{R}^{d-1}$, and such that $\|f\|_{L^2} = 1$, and consider the continuous real-valued function Φ defined on $G := \mathrm{PSO}(1,d)$ by

$$\Phi(\gamma) := \langle f \circ \gamma, f\rangle_{L^2}.$$

We know that $\Phi(\mathbf{1}) = 1$, that $\Phi(\gamma) = \Phi(\gamma^{-1})$ and that $\Phi(\gamma) = \Phi(\gamma\theta_u^+) = \Phi(\theta_u^+\gamma)$, for any real number u and any $\gamma \in G$; we want to show that $\Phi \equiv 1$ on G.

(2) We begin with $d = 2$, and use the Iwasawa decomposition in $\mathrm{PSO}(1,2)$ (recall Theorem 1.7.2, applied to γ^{-1}): any element can be written as $\mathcal{R}_\alpha\theta_r\theta_\sigma^+$ (writing θ_x^+, $x \in \mathbb{R}$, for $\theta_{xe_2}^+$) in a unique way. Here $\mathcal{R}_\alpha$ denotes a rotation by α in the oriented plane $(e_1, -e_2)$. From the hypothesis, we have for any reals α, r, x:

$$\Phi(\mathcal{R}_\alpha\theta_r\theta_x^+) = \Phi(\theta_x^+\mathcal{R}_\alpha\theta_r) = \Phi(\mathcal{R}_\alpha\theta_r) =: \varphi(\alpha, r)$$

for some continuous function φ on $(\mathbb{R}/2\pi\mathbb{Z}) \times \mathbb{R}$, such that $\varphi(0,0) = 1$.

We first show that for any $\alpha, \beta \notin \pi\mathbb{Z}$ and $t, r \in \mathbb{R}$, we have:

$$(1 - \cos\alpha)\,e^r = (1 - \cos\beta)\,e^t \implies \varphi(\alpha, r) = \varphi(\beta, t).$$

It is enough to show that

$$(1 - \cos\alpha)\,e^r = (1 - \cos\beta)\,e^t \implies (\exists\, x, x' \in \mathbb{R})\;\; \theta_x^+\mathcal{R}_\alpha = \mathcal{R}_\beta\,\theta_{x'}^+\theta_{t-r}\,,$$

since by the commutation relation of eqn (1.16), we have

$$\theta_x^+\mathcal{R}_\alpha = \mathcal{R}_\beta\,\theta_{x'}^+\theta_{t-r} \implies \theta_x^+\mathcal{R}_\alpha\theta_r = \mathcal{R}_\beta\,\theta_{x'}^+\theta_t = \mathcal{R}_\beta\theta_t\,\theta_{e^{-t}x'}^+$$

$$\implies \varphi(\alpha, r) = \varphi(\beta, t).$$

Now, for $\alpha, \beta \notin \pi\mathbb{Z}$ and $s \in \mathbb{R}$ such that $(1 - \cos\alpha)\,e^s = (1 - \cos\beta)$, the existence of real numbers x, x' such that $\theta_x^+\mathcal{R}_\alpha = \mathcal{R}_\beta\theta_{x'}^+\theta_s$ follows at once from Proposition 2.6.3(i). This proves the desired formula.

(3) We now deduce that, necessarily, $\Phi \equiv 1$ on $\mathrm{PSO}(1,2)$.

For any real $r < t$ and $\varepsilon > 0$, by continuity there exists $\alpha \in]0, \pi[$ such that $\beta \in]0, \alpha]$ implies $|\varphi(0, r) - \varphi(\beta, r)| < \varepsilon/2$ and $|\varphi(0, t) - \varphi(\beta, t)| < \varepsilon/2$. Therefore, choosing β such that $(1 - \cos \beta) = (1 - \cos \alpha)\, e^{r-t} < (1 - \cos \alpha)$, we have

$$|\varphi(0, r) - \varphi(0, t)| < \varepsilon + |\varphi(\alpha, r) - \varphi(\beta, t)| = \varepsilon$$

by (2) above. Hence for any real r, t we obtain $\varphi(0, r) = \varphi(0, t)$, whence $\varphi(0, r) = \varphi(0, 0) = 1$.

By the definition of φ in (2) above, this means that $\Phi(\theta_r) = 1$, and therefore that $f \circ \theta_r = f$ (for any $r \in \mathbb{R}$), and then that

$$\Phi(\mathcal{R}_\alpha\, \theta_r\, \theta_x^+) = \Phi(\mathcal{R}_\alpha\, \theta_r) = \langle f \circ \mathcal{R}_\alpha, f \circ \theta_{-r} \rangle_{L^2} = \Phi(\mathcal{R}_\alpha) =: \psi(\alpha),$$

for some even continuous function ψ on $(\mathbb{R}/2\pi\mathbb{Z})$ such that $\psi(0) = 1$. Since $\Phi(\gamma) = \Phi(\gamma^{-1})$ for any $\gamma \in G$ by the definition of Φ, we have $\Phi(\theta_r \theta_x^+ \mathcal{R}_\alpha) = \psi(\alpha)$ as well. Now, for any $\alpha \in]0, \pi[$, setting $r_\alpha := \operatorname{argsh}(\cot g\, \alpha)$, we have $\log \operatorname{ch} r_\alpha = -\log \sin \alpha$ and then, by Proposition 2.6.3(ii),

$$\theta_{-r_\alpha} \mathcal{R}_\alpha = \mathcal{R}_{\frac{\pi}{2}} \theta_{\log \sin \alpha} \theta_{-\cot g\, \alpha}^+.$$

By the above, this entails that

$$\psi(\alpha) = \Phi(\theta_{-r_\alpha} \mathcal{R}_\alpha) = \Phi\big(\mathcal{R}_{\frac{\pi}{2}} \theta_{\log \sin \alpha} \theta_{-\cot g\, \alpha}^+\big) = \psi(\pi/2).$$

By continuity, this holds for $\alpha \in [0, \pi]$. Therefore we obtain $\psi(\alpha) = \psi(0) = 1$ for any α, that is $\Phi \equiv 1$ on $\mathrm{PSO}(1, 2)$, as claimed.

(4) The above proves that any $(\theta_{x e_2}^+)$-invariant function $f \in L_0^2$ is also (θ_t)-invariant and $(\mathcal{R}_\alpha)$-invariant. Now we can do exactly the same for any $j \in \{3, \dots, d\}$, using $\{\theta_{x e_j}^+ | x \in \mathbb{R}\}$ instead of $\{\theta_{x e_2}^+ | x \in \mathbb{R}\}$, and considering the rotations $\mathcal{R}_\alpha^j$ in the plane (e_1, e_j) instead of the rotations $\mathcal{R}_\alpha$: the elements $\mathcal{R}_\alpha^j \theta_r \theta_{x e_j}^+$ clearly constitute a group isomorphic to $\mathrm{PSO}(1, 2)$, under which f is invariant.

Thus we see that any $\mathbb{T}_{d-1}$-invariant function $f \in L_0^2$ is also (θ_t)-invariant and invariant under any rotation $\mathcal{R}_\alpha^j$. Now, observe that these rotations $\mathcal{R}_\alpha^j$ (for $2 \le j \le d$ and $\alpha \in \mathbb{R}$) generate $\mathrm{SO}(d)$: in fact, the Lie algebra of the subgroup that they generate contains all matrices $E_{1,j}$ (recall Section 1.3), and then also $[E_{1,j}, E_{1,k}] = E_{k,j}$, and hence the whole of $\mathfrak{so}(d)$. Finally, any $\mathbb{T}_{d-1}$-invariant function $f \in L_0^2$ is indeed also G-invariant, owing to the Iwasawa decomposition (Theorem 1.7.2). $\diamond$

5.4   Poincaré inequalities

We shall need only the case of hyperbolic domains, i.e., of connected open subsets of $\mathbb{H}^d$, endowed with the restriction of the volume measure dp (recall Definition 3.6.2.1). However, it happens to be easier to begin with the Euclidean case, as long as the Lebesgue measure is provided with weights. Therefore we shall consider the Euclidean framework first.

5.4.1   The Euclidean case   We are interested in the following basic type of functional inequality.

Definition 5.4.1.1 *Consider a domain $D \subset \mathbb{R}^d$ and two positive measurable functions on D: φ which is bounded and ψ which is bounded away from 0. A **Poincaré inequality** $\mathcal{I}(D, \varphi, \psi)$ holds when there exists a $C = C(D, \varphi, \psi)$ such that for any integrable function f of class C^1 on D satisfying $\int_D f(x)\varphi(x)\, dx = 0$, we have*

$$\int_D f(x)^2 \varphi(x)\, dx \leq C \int_D |df|^2(x)\psi(x)\, dx, \tag{5.1}$$

where $|df|$ denotes the Euclidean norm of the differential of f (i.e., of its Euclidean gradient), and dx denotes the Lebesgue measure of $\mathbb{R}^d$.

Note that provided $\varphi \in L^1(D), \mathcal{I}(D, \varphi, \psi)$ is clearly equivalent to the existence of C such that for any $f \in L^1(D) \cap C^1(D)$,

$$\int_D f(x)^2 \varphi(x)\, dx \; \leq \; C \int_D |df|^2(x)\,\psi(x)\, dx + C \left(\int_D f(x)\varphi(x)\, dx \right)^2. \tag{5.2}$$

The following statement has the following interpretation: a spectral gap $\mathcal{I}(D, \varphi, \psi)$ yields a positive bottom for the spectrum of the induced theory obtained by a so-called killing procedure on a non-empty open set $U \subset D$ (provided $\varphi \in L^1(D)$).

Proposition 5.4.1.2 *Consider any Borelian subset U of a domain $D \subset \mathbb{R}^d$ on which a Poincaré inequality $\mathcal{I}(D, \varphi, \psi)$ holds. Then, for any function $h \in L^1\big(D, \varphi(x)\, dx\big) \cap C^1(D)$ which vanishes on U, we have*

$$\int_D h^2(x)\varphi(x)\, dx \; \leq \; \frac{\int_D \varphi(x)\, dx}{\int_U \varphi(x)\, dx} \times C(D, \varphi, \psi) \int_D |dh|^2(x)\,\psi(x)\, dx. \tag{5.3}$$

Proof   Suppose that $\varphi \in L^1(D)$. We denote by m the normalized measure $\frac{\mathbf{1}_D(x)\varphi(x)\, dx}{\int_D \varphi(x)\, dx}$, and set $V := D \smallsetminus U$ and $f := h - \int_D h\, dm$. Applying (eqn 5.1) yields

$$\frac{C}{\int_D \varphi} \int_D |dh|^2(x)\,\psi(x)\, dx = \frac{C}{\int_D \varphi} \int_D |df|^2(x)\,\psi(x)\, dx \geq \int f^2\, dm$$

$$= \int h^2\, dm - \left[\int h\, dm \right]^2$$

$$= \int_V h^2\, dm - \left[m(V) \int_V h\, \frac{dm}{m(V)} \right]^2 \geq \int_V h^2\, dm - m(V)^2 \int_V h^2\, \frac{dm}{m(V)}$$

$$= \int_V h^2\, dm - m(V) \int_V h^2\, dm = m(U) \int_D h^2\, dm. \qquad \diamond$$

Let us begin with the following example, which will prove to be sufficient for our purposes.

Lemma 5.4.1.3 *A Poincaré inequality* $\mathcal{I}(D, \varphi, \psi)$ *holds on a convex simplex* $D = D_a := \{(x^1, \ldots, x^d) \in (\mathbb{R}_+^*)^d \,|\, x^1 + \cdots + x^d < a\}$, *for all* φ *and* ψ *(and* $a > 0$).

Proof We fix a constant $c < \infty$ such that $\varphi \le c$ on D. We first have

$$2 \int_D \varphi \times \int_D f^2 \varphi - 2 \left(\int_D f\varphi \right)^2 = \int_{D^2} |f(x) - f(x')|^2 \varphi(x)\varphi(x') \, dx \, dx'.$$

Then, by the hypothesis on D, for any given $x, x' \in D$ we can find a piecewise linear path $x = x^0 \to x^1 \to \cdots \to x^d = x'$ in D such that up to some permutation (depending on (x, x')) of the d coordinates, we have $x^j \equiv (x'_1, \ldots, x'_j, x_{j+1}, \ldots, x_d)$ for $0 \le j \le d$, so that each segment $[x^{j-1}, x^j]$ is directed by the coordinate axis (x_j). Therefore

$$f(x) - f(x') = \sum_{j=1}^d \left[f(x^{j-1}) - f(x^j) \right]$$

$$= \sum_{j=1}^d \int_{x_j}^{x'_j} \partial_j f\left(x'_1, \ldots, x'_{j-1}, \xi_j, x_{j+1}, \ldots, x_d\right) d\xi_j$$

and

$$|f(x) - f(x')|^2 \le d \sum_{j=1}^d (x'_j - x_j) \int_{x_j}^{x'_j} \left| \partial_j f\left(\ldots, x'_{j-1}, \xi_j, x_{j+1}, \ldots\right) \right|^2 d\xi_j$$

$$\le d \times \operatorname{diam}(D) \sum_{j=1}^d \int \left[1_D \times |\partial_j f|^2 \right] \left(\ldots, x'_{j-1}, \xi_j, x_{j+1}, \ldots\right) d\xi_j.$$

Setting $k := d \times \operatorname{diam}(D) \times c^2$, we thus have

$$\int_{D^2} |f(x) - f(x')|^2 \, \varphi(x)\varphi(x') \, dx \, dx' \le c^2 \int_{D^2} |f(x) - f(x')|^2 \, dx \, dx'$$

$$\le k \int \sum_{j=1}^d \left[1_D |\partial_j f|^2 \right] (\ldots, x'_{j-1}, \xi_j, x_{j+1}, \ldots) 1_{D^2}(x, x') \, d\xi_j \, dx_1 \ldots dx_d \, dx'_1 \ldots dx'_d$$

$$\le k \operatorname{diam}(D)^{d+1} \int \sum_{j=1}^d \left[1_D |\partial_j f|^2 \right] (\ldots, x'_{j-1}, \xi_j, x_{j+1}, \ldots) \, dx'_1 \ldots dx'_{j-1} \, d\xi_j \, dx_{j+1} \ldots dx_d$$

$$= k' \int_D |df|^2(\xi) \, d\xi \le \frac{k'}{\inf \psi(D)} \int_D |df|^2(\xi) \, \psi(\xi) \, d\xi,$$

with $k' := d \times \operatorname{diam}(D)^{d+2} \times c^2 = k'(D, c)$. Finally, we obtain

$$\int_D f^2 \, \varphi \leq \frac{k'}{2 \int_D \varphi \times \inf_D \psi} \int_D |df|^2 \, \psi + \frac{1}{\int_D \varphi} \left(\int_D f \varphi \right)^2. \qquad \Diamond$$

Remark 5.4.1.4 The proof of Lemma 5.4.1.3 is valid also for a parallelepiped $\prod_{j=1}^{d}]a_j, a'_j[$ or an open ball in $\mathbb{R}^d$, and in fact for any bounded domain $D \subset \mathbb{R}^d$ for which there exists an $N(D) \in \mathbb{N}$ such that any pair of points $\{x, x'\} \subset D$ can be linked by a piecewise linear path made up of at most $N(D)$ segments (contained in D) directed by the coordinate axes.

We then have the following invariance under diffeomorphisms.

Proposition 5.4.1.5 *Consider two bounded domains D, D' in $\mathbb{R}^d$, with closures $\overline{D}, \overline{D'}$ that have C^1-diffeomorphic neighbourhoods, such that Poincaré inequalities $\mathcal{I}(D, \varphi, \psi)$ hold for any φ, ψ as in Definition 5.4.1.1. Then a Poincaré inequality $\mathcal{I}(D', \varphi, \psi)$ holds on D' as well, for all φ and ψ.*

Proof Let $H : \overline{D} \to \overline{D'}$ be a C^1 diffeomorphism (from a neighbourhood of $\overline{D}$ onto a neighbourhood of $\overline{D'}$). Note that the Jacobian $\det J_H$ does not vanish on $\overline{D}$, and is bounded and bounded away from 0 on D'. Consider φ, ψ (as in Definition 5.4.1.1) on D', and $f \in C^1 \cap L^1(D')$. Set $h := f \circ H, \Phi := \varphi \circ H \times |\det J_H|$ and $\Psi := \psi \circ H \times |\det J_H| \times \|J_H\|^{-2}$.

Applying the hypothesis $\mathcal{I}(D, \Phi, \Psi)$ to h (in the form of eqn (5.2)) yields

$$\int_{D'} f^2 \, \varphi = \int_D h^2 \, \Phi \leq C \int_D |dh|^2 \, \Psi + C \left(\int_D h \, \Phi \right)^2$$

$$\leq C \int_D |df|^2 \circ H \times \|J_H\|^2 \, \Psi + C \left(\int_D h \, \Phi \right)^2 = C \int_{D'} |df|^2 \, \psi + C \left(\int_{D'} f \varphi \right)^2. \Diamond$$

It is crucial to be able to consider different domains separately, in order to obtain a Poincaré inequality on their union. This is made possible by the following argument. Note that we relax the constraints on the functions φ, ψ here somewhat, in order to apply this statement, not only to prove Proposition 5.4.1.8 but also to the slightly more general context of Theorem 5.4.2.5.

Proposition 5.4.1.6 *Let D and D' be two intersecting domains in $\mathbb{R}^d$ and let φ, ψ be two positive measurable functions on $D \cup D'$, such that $\varphi \in L^1(D \cup D')$. If Poincaré inequalities $\mathcal{I}(D, \varphi, \psi)$ and $\mathcal{I}(D', \varphi, \psi)$ hold, then a Poincaré inequality $\mathcal{I}(D \cup D', \varphi, \psi)$ holds too.*

Proof We denote by μ the absolutely continuous measure with density φ, restricted to $D \cup D'$ and normalized to be a probability measure so that $\mu(dx) := 1_{D \cup D'}(x) \, \varphi(x) \, dx / \int_{D \cup D'} \varphi$, and consider any function $f \in C^1 \cap L^1(D \cup D', \mu)$. Let $\mu_D(f) := \mu(D)^{-1} \int_D f \, d\mu$ and $V_D(f) := \mu_D(f^2) - \mu_D(f)^2$ denote the mean and variance, respectively, of f on D. We write simply $\mu(f), V(f)$ for $\mu_{D \cup D'}(f), V_{D \cup D'}(f)$.

Applying Definition 5.4.1.1 to $(f - \mu_D(f))$ on D and to $(f - \mu_{D'}(f))$ on D', first we obtain

$$2 \int_{D \cup D'} |df|^2 \, \psi \, dx \geq \int_D |df|^2 \, \psi \, dx + \int_{D'} |df|^2 \, \psi \, dx$$

$$\geq C_D^{-1} \, \mu(D) \, V_D(f) + C_{D'}^{-1} \, \mu(D') \, V_{D'}(f). \qquad (5.4)$$

We then set $D_1 := D \smallsetminus D'$, $D_2 := D' \smallsetminus D$, $D_3 := D \cap D'$, and consider the conditional expectation $\tilde{f}$ of f with respect to the partition $D \cup D' = \bigsqcup_{j=1}^{3} D_j$: $\tilde{f} := \sum_{j=1}^{3} \mu_{D_j}(f) 1_{D_j}$. We have $\mu(f) = \mu(\tilde{f}) = \sum_{j=1}^{3} \mu(D_j) \mu_{D_j}(f)$, and

$$V(f) = V(f - \tilde{f}) + V(\tilde{f})$$

$$= V\left(\sum_{j=1}^{3} (f 1_{D_j} - \mu_{D_j}(f)) 1_{D_j} \right) + V\left(\sum_{j=1}^{3} \mu_{D_j}(f) 1_{D_j} \right)$$

$$= \mu\left[\sum_{j=1}^{3} (f - \mu_{D_j}(f))^2 \, 1_{D_j} \right] + \mu\left[\sum_{j=1}^{3} (\mu_{D_j}(f) - \mu(f))^2 \, 1_{D_j} \right]$$

$$= \sum_{j=1}^{3} \mu(D_j) \left[V_{D_j}(f) + (\mu_{D_j}(f) - \mu(f))^2 \right].$$

From this expression for $V(f)$, we then obtain

$$V(f) - \sum_{j=1}^{3} \mu(D_j) V_{D_j}(f) = \sum_{j=1}^{3} \mu(D_j) (\mu_{D_j}(f) - \mu_{D_3}(f) + \mu_{D_3}(f) - \mu(f))^2$$

$$= \sum_j \mu(D_j) (\mu_{D_j}(f) - \mu_{D_3}(f))^2 - (\mu_{D_3}(f) - \mu(f))^2$$

$$\left(\text{since } \sum_j \mu(D_j) (\mu_{D_j}(f) - \mu_{D_3}(f)) = \mu(f) - \mu_{D_3}(f) \text{ and } \sum_j \mu(D_j) = 1 \right)$$

$$\leq \sum_j \mu(D_j) (\mu_{D_j}(f) - \mu_{D_3}(f))^2. \qquad (5.5)$$

Moreover, writing $D = D_1 \sqcup D_3$ and $D' = D_2 \sqcup D_3$, we have

$$\mu(D) \, V_D(f) = \mu(D_1) \left[V_{D_1}(f) + (\mu_{D_1}(f) - \mu_D(f))^2 \right]$$

$$+ \mu(D_3) \left[V_{D_3}(f) + (\mu_{D_3}(f) - \mu_D(f))^2 \right]$$

$$= \mu(D_1) V_{D_1}(f) + \mu(D_3) V_{D_3}(f) + \mu(D)^{-1} \mu(D_1) \mu(D_3) (\mu_{D_1}(f) - \mu_{D_3}(f))^2,$$

since

$$\mu(D) = \mu(D_1) + \mu(D_3) \text{ and } \mu(D)\mu_D(f) = \mu(D_1)\mu_{D_1}(f) + \mu(D_3)\mu_{D_3}(f)$$

imply

$$\mu(D)\big(\mu_{D_1}(f) - \mu_D(f)\big) = \mu(D_3)\big(\mu_{D_1}(f) - \mu_{D_3}(f)\big)$$

and a similar relation with D_1, D_3 exchanged. In the same way, we have

$$\mu(D') V_{D'}(f) =$$
$$\mu(D_2)V_{D_2}(f) + \mu(D_3)V_{D_3}(f) + \mu(D')^{-1}\mu(D_2)\mu(D_3)\big(\mu_{D_2}(f) - \mu_{D_3}(f)\big)^2.$$

From these two expressions, we obtain the result that

$$\sum_j \mu(D_j)\, V_{D_j}(f) \leq \mu(D)\, V_D(f) + \mu(D')\, V_{D'}(f) \tag{5.6}$$

and $\big($setting $C := \max\{\mu(D), \mu(D')\}/\mu(D_3)\big)$

$$\sum_j \mu(D_j)\big(\mu_{D_j}(f) - \mu_{D_3}(f)\big)^2 \leq C\big[\mu(D)\, V_D(f) + \mu(D')\, V_{D'}(f)\big]. \tag{5.7}$$

The conclusion now follows at once from the inequalities (5.4), (5.5), (5.6), and (5.7):

$$V(f) \leq (C+1)\big[\mu(D)V_D(f) + \mu(D')V_{D'}(f)\big] \leq C' \int_{D \cup D'} |df|^2\, \psi\, dx,$$

which is a Poincaré inequality $\mathcal{I}(D \cup D', \varphi, \psi)$. $\diamondsuit$

Remark 5.4.1.7 Some more care in the above proof yields the following estimate for the best constant in the Poincaré inequality under consideration:

$$C_{D \cup D'} \leq 2\, \frac{\max\{C_D, C_{D'}\}}{\mu(D \cap D')}.$$

In fact, a careful computation yields

$$\mu(D_3)^{-1}\big[\mu(D)V_D(f) + \mu(D')V_{D'}(f)\big] - V(f) - V_{D_3}(f) - \big(\mu_{D_3}(f) - \mu(f)\big)^2$$

$$= \Big[\tfrac{1}{\mu(D_3)} - 1\Big] \sum_j \mu(D_j)V_{D_j}(f) + \mu(D_1)\mu(D_2)\Big[\frac{[\mu_{D_1}(f) - \mu_{D_3}(f)]^2}{\mu(D)}$$

$$+ \frac{[\mu_{D_2}(f) - \mu_{D_3}(f)]^2}{\mu(D')}\Big] \geq 0,$$

whence

$$V(f) \le \frac{\mu(D)V_D(f) + \mu(D')V_{D'}(f)}{\mu(D_3)} \le 2\,\frac{\max\{C_D, C_{D'}\}}{\mu(D \cap D')} \int_{D \cup D'} |df|^2\, \psi\, dx.$$

Proposition 5.4.1.8 *A Poincaré inequality $\mathcal{I}(D, \varphi, \psi)$ holds on a domain $D \subset \mathbb{R}^d$ whose closure is C^1-diffeomorphic to a convex polyhedron, for all φ and ψ.*

Proof By Proposition 5.4.1.5, we just have to consider the case of a convex polyhedron. Now, this is a finite union of adjacent simplexes, C^1-diffeomorphic to the simplexes considered in Lemma 5.4.1.3. Moreover, we obtain an overlap between adjacent simplexes by means of diffeomorphisms acting non-trivially in a small neighbourhood of adjacent faces. The proof is concluded by applying Propositions 5.4.1.5 and 5.4.1.6. ◇

5.4.2 The case of a fundamental domain in $\mathbb{H}^d$

We now apply Proposition 5.4.1.8 to our main concern: hyperbolic domains, i.e., connected open subsets $D \subset \mathbb{H}^d$, although most of the following is valid in a more general setting. Any hyperbolic domain D is endowed with a restriction of the volume measure dp (recall Definition 3.6.2.1 and Proposition 3.6.2.2(i)), and with a hyperbolic gradient, as appears implicitly in, for example, Proposition 3.2.3. Namely, with any $f \in C^2(D)$ we associate its squared hyperbolic gradient

$$\gamma(f) := \frac{1}{2}\Delta(f^2) - f\,\Delta f = y^2 \left(\frac{\partial f}{\partial y}\right)^2 + y^2 \sum_{j=2}^{d} \left(\frac{\partial f}{\partial x_j}\right)^2, \qquad (5.8)$$

in Poincaré coordinates (recall Proposition 1.6.2) according to eqn (3.10). We shall henceforth be concerned with the following notion of a Poincaré inequality.

Definition 5.4.2.1 *A **Poincaré inequality** holds on a hyperbolic domain D when there exists a $C_D > 0$ such that, for any integrable function f of class C^1 on D such that $\int_D f(p)\, dp = 0$, we have*

$$\int_D f^2(p)\, dp \le C_D \int_D \gamma(f)(p)\, dp. \qquad (5.9)$$

Using Poincaré coordinates, Proposition 3.6.2.2(i) and eqn (5.8), this reads

$$\int_D f^2(x, y)\, y^{-d}\, dx\, dy \le C_D \int_D |df|^2(x, y)\, y^{2-d}\, dx\, dy \qquad (5.10)$$

provided $\int_D f(x, y)\, y^{-d}\, dx\, dy = 0$.

This explains why we needed to consider weights φ, ψ in Section 5.4.1. We particularize now to $\varphi \equiv y^{-d}$ and $\psi \equiv y^{2-d}$.

By applying Proposition 5.4.1.8 to a bounded hyperbolic domain D, we obtain the following corollary.

Corollary 5.4.2.2 *A Poincaré inequality as in eqn (5.9) holds on any bounded domain $D \subset \mathbb{H}^d$ whose closure is C^1-diffeomorphic to a convex polyhedron.*

To handle unbounded fundamental domains, we need to address their unbounded ends in particular. We use the term **solid cusp** to denote the image in the orbit space $\Gamma\backslash\mathbb{H}^d$ of a horoball $\mathcal{H}^+$ that is based at some parabolic point η and small enough in order that $\Gamma\backslash\mathcal{H}^+$ can be identified with $\Gamma_\eta\backslash\mathcal{H}^+$, where Γ_η denotes the maximal parabolic subgroup of Γ that fixes η (see Section A.4). We shall need to consider only the case of an isolated solid cusp that has a finite volume and maximal rank, which means that the stabilizer Γ_η contains a free Abelian subgroup (of finite index) isomorphic to $\mathbb{Z}^{d-1}$. Thus we need to handle finite quotients (in dimension $d \geq 3$ only), as follows.

Lemma 5.4.2.3 *Consider the disjoint union $D = D_1 \sqcup \ldots \sqcup D_k \subset \mathbb{H}^d$ of k bounded isometric domains $D_1, \ldots, D_k$. If a Poincaré inequality of type (5.9) holds on D, then a Poincaré inequality of type (5.9) holds on D_1 as well.*

Proof We denote by φ_j an isometry from D_j onto D_1, and consider an $f \in C^1(D_1)$ such that $\int_{D_1} f = 0$. Extending this function to D by setting $\tilde{f} := f \circ \varphi_j$ on D_j, we first have

$$\int_D \tilde{f} = \sum_{j=1}^k \int_{D_j} f \circ \varphi_j = k \int_{D_1} f = 0,$$

since any isometry preserves the volume measure. Therefore (and since the Jacobian matrix J_{φ_j} is bounded on D_j) we have

$$\int_{D_1} f^2 = k^{-1} \int_D \tilde{f}^2 \leq \frac{C_D}{k} \int_D \gamma(\tilde{f}) = \frac{C_D}{k} \sum_{j=1}^k \int_{D_j} \gamma(f \circ \varphi_j)$$

$$\leq \frac{C_D}{k} \sum_{j=1}^k \int_{D_1} \gamma(f) \times (\|J_{\varphi_j}\|^2 \circ \varphi_j^{-1}) \leq C_{D_1} \int_{D_1} \gamma(f). \qquad \diamond$$

We come now to the Poincaré inequality relative to a solid cusp.

Proposition 5.4.2.4 *A Poincaré inequality of type (5.9) holds on any solid cusp of finite volume.*

Proof

(1) We take a solid cusp D, and use a Poincaré model $\mathbb{R}^{d-1} \times \mathbb{R}_+^*$ such that the cusp associated with D is ∞, and, according to Theorem A.4.5, $D = \mathcal{P} \times]1, \infty[$ for some $\mathcal{P} \subset \mathbb{R}^{d-1}$, a relatively compact fundamental domain of the maximal parabolic subgroup Γ_∞ (which stabilizes the cusp ∞ and contains a finite-index subgroup isomorphic to $\mathbb{Z}^{d-1}$) acting on $\mathbb{R}^{d-1}$. We must establish eqn (5.10) or, equivalently,

$$\int_D \left| f(x,y) - \frac{\int_D f(x,y)y^{-d} dx\, dy}{\int_D y^{-d}\, dx\, dy} \right|^2 y^{-d} dx\, dy \le C_D \int_D |df|^2(x,y)y^{2-d}\, dx\, dy,$$

for any function f of class C^1 on D such that $\int_D f^2(x,y)y^{-d}\, dx\, dy < \infty$.

(2) We first consider functions of the vertical coordinate y alone or, more precisely, the space $\mathcal{H}$ of real C^1 functions h on $[1,\infty[$ such that

$$\int_1^\infty h(y)\, y^{-d}\, dy = 0 \qquad \text{and}$$

$$\|h\|^2 := (d-1)^2 \int_1^\infty h(y)^2\, y^{-d}\, dy + \int_1^\infty |h'(y)|^2\, y^{2-d}\, dy < \infty.$$

We also set $\mathcal{H}' := \{ h \in \mathcal{H} \,|\, h(1) = 0 \}$, and $\varphi(y) := (d-1)^2 \log y - 1$.

It is easily checked that the function φ belongs to $\mathcal{H}$. Furthermore, it is orthogonal to the space $\mathcal{H}'$ with respect to the underlying scalar product, since for any $h \in \mathcal{H}'$, we obtain the following by integrating by parts:

$$(d-1)^{-2} \int_1^\infty h'(y)\varphi'(y)y^{2-d}\, dy = \left[h(y)\, y^{1-d} \right]_1^\infty - (1-d) \int_1^\infty h(y) y^{-d}\, dy = 0.$$

(The boundary term vanishes since, for any $h \in \mathcal{H}'$, $Y \ge 1$ and $\varepsilon \in]0, \tfrac{1}{4}]$, we have

$$\left| h(Y)Y^{1+\varepsilon-d} \right|^2 = \left| (1+\varepsilon-d) \int_1^Y h(y)y^{\varepsilon-d}\, dy + \int_1^Y h'(y)y^{1+\varepsilon-d}\, dy \right|^2 \le c\, \|h\|^2$$

by Schwarz's inequality.) We then set

$$c_d := \int_1^\infty \varphi(y)^2\, y^{-d}\, dy \Big/ \int_1^\infty [\varphi'(y)]^2\, y^{2-d}\, dy.$$

Integrating by parts and applying Schwarz's inequality, for any $h \in \mathcal{H}'$ we obtain

$$(d-1) \int_1^\infty h^2(y)\, y^{-d}\, dy \le 2 \int_1^\infty h(y)\, h'(y)\, y^{1-d}\, dy$$

$$\le 2\sqrt{ \int_1^\infty h^2(y)\, y^{-d}\, dy \int_1^\infty [h'(y)]^2\, y^{2-d}\, dy }$$

and therefore

$$(d-1)^2 \int_1^\infty h^2(y)\, y^{-d}\, dy \le 4 \int_1^\infty [h'(y)]^2\, y^{2-d}\, dy.$$

We thus obtain for any $h \in \mathcal{H}'$ and $t \in \mathbb{R}$,

$$\int_1^\infty \left[h(y) + t\,\varphi(y)\right]^2 y^{-d}\,dy \;\le\; 2\int_1^\infty h(y)^2\,y^{-d}\,dy + 2\,t^2\int_1^\infty \varphi(y)^2\,y^{-d}dy$$

$$\le 8(d-1)^{-2}\int_1^\infty [h'(y)]^2\,y^{2-d}\,dy + 2c_d\,t^2\int_1^\infty [\varphi'(y)]^2\,y^{2-d}\,dy$$

$$\le c_d'\int_1^\infty \left[h' + t\,\varphi'\right]^2(y)\,y^{2-d}\,dy\,,$$

which shows that the Poincaré inequality holds in $\mathcal{H}$.

(3) We now take a function $f \in C^1(D)$ which is square-integrable with respect to the volume measure of D, i.e., $\int_D f(x,y)^2 y^{-d}\,dx\,dy < \infty$, denote by m its average with respect to the normalized volume measure, i.e.,

$$m := \frac{\int_D f(x,y)\,y^{-d}\,dx\,dy}{\int_D y^{-d}\,dx\,dy}\,,$$

and set for any $y \ge 1$

$$h(y) := \frac{\int_\mathcal{P} f(x,y)\,dx}{\int_\mathcal{P} dx} \quad \text{and} \quad H(y) := h(y) - m.$$

Note that $(d-1)\int_1^\infty h(y)y^{-d}\,dy = m$, $\int_1^\infty H(y)^2\,y^{-d}\,dy < \infty$ and $\int_1^\infty H(y)y^{-d}dy = 0$, so that we can apply the Poincaré inequality obtained in (2) above to H. Moreover, by Theorem A.4.5, the bounded polyhedron $\mathcal{P}$, which is the base of the solid cusp D, is a quotient (by a finite group of isometries) of a bounded polyhedron $\mathcal{P}'$ diffeomorphic to a hypercube. Hence, by Proposition 5.4.2.2 and Lemma 5.4.2.3, we have for any $y \ge 1$

$$\int_\mathcal{P} \left[f(x,y) - h(y)\right]^2 dx \;\le\; C_\mathcal{P}\int_\mathcal{P} \left|d_x f(x,y)\right|^2 dx \quad \text{for some } C_\mathcal{P} > 0.$$

Finally, we obtain

$$\int_D \left|f(x,y) - m\right|^2 y^{-d}\,dx\,dy = \int_D \left(\left[f(x,y) - h(y)\right] + H(y)\right)^2 y^{-d}\,dx\,dy$$

$$\le 2\int_1^\infty \int_\mathcal{P} \left[f(x,y) - h(y)\right]^2 dx\,y^{-d}\,dy + 2\int_\mathcal{P} dx \int_1^\infty H(y)^2\,y^{-d}\,dy$$

$$\le 2\,C_\mathcal{P}\int_1^\infty \int_\mathcal{P} \left|d_x f(x,y)\right|^2 dx\,y^{-d}\,dy + 2\,c_d'\int_\mathcal{P} dx \int_1^\infty \left|H'(y)\right|^2 y^{2-d}\,dy$$

$$\le 2\,C_\mathcal{P}\int_D \left|d_x f(x,y)\right|^2 y^{2-d}\,dx\,dy + 2\,c_d'\int_D \left|d_y f(x,y)\right|^2 y^{2-d}\,dx\,dy,$$

and hence the Poincaré inequality on D, with $C_D := 2\max\{C_\mathcal{P}, c_d'\}$.   ◇

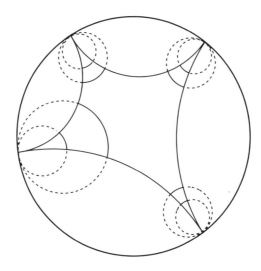

Fig. 5.1 Ideal core–cusp decomposition for Theorem 5.4.2.5

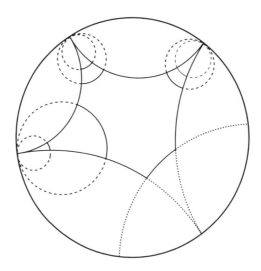

Fig. 5.2 Non-ideal core–cusp decomposition for Theorem 5.4.2.5

We can now conclude with the main result of the present section, which yields the spectral-gap result we need for Chapter 8.

Theorem 5.4.2.5 *A Poincaré inequality of type (5.9) holds on a fundamental domain D of a geometrically finite and cofinite Kleinian group Γ.*

Proof We use Corollary 5.4.2.2, and Propositions 5.4.2.4 and 5.4.1.6. We note that in the cofinite examples in Chapter 4, we can cover a fundamental polyhedron with a finite union of solid cusps and a so-called core, which is compact and diffeomorphic to a convex polyhedron.

Indeed, the solid cusps are obtained by cutting each of the unbounded ends of the fundamental polyhedron with a sufficiently small horoball (based at the corresponding infinite end, i.e., the cusp). The remaining core is diffeomorphic to a convex polyhedron. This core can then be slightly enlarged, by cutting each unbounded end with a strictly smaller horoball, in order to intersect the interior of every solid cusp (in such a way that the core is still diffeomorphic to a convex polyhedron); see Figures 5.1 and 5.2.

Finally, the core–cusp decomposition of Theorem A.4.5 (again with the core slightly enlarged, so that it intersects the solid cusps) states that the above picture holds generally. $\Diamond$

5.5   Notes and comments

The ergodic theorem on hyperbolic surfaces is due to Hopf [Hop]. A fundamental reference on this subject is [CFS]; see also [Rag], [Ni], [Mar], [Gu], [HK] and [Da]. Also, [N2] and [St] prove the basic subadditive ergodic theorem elegantly.

The ergodic theorem extends to general manifolds with negative curvature and a finite volume. In the case of an infinite volume and constant negative curvature, an invariant measure can be defined, namely the Patterson–Sullivan measure, which is related to the invariant measure of the image of the Laplacian under the ground state transformation $f \mapsto \frac{1}{h}\Delta(hf)$; see in particular [P1], [P2], [PS], [Su], and [Y].

The spectral-gap property (equivalent to a Poincaré inequality) has an important meaning in quantum mechanics, as it shows the existence of a first excited state for the Hamiltonian associated with the energy.

Proposition 5.4.1.2 means that a spectral gap $\mathcal{I}(D, \psi, \psi)$ yields a positive bottom for the spectrum of the induced theory obtained by a so-called killing procedure on a non-empty open set $U \subset D$ (provided $\varphi \in L^1(D)$). It can be applied as follows: let $T_U = \inf\{s \geq 0 | z_s \in U\}$ be the hitting time of U for the drifted (by $\nabla \log \psi$) Brownian motion (z_s), and let (Q_s^U) be the **killed** semigroup, defined on $C_b(D)$ by $Q_s^U f(z) := \mathbb{E}_z\big[f(z_s) 1_{\{T_U > t\}}\big]$. It can be shown that the above positivity of the bottom of the spectrum entails an exponential decay of (Q_s^U), which, when applied in particular to the constant function $f \equiv 1$, yields an exponential decay of the tail $\mathbb{P}(T_U > s)$ as $s \to \infty$ or, equivalently, some exponential integrability of the hitting time T_U; see also [CK].

Poincaré inequalities pertain to a family of functional inequalities which have been extensively developed in recent years, in particular Log–Sobolev inequalities (due to L. Gross); see for example [By].

Chapter Six
Basic Itô calculus

The passage from the discrete case to the continuum is fundamental in probability theory. The study of this topic originated with De Moivre's theorem on the convergence of the rescaled binomial distribution towards a bell-shaped curve. But it is only during the twentieth century that the scaling limit of a simple random walk, known as a Wiener process or Brownian motion, was mathematically defined and studied. Itô's calculus is the continuous limit of elementary calculations that can be done in discrete time. It turns out that a second-order correction to the usual rules of calculus is needed, arising from the fact that Brownian paths have a non-vanishing quadratic variation.

Fundamental notions such as predictability, martingales and stopping times, are introduced in Section 1 in the discrete case. We then give a short introduction to Brownian motion and finally deal with the basic tools of Itô's calculus: the stochastic integral and the Itô change-of-variable formula. We have tried to avoid unnecessary technicalities, while at the same time providing complete proofs.

6.1 Discrete martingales and stochastic integrals

We assume that the reader is familiar with the notion of a probability space. Random variables will always be defined almost surely.

Consider a probability space $(\Omega, \mathcal{F}, \mathbb{P})$, endowed with a **filtration** $(\mathcal{F}_n)$, i.e., an increasing sequence of sub-σ-fields of $\mathcal{F}$, which will often be generated by a process.

Definition 6.1.1 *An* $(\mathcal{F}_n)$-**martingale** *is a sequence of real random variables* $\{X_n | n \in \mathbb{N}\}$ *such that for any* $n \in \mathbb{N}$, *(i)* X_n *is* $\mathcal{F}_n$-*measurable (defining a non-anticipating sequence) and integrable, i.e.,* $X_n \in L^1(\mathcal{F}_n, \mathbb{P})$; *and (ii)* $X_n = \mathbb{E}(X_{n+1} | \mathcal{F}_n)$ *almost surely. If in (ii),* $=$ *is replaced by* $\leq$ *or by* $\geq$, *the process* (X_n) *is said to be a* **submartingale** *or a* **supermartingale**, *respectively.*

Note that for a submartingale, $\mathbb{E}(X_n)$ is non-decreasing and that for a supermartingale, $\mathbb{E}(X_n)$ is non-increasing.

Example

- If Z is integrable, then $\mathbb{E}(Z | \mathcal{F}_n)$ defines a martingale.
- If $S_n := \sum_{j=1}^{n} X_j$ denotes a simple random walk on $\mathbb{Z}$, with independent variables (X_j) such that $\mathbb{P}(X_j = 1) = p = 1 - \mathbb{P}(X_j = -1)$, then this is a martingale if and only if $p = \frac{1}{2}$, a submartingale if and only if $p \geq \frac{1}{2}$, and a supermartingale if and only if $p \leq \frac{1}{2}$.

By Jensen's inequality for conditional expectations, if (X_n) is a martingale and ϕ is a convex real function such that $\mathbb{E}\big(|\phi \circ X_n|\big) < \infty$, then $(\phi \circ X_n)$ is a submartingale. Moreover, if (X_n) is a submartingale and ϕ is a convex non-decreasing real function such that $\mathbb{E}\big(|\phi \circ X_n|\big) < \infty$, then $(\phi \circ X_n)$ is a submartingale. For example, $(X_n - a)^+$ and $\max\{X_n, a\}$ are submartingales, for any real constant a. Consequently, if (X_n) is a supermartingale, then $\min\{X_n, a\}$ is a supermartingale.

Definition 6.1.2 A **predictable** process in discrete time is a process $\{H_n | n \in \mathbb{N}^*\}$ such that X_n is $\mathcal{F}_{n-1}$-measurable for all $n \geq 1$.

We can think of H_n as the amount a gambler bets in the n-th game, knowing the outcome of the previous games.

Proposition 6.1.3 Let H be bounded and predictable and let X be a martingale, and define $(H \cdot X)_n := \sum_{m=1}^{n} H_m(X_m - X_{m-1})$. Then $H \cdot X$ is a martingale. If H is non-negative bounded and predictable and if X is a submartingale or supermartingale, then so is $H \cdot X$.

This is an essential property. $H \cdot X$ is called the **stochastic integral** of H against X. The proof is straightforward from the definitions.

Recall that an $(\mathcal{F}_n)$-**stopping time** is a random variable S taking values in $\overline{\mathbb{N}} = \mathbb{N} \cup \{\infty\}$ such that $\{S = n\} \in \mathcal{F}_n$ for all n (e.g., hitting times for random walks or, more generally, for any non-anticipating process).

The process $H_n := 1_{\{S \geq n\}} = 1 - \sum_{k=0}^{n-1} 1_{\{S=k\}}$ is predictable, and $(H \cdot X)_n = X_{n \wedge S} - X_0$. Moreover, $((1 - H) \cdot X)_n = X_n - X_{n \wedge S}$. Therefore, we have the following corollary.

Corollary 6.1.4 If X is a submartingale and if S is a stopping time, then $\{X_{n \wedge S} | n \in \mathbb{N}\}$ and $\{X_n - X_{n \wedge S} | n \in \mathbb{N}\}$ are submartingales too. In particular, $\mathbb{E}(X_{S \wedge n}) \leq \mathbb{E}(X_n)$ for any $n \in \mathbb{N}$.

Note that the last point can be seen directly: in fact,

$$\mathbb{E}(X_{S \wedge n}) = \sum_{k=0}^{n} \mathbb{E}\Big(X_k 1_{\{S \wedge n = k\}}\Big) \leq \sum_{k=0}^{n} \mathbb{E}\Big(\mathbb{E}(X_n | \mathcal{F}_k) 1_{\{S \wedge n = k\}}\Big) = \mathbb{E}(X_n).$$

Example Doubling strategy in a coin-tossing game: double your bets until you win! Mathematically, this is modelled by a simple random walk (S_n). We set $\tau := \inf\{n | X_n = 1\}$, and take $H_n := 2^{n-1} 1_{\{\tau \geq n\}}$. You win 1 at time τ. But beware that, in practice, you cannot play that way infinitely many times, so your risk of losing a lot is significant.

As a consequence of the previous corollary, we obtain the following theorem.

Theorem 6.1.5 (Doob's inequalities) Let (X_n) be a submartingale. We then have the following.

(*i*) *For any* $\lambda > 0$, *we have*

$$\lambda \mathbb{P}\left[\sup_{m \leq n} X_m^+ \geq \lambda\right] \leq \mathbb{E}\left[X_n 1_{\left\{\sup_{m \leq n} X_m^+ \geq \lambda\right\}}\right] \leq \mathbb{E}(X_n^+).$$

(*ii*) *If* (X_n) *is non-negative or is a martingale, then for any* $p > 1$,

$$\left\|\sup_{m \leq n} |X_m|\right\|_p \leq \frac{p}{p-1} \|X_n\|_p \text{ for all } n,$$

and

$$\left\|\sup_n |X_n|\right\|_p \leq \frac{p}{p-1} \sup_n \|X_n\|_p.$$

Proof (*i*) For the first inequality, denoting by T the hitting time for $[\lambda, \infty]$ by X, we have $\{T \leq n\} = \left\{\sup_{m \leq n} X_m^+ \geq \lambda\right\} = \{X_{T \wedge n} \geq \lambda\}$, and then

$$\lambda \mathbb{P}\left[\sup_{m \leq n} X_m^+ \geq \lambda\right] \leq \mathbb{E}\left[X_{T \wedge n} 1_{\{T \leq n\}}\right] = \sum_{k=1}^{n} \mathbb{E}\left[X_k 1_{\{T=k\}}\right] \leq \mathbb{E}\left[X_n 1_{\{T \leq n\}}\right].$$

The second inequality is obvious.

(*ii*) The absolute value of a martingale is a non-negative submartingale. It is thus enough to consider a non-negative submartingale (X_n). Note that $\|X_n\|_p$ increases, since X^p is also a submartingale. Set $Y_n := \sup_{m \leq n} X_m$. From (*i*), by Fubini's theorem we obtain

$$\|Y_n\|_p^p = \int_0^\infty p\lambda^{p-1} \, \mathbb{P}(Y_n \geq \lambda) \, d\lambda \leq p \int_0^\infty \lambda^{p-2} \, \mathbb{E}\left[X_n 1_{\{Y_n \geq \lambda\}}\right] \, d\lambda$$

$$= p \, \mathbb{E}\left[X_n \int_0^{Y_n} \lambda^{p-2} \, d\lambda\right] = \frac{p}{p-1} \mathbb{E}\left(X_n Y_n^{p-1}\right)$$

$$\leq \frac{p}{p-1} \|X_n\|_p \, \|Y_n^{p-1}\|_{(p/p-1)} = \frac{p}{p-1} \|X_n\|_p \, \|Y_n\|_p^{p-1}$$

by Hölder's inequality. This yields the required result if $0 < \|Y_n\|_p < \infty$. There is nothing to prove if $\|Y_n\|_p = 0$, and if $\|Y_n\|_p = \infty$, then we must have $\|X_n\|_p = \infty$ too, since $\|Y_n\|_p \leq \|X_0\|_p + \cdots + \|X_n\|_p \leq (n+1) \|X_n\|_p$. Finally, this implies at once the inequality relating to $\|\sup_n Y_n\|_p$, merely by applying the monotone convergence theorem.   ◇

A martingale (M_n) which is bounded in L^2 can easily be seen to converge in L^2: since its L^2 squared norm is the sum of the squared norms of its increments

$$\sum_{k=1}^{n} \|M_k - M_{k-1}\|_2^2 = \|M_n\|_2^2 - \|M_0\|_2^2 \le C < \infty,$$

the Cauchy criterion applies at once. Letting m go to infinity in $\mathbb{E}(M_{n+m}|\mathcal{F}_n) = M_n$, we deduce that $\mathbb{E}(M_\infty|\mathcal{F}_n) = M_n$ for any n, almost surely.

6.2 Brownian motion

In the case of continuous time, we will consider only continuous or **càdlàg** (i.e., left-limited and right-continuous) matrix-valued processes, i.e., random variables taking values in the space of continuous or càdlàg functions X_t of the time coordinate $t \in \mathbb{R}_+$, equipped with its natural σ-field $\sigma\{X_s | s \in \mathbb{R}_+\} = \sigma\{X_s | s \in \mathbb{D}\}$, $\mathbb{D}$ denoting the set of non-negative dyadic numbers.

Definition 6.2.1 *A real Brownian motion (or Wiener process) is a real-valued continuous process $(B_t)_{t \ge 0}$ such that $B_0 = 0$ and for any $n \in \mathbb{N}^*$ and $0 = t_0 < \cdots < t_n$, the random variables $(B_{t_j} - B_{t_{j-1}})$ are independent, and the law of $(B_{t_j} - B_{t_{j-1}})$ is $\mathcal{N}(0, t_j - t_{j-1})$, i.e., it is centred Gaussian with variance $(t_j - t_{j-1})$.*

The above Brownian motion (B_t) is sometimes called **standard** to underline the fact that it starts from 0 and has unit variance (at time 1). The generic real Brownian motion has the law of $(a + c\,B_t)$, with real constants a, c.

A slightly different formulation of the second part of the definition is that the increments of (B_t) are independent and stationary, i.e., for all $0 \le s \le t$, $(B_t - B_s) \overset{\text{law}}{\equiv} B_{t-s}$ and the law of B_t is $\mathcal{N}(0, t)$. A simple construction of (B_t), by means of a multi-scale series, is given in the appendices; see Section B.1.

The (probability) law of such a process is clearly unique, and is known as the **Wiener measure** on the space $C_0(\mathbb{R}_+, \mathbb{R})$ of real continuous functions indexed by $\mathbb{R}_+$ (and vanishing at 0).

The following property is obtained straightforwardly from the definition, since the law of a Gaussian process is prescribed by its mean and its covariance.

Proposition 6.2.2 *The standard real Brownian motion (B_t) is the unique real process which is Gaussian and centred with covariance function $\mathbb{R}_+^2 \ni (s, t) \longmapsto \mathbb{E}(B_s B_t) = \min\{s, t\}$.*

The processes $t \mapsto B_{a+t} - B_a$, $t \mapsto c^{-1} B_{c^2 t}$, $t \mapsto t B_{1/t}$ and $t \mapsto (B_T - B_{T-t})$ (for $0 \le t \le T$) satisfy the same conditions. We therefore deduce the following fundamental properties.

Corollary 6.2.3 *The standard real Brownian motion (B_t) satisfies the following conditions:*

(1) *The **Markov property**: for all $a \in \mathbb{R}_+$, $(B_{a+t} - B_a)$ is also a standard Brownian motion, and is independent of $\mathcal{F}_a := \sigma\{B_s \,|\, 0 \le s \le a\}$.*

(2) *Self-similarity: for any $c > 0$, $(c^{-1} B_{c^2 t})$ is also a standard real Brownian motion.*

(3) $(-B_t)$ and $(t\,B_{1/t})$ are also standard real Brownian motions.

(4) For any fixed $T > 0$, $(B_T - B_{T-t})_{0 \le t \le T}$ is also a standard real Brownian motion (on $[0,T]$).

An $\mathbb{R}^d$-valued process $(B_t^1, \ldots, B_t^d)$ made up of d independent standard Brownian motions (B_t^j) is called a d-**dimensional Brownian motion**.

Of course, for any $v \in \mathbb{R}^d$, $v + (B_t^1, \ldots, B_t^d)$ is called a d-dimensional Brownian motion starting at v.

Exercise   Prove that the law of a d-dimensional Brownian motion (starting at 0) is preserved by Euclidean rotations (of the vector space $\mathbb{R}^d$).

Proposition 6.2.4 *(quadratic variation) Take a $t > 0$, and denote by $\mathcal{P} := \{0 = t_0 < t_1 < \ldots < t_N = t\}$ a subdivision of $[0,t]$, with mesh $|\mathcal{P}|$. Set $V_{\mathcal{P}} = \sum_{j=0}^{N-1} (B_{t_{j+1}} - B_{t_j})^2$, for some standard real Brownian motion (B_t). Then as $|\mathcal{P}|$ goes to 0, $V_{\mathcal{P}}$ converges to t in the L^2-norm.*

Proof   We have $E(V_{\mathcal{P}}) = t$, and by the independence of increments,

$$\mathbb{E}\big[(V_{\mathcal{P}} - t)^2\big] = \sum_{j=0}^{N-1} (t_{j+1} - t_j)^2 \times \mathbb{E}\left(\left[\frac{(B_{t_{j+1}} - B_{t_j})^2}{t_{j+1} - t_j} - 1\right]^2\right)$$

$$= 3 \sum_{j=0}^{N-1} (t_{j+1} - t_j)^2 \le 3t \ |\mathcal{P}|. \qquad \diamond$$

6.3   Martingales in continuous time

Consider a probability space $(\Omega, \mathcal{F}, \mathbb{P})$ endowed with a **filtration** $(\mathcal{F}_t)$, i.e., an increasing family of sub-σ-fields of $\mathcal{F}$. We suppose that $\mathcal{F}_0$ contains the set $\mathcal{N}(\mathbb{P})$ of all $\mathbb{P}$-negligible subsets of Ω.

Definition 6.3.1   *A continuous $(\mathcal{F}_t)$-martingale is a continuous real-valued process $\{X_t | t \in \mathbb{R}_+\}$ such that (i) X_t is $\mathcal{F}_t$-measurable (i.e., a **non-anticipating** or **adapted** process) and integrable, i.e., $X_t \in L^1(\mathcal{F}_t, \mathbb{P})$; and (ii) $X_t = \mathbb{E}(X_{t+s} | \mathcal{F}_t)$ almost surely, for all $s, t \ge 0$. If in (ii), $=$ is replaced by $\le$ or by $\ge$, the process (X_t) is said to be a **submartingale** or a **supermartingale**, respectively.*

The properties of discrete martingales extend to martingales in continuous time. In particular, if (X_t) is a submartingale, then $(X_t - a)^+$ is a submartingale. If (X_t) is a supermartingale, then $(t \mapsto \min\{X_t, a\})$ is a supermartingale.

Example

- A Brownian motion (B_t) is a square-integrable continuous martingale, with respect to its completed proper filtration $\overline{\mathcal{F}}_t = \sigma\{B_s | s \le t\} \vee \mathcal{N}(\mathbb{P})$. This holds for $(B_t^2 - t)$ as well.

- For any real α, the '**exponential martingale**' $(e^{\alpha B_t - \alpha^2 t/2})$ is a (continuous) martingale.

The following is easily deduced from Theorem 6.1.5 by using dyadic approximation.

Theorem 6.3.2 *(Doob's inequalities) Let (X_t) be a continuous submartingale. We then have the following.*

(i) for any $\lambda, t > 0$, we have

$$\lambda \, \mathbb{P}\left[\sup_{s \le t} X_s^+ \ge \lambda\right] \le \mathbb{E}\left[X_t 1_{\left\{\sup_{s \le t} X_s^+ \ge \lambda\right\}}\right] \le \mathbb{E}(X_t^+).$$

(ii) If (X_t) is non-negative or is a martingale, then for any $p > 1$,

$$\left\|\sup_{s \le t} |X_s|\right\|_p \le \frac{p}{p-1} \, \|X_t\|_p \text{ for any time } t,$$

and then

$$\left\|\sup_s |X_s|\right\|_p \le \frac{p}{p-1} \sup_s \|X_s\|_p.$$

We shall use the following asymptotic property, known as the Khinchin law of the iterated logarithm. The elegant proof that we give (taken from [MK]) uses essentially Theorem 6.3.2(i), applied to the exponential martingale given above as an example.

Proposition 6.3.3 *For any real Brownian motion (B_t), we almost surely have*

$$\limsup_{t \to \infty} \frac{B_t}{\sqrt{2t \, \log(\log t)}} = 1 = -\liminf_{t \to \infty} \frac{B_t}{\sqrt{2t \, \log(\log t)}}.$$

Proof We have to prove that

$$\limsup_{t \searrow 0} \frac{B_t}{\sqrt{2t \, \log \, \log(1/t)}} = 1.$$

This is equivalent to the claim via the transforms $(B_t) \mapsto (-B_t)$ and $(B_t) \mapsto (t\, B_{1/t})$ of Corollary 6.2.3(3).

We set $h(t) := \sqrt{2t \, \log \, \log(1/t)}$, and fix $\theta, \delta \in]0, 1[$. We also set $\alpha_n := (1+\delta)\theta^{-n}h(\theta^n)$ and $\beta_n := h(\theta^n)/2$, for any $n \in \mathbb{N}^*$. Doob's inequality (more precisely, Theorem 6.3.2(i)) applied to the exponential martingale $X_s := e^{\alpha_n B_s - \alpha_n^2 s/2}$ ensures that

$$\mathbb{P}\left[\sup_{0 \le s \le 1}(B_s - \frac{\alpha_n s}{2}) \ge \beta_n\right] = \mathbb{P}\left[\sup_{0 \le s \le 1} X_s \ge e^{\alpha_n \beta_n}\right] \le e^{-\alpha_n \beta_n} = \left(n \, |\log \, \theta|\right)^{-1-\delta},$$

whence

$$\mathbb{P}\left(\liminf_n \left\{ \sup_{0 \le s \le 1} (B_s - \alpha_n s/2) < \beta_n \right\} \right) = 1$$

by the Borel–Cantelli lemma. Hence, we almost surely have the following: for any large enough n and for $\theta^n \le s < \theta^{n-1}$,

$$B_s \le \beta_n + \frac{\alpha_n s}{2} \le \beta_n + \frac{\alpha_n \theta^{n-1}}{2} = \left(\tfrac{1}{2} + \frac{1+\delta}{2\theta} \right) h(\theta^n) \le \left(\tfrac{1}{2} + \frac{1+\delta}{2\theta} \right) h(s).$$

This proves that almost surely

$$\limsup_{s \searrow 0} B_s/h(s) \le \left(\tfrac{1}{2} + \frac{1+\delta}{2\theta} \right),$$

whence $\limsup_{s \searrow 0} B_s/h(s) \le 1$, letting $(1-\theta)$ and δ go to 0.

As we shall use only this upper bound, we refer to [MK] for the analogous proof of the lower bound. ◇

Note that the above law of the iterated logarithm, in the form

$$\limsup_{t \searrow 0} \frac{B_t}{\sqrt{2t \, \log \, \log(1/t)}} = 1,$$

shows the almost sure non-differentiability of the Brownian path at 0, and then at any $a \in \mathbb{R}_+$ by the Markov property (of Corollary 6.2.3). In fact, this non-differentiability occurs in a stronger sense, see for example [RY], Chapter I, eqn (2.9).

Definition 6.3.4 An $(\mathcal{F}_t)$-***stopping time*** *is a random variable S taking values in $\overline{\mathbb{R}_+} = \mathbb{R}_+ \cup \{\infty\}$ such that $\{S \le t\} \in \mathcal{F}_t$ for all $t \ge 0$. The σ-field associated with a stopping time S is*

$$\mathcal{F}_S := \{ F \in \mathcal{F} \mid F \cap \{S \le t\} \in \mathcal{F}_t \quad \text{for any } t \ge 0 \}.$$

This implies that $\{S < t\} \in \mathcal{F}_t$ for all t, and this is actually equivalent if the filtration is right-continuous, which means that $\mathcal{F}_{s+} := \bigcap_{t>s} \mathcal{F}_t$ equals $\mathcal{F}_s$ for any $s > 0$. The classical examples of stopping times are the hitting times of open sets by a right-continuous non-anticipating process.

It is easy to check that if S and T are stopping times, their supremum or infimum is also a stopping time. Almost sure limits of non-decreasing sequences of stopping times are stopping times, and the same holds for non-increasing sequences if the filtration is right-continuous.

If S, T are stopping times then the stochastic interval $[S, T[:= \{(s, \omega) \in \mathbb{R}_+ \times \Omega \mid S(\omega) \le s < T(\omega)\}$ is adapted and càdlàg. If $S \le T$, then $\mathcal{F}_S \subset \mathcal{F}_T$.

A stopping time can always be approximated uniformly from above by a non-increasing sequence of discrete-valued stopping times, by taking $S_n = k + 1/2^n$ on $\{k/2^n \le S < k+1/2^n\}$ to approach S.

This approximation procedure easily implies that the Markov property of Corollary 6.2.3 extends to stopping times, as follows.

Corollary 6.3.5 *(Strong Markov property) A Brownian motion (B_t) satisfies the strong Markov property: for any finite stopping time S, $(B_{S+t} - B_S)$ is also a Brownian motion, and is independent of $\mathcal{F}_S$ (if $(\mathcal{F}_t)$ denotes the natural Brownian filtration of Corollary 6.2.3 or at least, if for all $0 < s < t$, $(B_t - B_s)$ is independent of $\mathcal{F}_s$).*

Proof This is straightforward from the weak Markov property (Corollary 6.2.3) if $T = \sum_{j \in \mathbb{N}} \alpha_j 1_{A_j}$, with $A_j \in \mathcal{F}_{\alpha_j}$ for each j. For the general case, consider a sequence T_n of stopping times having the preceding form, which decreases to T: we obtain a sequence of standard Brownian motions, which are independent of $\mathcal{F}_T$ and which converge (almost surely uniformly on any $\{T \leq N, t \leq N\}$) to $(B_{T+t} - B_T)$. $\diamond$

Proposition 6.3.6 *(hitting times) For all $t, x > 0$, the law of the hitting time $T_x := \min\{s > 0 | B_s = x\}$ of a real Brownian motion (B_s) is given by*

$$\mathbb{P}(T_x < t) = \mathbb{P}(\max\{B_s | s \leq t\} > x) = 2\mathbb{P}(B_t > x) = \mathbb{P}(|B_t| > x)$$

$$= \int_0^t \frac{x}{\sqrt{2\pi s^3}} e^{-x^2/2s} \, ds.$$

Proof The second equality is due to the reflection principle between times T_x and t, given the event $\{T_x < t\}$: setting $T_x^t := \min\{T_x, t\}$, so that $\{T_x < t\} = \{T_x^t < t\}$, and using the strong Markov property (Corollary 6.3.5), we know that the random variable $B_{T_x^t + (t - T_x^t)} - B_{T_x^t}$ is centred Gaussian (null on $\{T_x^t = t\}$) conditionally on $\mathcal{F}_{T_x^t}$. So, we have

$$\mathbb{P}(B_t > x) = \mathbb{P}(T_x < t, B_t - B_{T_x} > 0) = \mathbb{P}\big(T_x^t < t, B_{T_x^t + (t - T_x^t)} - B_{T_x^t} > 0\big)$$

$$= \mathbb{E}\Big(\mathbb{P}\big(B_{T_x^t + (t - T_x^t)} - B_{T_x^t} > 0 | \mathcal{F}_{T_x^t}\big) 1_{\{T_x^t < t\}}\Big) = \mathbb{E}\big(\tfrac{1}{2} 1_{\{T_x^t < t\}}\big) = \tfrac{1}{2}\mathbb{P}(T_x < t).$$

For the last equality, we differentiate $2\mathbb{P}(B_t > x) = \sqrt{2/\pi t} \int_x^\infty e^{-y^2/2t} \, dy$ with respect to t. $\diamond$

Note Note in particular that the hitting times T_x are not integrable (for any $x \neq 0$).

Proposition 6.3.7 *Denote by $\mathcal{M}_c$ the space of continuous square-integrable martingales, and by $\mathcal{M}_c^\infty$ the subspace of those which are bounded in L^2. We have the following.*
   (i) Any $(M_t) \in \mathcal{M}_c^\infty$ converges in L^2 to some random variable $M_\infty \in L^2$. Moreover $\mathbb{E}(M_\infty | \mathcal{F}_t) = M_t$ for any t, almost surely.
   (ii) $\mathcal{M}_c^\infty$ is a Hilbert space for the norm $\|M\| := \|M_\infty\|_2$.

Proof The discrete-time case (recall the end of Section 6.1) provides a random variable $M_\infty \in L^2$, which is the limit in L^2 of M_s along any sequence $\{n/2^m | n \in \mathbb{N}\}$. As above, we have

$$\left\| M_\infty - M_{n/2^m} \right\|_2^2 \leq \sum_{k=n+1}^{\infty} \left\| M_{k/2^m} - M_{(k-1)/2^m} \right\|_2^2 = \left\| M_\infty \right\|_2^2 - \left\| M_{n/2^m} \right\|_2^2 .$$

Now, for any continuous martingale, continuity holds also in the L^2-norm, by Doob's second inequality (i.e., Theorem 6.1.5(ii)). Thus we obtain $\left\| M_\infty - M_t \right\|_2^2 \leq \left\| M_\infty \right\|_2^2 - \left\| M_t \right\|_2^2$ for any positive t, and the right-hand side decreases to 0 as t increases to infinity. This yields the statement (i). Moreover, the same Doob's inequality also implies straightforwardly statement (ii).   ◇

Remark 6.3.8 (i) The convergence $M_t \to M_\infty$ of Proposition 6.3.7 occurs almost surely also. Equivalently, L^2 convergence of martingales entails almost sure convergence; see for example [Do], [N1] or [RY], Theorem II.2.10.

(ii) If (M_s) belongs to $\mathcal{M}_c$, it is clear by Corollary 6.1.4 that for any bounded stopping time T, $(s \mapsto M_{s \wedge T})$ belongs to $\mathcal{M}_c^\infty$.

6.4   The Itô integral

We will now define the notion of a stochastic integral in continuous time. This is the Itô integral. The problem is to make sense of expressions of the form $\int_0^t H_s \, dB_s$, where (B_s) is a Brownian motion which is not differentiable and has infinite variation on any non-empty open interval. The solution is to restrict in a suitable way the class of processes that we integrate.

The simplest càdlàg adapted processes are **the step processes**, i.e., those processes (H_s) which can be expressed as

$$H_s(\omega) := \sum_{j=0}^{\infty} U_j(\omega) 1_{[T_j, T_{j+1}[}(s), \qquad (6.1)$$

with a non-decreasing sequence $0 = T_0 \leq T_1 \leq \ldots$ of stopping times going almost surely to infinity, with $U_j \in b\mathcal{F}_{T_j}$ (i.e., U_j is bounded and $\mathcal{F}_{T_j}$-measurable), and such that $\mathbb{E}\left[\int_0^\infty H_s^2 \, ds \right] < \infty$. Let Λ^∞ denote the space of all càdlàg adapted processes such that $\mathbb{E}\left[\int_0^\infty H_s^2 \, ds \right] < \infty$. It clearly contains the step processes.

Note for future reference that the integrals of these processes are almost surely continuous, since $\left| \int_{t-\varepsilon}^t H_s \, ds \right|^2 \leq \varepsilon \int_0^t H_s^2 ds$.

Lemma 6.4.1 *Step processes are dense in Λ^∞ in the sense that for all H in Λ^∞, there exists a sequence of step processes $H^{(n)}$ such that*

$$\lim_{n \to \infty} \int_0^\infty \mathbb{E}\left[|H_s - H_s^{(n)}|^2 \right] ds = 0.$$

*This holds even for **simple** step processes, i.e., step processes of the form of eqn (6.1) but with constant times T_j.*

Proof Take for example $\tilde{H}_s^{(n)} := \max\{-n, \min\{n, H_s\}\}$ and

$$H_s^{(n)} := \sum_{j \geq 0} \tilde{H}_{T_j^{(n)}}^{(n)} 1_{[T_j^{(n)}, T_{j+1}^{(n)}[}(s),$$

where $T_0^{(n)} := 0$ and $T_{j+1}^{(n)} := \inf\{s > T_j^{(n)} \mid |\tilde{H}_s^{(n)} - \tilde{H}_{T_j^{(n)}}^{(n)}| > 2^{-n}\}$. The second claim follows immediately from the first and from the uniform approximation of stopping times by discrete-valued ones. $\diamond$

Similarly, for any $t \geq 0$, let Λ^t denote the space of càdlàg adapted processes such that $\mathbb{E}\left[\int_0^t H_s^2 \, ds\right] < \infty$.

Assume we are given an $(\mathcal{F}_t)$-**Brownian motion** B, that is to say, an $(\mathcal{F}_t)$-adapted Brownian motion such that for all $0 < s < t$, $(B_t - B_s)$ is independent of $\mathcal{F}_s$. The Itô integral is defined naturally on step processes H written as in eqn (6.1), by

$$\int_0^t H \, dB \equiv \int_0^t H_s \, dB_s := \sum_{j \geq 0} U_j (B_{T_{j+1} \wedge t} - B_{T_j \wedge t}).$$

Equivalently, denoting by $N(t)$ the largest integer j such that $T_j \leq t$ and setting $U_{N(t)} = 0$,

$$\int_0^t H \, dB = \sum_{j=0}^{N(t)-1} U_j (B_{T_{j+1}} - B_{T_j}) + U_{N(t)} (B_T - B_{T_{N(t)}}).$$

Lemma 6.4.2 *(i) $\int_0^t H \, dB$ is a continuous martingale; (ii) $\int_0^t H \, dB$ does not depend on the representation (6.1) of H; (iii) for any step process, we have for all t, positive or infinite,*

$$\mathbb{E}\left[\left|\int_0^t H_s \, dB_s\right|^2\right] = \mathbb{E}\left[\int_0^t H_s^2 \, ds\right].$$

Proof This is left as an exercise. (For (i), consider the case $N(t) = 2$ first and $s < T_1 < t < T_2$, to understand why $\mathbb{E}\left[\int_0^t H \, dB \mid \mathcal{F}_s\right] = \int_0^s H \, dB$. (iii) follows very easily from a simple calculation, in which all cross terms vanish under the expectation, by the definition of an $(\mathcal{F}_t)$-Brownian motion.) $\diamond$

It is now clear (using Proposition 6.3.7) that the Itô integral extends by density to a linear map $\mathcal{I}s$ from Λ^∞ into the space $\mathcal{M}_c^\infty$ of square-integrable continuous martingales bounded in L^2, and that we have

$$\mathbb{E}\left[\int_0^\infty H_s^2\,ds\right] = \mathbb{E}\left[\mathcal{I}s(H)_\infty^2\right] = \|\mathcal{I}s(H)\|^2.$$

$\mathcal{I}s(H)_t$ is usually denoted by $\int_0^t H_s\,dB_s$ or $\displaystyle\int_0^t H\,dB$, for all $t \leq \infty$. This notation is coherent with the following natural property.

Lemma 6.4.3 *For any stopping time τ and any $H \in \Lambda^\infty$, $t \mapsto H_t\,1_{[0,\tau[}(t)$ belongs to Λ^∞,*

$$\int_0^\infty 1_{\{s<\tau\}} H_s\,dB_s = \mathcal{I}s(H)_\tau =: \int_0^\tau H\,dB$$

and

$$\mathbb{E}\left[\left|\int_0^\tau H\,dB\right|^2\right] = \mathbb{E}\left[\int_0^\tau H_s^2\,ds\right].$$

Proof This is obvious for step processes, and if H_n converges towards H in Λ^∞, then $H_n\,1_{[0,\tau]}$ also converges towards $H\,1_{[0,\tau]}$ in Λ^∞, and we have (by Doob's inequality) the almost sure uniform convergence of the two stochastic integrals. ◇

Remark 6.4.4 In exactly the same way, we can see that for any pair of stopping times $\sigma \leq \tau$ and any (H_t) in Λ^∞, $(t \mapsto H_t\,1_{[\sigma,\tau[}(t))$ belongs to Λ^∞, and

$$\int_0^\infty 1_{\{\sigma\leq s<\tau\}} H_s\,dB_s = \mathcal{I}s(H)_\tau - \mathcal{I}s(H)_\sigma =: \int_\sigma^\tau H_s\,dB_s \equiv \int_\sigma^\tau H\,dB.$$

Moreover, the Itô isometric identity holds:

$$\mathbb{E}\left[\left|\int_\sigma^\tau H\,dB\right|^2\right] = \mathbb{E}\left[\int_\sigma^\tau H_s^2\,ds\right].$$

The following is known as Doob's optional sampling theorem (written here for stochastic integrals).

Proposition 6.4.5 *For any stopping time T, any $H \in \Lambda^\infty$ and any $(\mathcal{F}_t)$-Brownian motion (B_s), we have almost surely*

$$\mathbb{E}\left[\int_0^\infty H\,dB \,\Big|\, \mathcal{F}_T\right] = \int_0^T H\,dB.$$

Proof By the definition of $\mathcal{I}s(H)$, we have to verify that for any $A \in \mathcal{F}_T$, $\mathbb{E}\left[1_A \int_T^\infty H\,dB\right] = 0$. This holds for constant times T by Lemma 6.4.2(i). Therefore if $T = \sum_{k\in\mathbb{N}} 1_{A_k} k2^{-n}$ (with pairwise disjoint $A_k \in \mathcal{F}_{k2^{-n}}$) is a discrete-valued stopping time, we have

$$\mathbb{E}\left[1_A \int_T^\infty H\,dB\right] = \sum_{k\in\mathbb{N}} \mathbb{E}\left[1_A \int_{k2^{-n}}^\infty 1_{A_k} H\,dB\right] = \sum_{k\in\mathbb{N}} \mathbb{E}\left[1_{A\cap A_k} \int_{k2^{-n}}^\infty H\,dB\right] = 0.$$

Hence if (T_n) is a non-increasing sequence of stopping times converging uniformly to T, we have $A \in \mathcal{F}_{T_n}$, and then

$$\mathbb{E}\left[1_A \int_T^\infty H\,dB\right] = \lim_{n\to\infty} \mathbb{E}\left[1_A \int_{T_n}^\infty H\,dB\right] = 0. \qquad \diamondsuit$$

We can extend the definition of the Itô integral slightly by considering the space $\Lambda := \bigcap_{t>0} \Lambda^t$ of càdlàg adapted processes such that $\mathbb{E}\left[\int_0^t H_s^2\,ds\right] < \infty$ for all t. It is obvious that the Itô integral $\int_0^t H\,dB$ extends to a linear map from Λ into the space $\mathcal{M}_c$ of square-integrable continuous martingales, and that we still have the **Itô isometric identity**

$$\mathbb{E}\left[\left|\int_0^t H\,dB\right|^2\right] = \mathbb{E}\left[\int_0^t H_s^2\,ds\right].$$

Note that this is equivalent to

$$\mathbb{E}\left[\left(\int_0^t H\,dB\right)\times\left(\int_0^t K\,dB\right)\right] = \mathbb{E}\left[\int_0^t H_s K_s\,ds\right], \ \forall\, H, K \in \Lambda. \qquad (6.2)$$

As before, for any H in Λ and for any stopping times $\sigma \le \tau$, $H\,1_{[\sigma,\tau[}$ belongs to Λ and $\int_0^t 1_{\{\sigma \le s < \tau\}} H_s\,dB_s = \int_{t\wedge\sigma}^{t\wedge\tau} H\,dB$. Moreover, for any $H \in \Lambda$, any $t \ge 0$ and any stopping times $\sigma \le \tau$, we have

$$\mathbb{E}\left[\left|\int_{t\wedge\sigma}^{t\wedge\tau} H\,dB\right|^2\right] = \mathbb{E}\left[\int_{t\wedge\sigma}^{t\wedge\tau} H_s^2\,ds\right]. \qquad (6.3)$$

In particular, $(t \mapsto B_{t\wedge\tau})$ is a square-integrable martingale.

Exercise Show that if $\mathbb{E}(\tau)$ is finite, then $\mathbb{E}(B_\tau) = 0$ and $\mathbb{E}(B_\tau^2) = \mathbb{E}(\tau)$. Deduce that the expectation of the hitting time for 1 by B is infinite.

6.5 Itô's formula

To prove Itô's formula, which is the fundamental result of stochastic calculus, we begin with two lemmas. We first observe the following.

Lemma 6.5.1 *For any càdlàg adapted process H bounded by C and for any $s, t \ge 0$, we have*

$$\mathbb{E}\left[\left|\int_s^t H\,dB\right|^4\right] \le 9\,C^4(t - s)^2.$$

Proof We can take $s = 0$ for simplicity. Let us consider simple step processes first. If $H_t(\omega) = \sum_{j\geq0} U_j(\omega)\, 1_{]t_j,t_{j+1}]}(t)$, with $\max_j |U_j| \leq C$, then we have

$$
\mathbb{E}\left[\left|\int_0^t H\, dB\right|^4\right] = \mathbb{E}\left[\sum_{j\geq0} U_j^4\, (B_{t\wedge t_{j+1}} - B_{t\wedge t_j})^4\right]
$$

$$
+ 6\,\mathbb{E}\left[\sum_{0\leq j<k<N} U_j^2 U_k^2\, (B_{t\wedge t_{j+1}} - B_{t\wedge t_j})^2 (B_{t\wedge t_{k+1}} - B_{t\wedge t_k})^2\right]
$$

$$
+ 12\,\mathbb{E}\left[\sum_{0\leq i<j<k<N} U_i U_j U_k^2\, [B_{t\wedge t_{i+1}} - B_{t\wedge t_i}][B_{t\wedge t_{j+1}} - B_{t\wedge t_j}][B_{t\wedge t_{k+1}} - B_{t\wedge t_k}]^2\right],
$$

as all other terms vanish. The first term is clearly bounded by $3\,C^4\,t^2$, and the sum of the two last terms equals

$$
6\sum_{k\geq1}\mathbb{E}\left[\left(\sum_{j=0}^{k-1} U_j\,(B_{t\wedge t_{j+1}} - B_{t\wedge t_j})\right)^2 U_k^2(B_{t\wedge t_{k+1}} - B_{t\wedge t_k})^2\right],
$$

Then, by conditioning first with respect to $\mathcal{F}_{t_k}$, the latter sum is bounded by

$$
6\,C^2\sum_{k\geq1}(t\wedge t_{k+1} - t\wedge t_k)\,\mathbb{E}\left[\left|\int_0^{t\wedge t_k} H\, dB\right|^2\right] \leq 6\,C^2\,t\,\mathbb{E}\left[\int_0^t H_s^2\, ds\right] \leq 6\,C^4\,t^2.
$$

This gives the required result for simple step processes. The general case follows immediately by Fatou's lemma and an approximation by a bounded sequence of simple step processes (recall Lemma 6.4.1), from which we can extract a subsequence to obtain almost sure convergence.   ◇

The following lemma is the essential step in the proof of Itô's formula.

Lemma 6.5.2 *For any bounded càdlàg adapted process H in Λ, we have:*
(i) $\sum_{k=1}^n \left|\int_{t(k-1)/n}^{tk/n} H_s\, dB_s\right|^2$ *converges to* $\int_0^t H_s^2\, ds$ *in L^2 as $n\to\infty$; and*
(ii) $\sum_{k=1}^n Y_{t(k-1)/n}\int_{t(k-1)/n}^{tk/n} H_s\, dB_s$ *converges to* $\int_0^t Y_s H_s\, dB_s$ *in L^2 as $n\to\infty$,*
for any bounded continuous càdlàg adapted process Y.

Proof *(i)* First, note that for any positive s and t,

$$
\mathbb{E}\left[\left(\int_t^{t+s} H_u\, dB_u\right)^2 - \int_t^{t+s} H_u^2\, du\,\Big|\,\mathcal{F}_t\right] = 0.
$$

In fact, for any $\mathcal{F}_t$-measurable set A, $\tau = 1_{A(\omega)}t + 1_{A^c(\omega)}(t+s)$ is a stopping time, and Remark 6.4.4 (with $\sigma \equiv t$) shows that

$$\mathbb{E}\left[\left(\left|\int_t^{t+s} H_u \, dB_u\right|^2 - \int_t^{t+s} H_u^2 \, du\right) 1_A\right]$$

vanishes. From this, we find easily that

$$\mathbb{E}\left[\left(\sum_{k=1}^n \left|\int_{t(k-1)/n}^{tk/n} H \, dB\right|^2 - \int_0^t H_s^2 \, ds\right)^2\right]$$

$$= \sum_{k=1}^n \mathbb{E}\left[\left(\left|\int_{t(k-1)/n}^{tk/n} H \, dB\right|^2 - \int_{t(k-1)/n}^{tk/n} H_s^2 \, ds\right)^2\right],$$

since the expectations of all cross terms vanish. By Lemma 6.5.1, this quantity is dominated by

$$2\sum_{k=1}^n \mathbb{E}\left[\left|\int_{t(k-1)/n}^{tk/n} H \, dB\right|^4 + \left|\int_{t(k-1)/n}^{tk/n} H_s^2 \, ds\right|^2\right] \leq \frac{20\,C^4\,t^2}{n}.$$

(ii) Set $Y^{n,t} := \sum_{k=0}^{n-1} Y_{t(k-1)/n} 1_{[tk/n, t(k+1)/n[}$, so that

$$\sum_{k=1}^n Y_{t(k-1)/n} \int_{t(k-1)/n}^{tk/n} H_s \, dB_s = \int_0^t Y_s^{n,t} H_s \, dB_s.$$

This is obvious for any simple step process H, and this extends by continuity to every $H \in \Lambda$. Then (ii) follows from the convergence of $Y^{n,t}H$ to $YH1_{[0,t[}$ in Λ^∞, by dominated convergence. ◇

Let us now define the spaces $\mathcal{S}_b$ and $\mathcal{S}$ of **semimartingales** of bounded and L^2 type, respectively as the space of square-integrable continuous processes which can be written as a sum

$$x_0 + \int_0^t H_s \, dB_s + \int_0^t K_s \, ds$$

where $x_0 \in \mathbb{R}$, and H, K are bounded càdlàg adapted processes and processes belonging to Λ respectively.

Note $\mathcal{S}$ is made up of continuous semimartingales and is included in Λ, since

$$\mathbb{E}\left[\int_0^t \left|\int_0^u K_s \, ds\right|^2 du\right] \leq \frac{t^2}{2} \mathbb{E}\left[\int_0^t K_s^2 \, ds\right]$$

and

$$\mathbb{E}\left[\int_0^t \left|\int_0^u H\, dB\right|^2 du\right] \le t\, \mathbb{E}\left[\int_0^t H_s^2\, ds\right].$$

Lemma 6.5.2 can be extended easily to $\mathcal{S}_b$, since the treatment of the terms $\int_0^t K_s ds$ is straightforward, with obvious upper bounds. We thus get the following lemma.

Lemma 6.5.3 *For any* $X_t = x_0 + \int_0^t H_s\, dB_s + \int_0^t K_s\, ds$ *in* $\mathcal{S}_b$, *we have*
(i) $\sum_{k=1}^n (X_{tk/n} - X_{t(k-1)/n})^2$ *converges to* $\int_0^t H_s^2\, ds$ *in* L^2 *as* $n \to \infty$; *and*
(ii) $\sum_{k=1}^n Y_{t(k-1)/n}(X_{tk/n} - X_{t(k-1)/n})$ *converges to* $\int_0^t Y_s H_s\, dB_s + \int_0^t Y_s K_s\, ds$ *in* L^2 *as* $n \to \infty$, *for any bounded continuous càdlàg adapted process* Y.

Theorem 6.5.4 *(Itô's formula) Consider a semimartingale in* $\mathcal{S}_b$,

$$X_t := x_0 + \int_0^t H_s\, dB_s + \int_0^t K_s\, ds,$$

and a C^2 *function* Φ *having bounded derivatives. Then* $\Phi(X_t)$ *is in* $\mathcal{S}_b$, *and*

$$\Phi(X_t) - \Phi(x_0) = \int_0^t \Phi'(X_s)H_s\, dB_s + \int_0^t \Phi'(X_s)K_s\, ds + \frac{1}{2}\int_0^t \Phi''(X_s)H_s^2\, ds.$$
(6.4)

The last term is known as the **Itô correction**, by comparison with the usual calculus (chain rule) formula, which we recover by taking $H = 0$. Note that formally, one can write

$$d\,\Phi(X_t) = \Phi'(X_s)H_s\, dB_s + \Phi'(X_s)K_s\, ds + \frac{1}{2}\,\Phi''(X_s)H_s^2\, ds$$

$$= \Phi'(X_s)\, dX_s + \frac{1}{2}\,\Phi''(X_s)H_s^2\, ds.$$

The formula looks like a second-order Taylor formula with '$(dB_t)^2 = dt$'. For example, for any real a, b and any $t \ge 0$ (by linearity, Φ can be $\mathbb{C}$-valued as well),

$$e^{\sqrt{-1}\, a\, (B_t + b\, t)} = 1 + \sqrt{-1}\, a\int_0^t e^{\sqrt{-1}\, a\, (B_s + b\, s)}\, dB_s$$

$$+ \left(\sqrt{-1}\, ab - \frac{a^2}{2}\right)\int_0^t e^{\sqrt{-1}\, a\, (B_s + b\, s)}\, ds.$$

Before proving the Itô formula, let us notice that it can be extended by an important method called **localization**.

Corollary 6.5.5 *Itô's formula (6.4) holds in* $\mathcal{S}$ *for any* C^2 *function* Φ *and any semimartingale* $X \in \mathcal{S}$ *such that* $(\Phi'(X_s)H_s)$, $(\Phi'(X_s)K_s)$ *and* $(\Phi''(X_s)H_s^2)$ *are in* Λ.

Proof If H, K are bounded, we can define a sequence of stopping times $T_n \leq n$, increasing to infinity and such that for $0 \leq s \leq T_n$, X_s, $\Phi'(X_s)H_s$, $\Phi'(X_s)K_s$ and $\Phi''(X_s)H_s^2$ are bounded by n. We then apply Itô's formula in $\mathcal{S}_b$ to $(X_{t \wedge T_n})$, and let n go to infinity. Convergence follows straightforwardly from the Itô isometric identity (6.3). The general case is obtained by approaching H, K by bounded processes. $\diamond$

Proof of Theorem 6.5.4 Using the localization argument given in the proof of Corollary 6.5.5, we can assume that X is bounded. Since, on any bounded interval, we can find a sequence of polynomials (Φ_n) such that Φ_n and its first two derivatives approximate Φ and its first two derivatives uniformly, we can, see that it is enough to show the formula when Φ is a polynomial. Therefore, by induction on the degree of Φ, it is enough to show that for two processes (X_t) and $(\overline{X}_t)$ in $\mathcal{S}_b$, the following integration-by-parts formula holds:

$$X_t \overline{X}_t - x_0 \overline{x}_0 = \int_0^t (\overline{X}_s H_s + X_s \overline{H}_s) \, dB_s + \int_0^t (\overline{X}_s K_s + X_s \overline{K}_s + H_s \overline{H}_s) \, ds.$$

In fact, applying this to $\overline{X} = \Phi \circ X$ completes the induction. Now, using bilinearity, it is enough to consider the case $X = \overline{X}$. Finally, writing

$$X_t^2 - x_0^2 = 2 \sum_{k=1}^n X_{t(k-1)/n} \left(X_{tk/n} - X_{t(k-1)/n} \right) + \sum_{k=1}^n \left(X_{tk/n} - X_{t(k-1)/n} \right)^2,$$

we see that the result is a direct consequence of Lemma 6.5.3. $\diamond$

Example We get

$$e^{a \, B_t - (a^2/2t)} = 1 + a \int_0^t e^{a \, B_s - (a^2/2s)} \, dB_s$$

for any $a \in \mathbb{R}$ and $t \geq 0$. This shows that Brownian exponential martingales are stochastic integrals (added to 1).

In particular, for any stopping time τ, $e^{a \, B_{\tau \wedge t} - (a^2/2)\tau \wedge t}$ is a square-integrable martingale. As an application, we can compute the law of the hitting time T_x for the level x by a standard Brownian motion starting at 0. Then $\mathbb{E}\left(e^{a \, B_{T_x \wedge t} - (a^2/2)T_x \wedge t}\right) = 1$. By symmetry, we can take $ax > 0$, let $t \to \infty$ and use dominated convergence to conclude that $\mathbb{E}\left(e^{-(a^2/2) \, T_x}\right) = e^{-|a \, x|}$.

Exercise Find the law of the hitting point for a given line by a planar Brownian motion.

The Itô formula can easily be generalized to functions of d variables and to the case where we consider r independent Brownian motions B^p. The spaces $\mathcal{S}_b$ and $\mathcal{S}$ of semimartingales of bounded and L^2 type, respectively, become the spaces of square-integrable continuous processes which can be written as a sum $x_0 + \sum_{p=1}^{r} \int_0^t H_s^p \, dB_s^p + \int_0^t K_s \, ds$, with bounded càdlàg adapted processes and processes in Λ, respectively, H^p and K. Thus we have the following theorem.

Theorem 6.5.6 *(Itô's formula) Consider d semimartingales in $\mathcal{S}_b$,*

$$X_t^j := x_0^j + \sum_{p=1}^{r} \int_0^t H_s^{j,p} \, dB_s^p + \int_0^t K_s^j \, ds\,, \quad 1 \le j \le d,$$

and a C^2 function Φ with bounded derivatives on $\mathbb{R}^d$. Then $\Phi(X_t) \equiv \Phi(X_t^1, \ldots, X_t^d)$ belongs to $\mathcal{S}_b$, and we have

$$\Phi(X_t) - \Phi(x_0) = \sum_{j=1}^{d} \sum_{p=1}^{r} \int_0^t \Phi_j'(X_s) H_s^{j,p} \, dB_s^p + \sum_{j=1}^{d} \int_0^t \Phi_j'(X_s) K_s^j \, ds$$

$$+ \frac{1}{2} \sum_{1 \le i,j \le d} \sum_{p=1}^{r} \int_0^t \Phi_{i,j}''(X_s) H_s^{i,p} H_s^{j,p} \, ds. \tag{6.5}$$

Proof The proof is the same as for Theorem 6.5.4, except that we have to consider also terms of the form $\sum_{k=1}^{n} \left(\int_{t(k-1)/n}^{tk/n} H_s^p \, dB_s^p \right) \left(\int_{t(k-1)/n}^{tk/n} H_s^q \, dB_s^q \right)$ for $1 \le p < q \le r$, to establish that they converge to 0 in L^2. To prove this, as in the proof of Lemma 6.5.2, we first check that $\mathbb{E}\left[\left(\int_s^t H_u^p \, dB_u^p \right) \left(\int_s^t H_u^q \, dB_u^q \right) \Big| \mathcal{F}_s \right] = 0$, which is obvious for step processes. Then we observe that by the Schwarz inequality and Lemma 6.5.1, if H^p and H^q are uniformly bounded by C, then

$$\mathbb{E}\left[\left| \int_s^t H_u^p \, dB_u^p \right|^2 \times \left| \int_s^t H_u^q \, dB_u^q \right|^2 \right] \le 9\,C^4 \,(t-s)^2. \qquad \diamond$$

A statement similar to Corollary 6.5.5 holds for $\mathcal{S}$.

Notation 6.5.7 If

$$X_t := x_0 + \sum_{p=1}^{r} \int_0^t H_s^p \, dB_s^p + \int_0^t K_s \, ds$$

is in $\mathcal{S}$, we use the notation

$$\langle X, X \rangle_t := \sum_{p=1}^{r} \int_0^t |H_s^p|^2 \, ds.$$

If

$$\overline{X}_t := \overline{x}_0 + \sum_{p=1}^{r} \int_0^t \overline{H_s^p} \, dB_s^p + \int_0^t \overline{K}_s \, ds$$

is another element of $\mathcal{S}$, we use the notation

$$\langle X, \overline{X} \rangle_t := \sum_{p=1}^{r} \int_0^t H_s^p \, \overline{H_s^p} \, ds.$$

If J is a càdlàg adapted process such that $(H_s^p J_s)$ and $(K_s J_s)$ belong to Λ, we use the notation

$$\int_0^t J_s \, dX_s := \sum_{p=1}^{r} \int_0^t J_s \, H_s^p \, dB_s^p + \int_0^t J_s \, K_s \, ds.$$

In this way, the multidimensional Itô formula in $\mathcal{S}_b$ (or in $\mathcal{S}$, when it holds) can be written as

$$\Phi(X_t) - \Phi(x_0) = \sum_i \int_0^t \Phi'_i(X_s) \, dX_s^i + \tfrac{1}{2} \sum_{i,j} \int_0^t \Phi''_{i,j}(X_s) \, d\langle X^i, X^j \rangle_s. \quad (6.6)$$

Note in particular that the **quadratic covariation** $\langle X, \overline{X} \rangle_t$ is bilinear, and note also the following integration-by-parts formula: for any X in $\mathcal{S}$ and Y in $\mathcal{S}_b$, $X\,Y$ belongs to $\mathcal{S}$, and we have

$$X_t \, Y_t = x_0 \, y_0 + \int_0^t Y_t \, dX_t + \int_0^t Y_t \, dX_t + \langle X, Y \rangle_t. \quad (6.7)$$

Observe that by eqn (6.7) uniqueness of the decomposition in the space $\mathcal{S}$ holds, since if

$$X_t = \sum_{p=1}^{r} \int_0^t H_s^p \, dB_s^p + \int_0^t K_s \, ds = 0,$$

then we get

$$\sum_{p=1}^{r} \int_0^t \left(H_s^p \right)^2 \, ds = \langle X, X \rangle_t = 0$$

and then

$$\sum_{p=1}^{r} \int_0^t H_s^p \, dB_s^p = \int_0^t K_s \, ds = 0.$$

The following application of Itô's formula provides a characterization of d-dimensional Brownian motion.

Theorem 6.5.7 (P. Lévy) Let

$$M_t^j := x_0^j + \sum_{p=1}^{r} \int_0^t H_s^{j,p} \, dB_s^p,$$

$1 \leq j \leq d$, be d martingales in $\mathcal{S}$ such that $\langle M^j, M^k \rangle_t = t 1_{\{j=k\}}$ for all $1 \leq j, k \leq d$ and $t \in \mathbb{R}_+$.

Then $M^1, \ldots, M^d$ are d independent Brownian motions (in other words, $(M^1, \ldots, M^d)$ is a d-dimensional Brownian motion).

Proof Itô's formula (6.6) (extended as in Corollary 6.5.5) ensures that

$$Z_t := \exp\left[\sum_{j=1}^d \lambda_j M_t^j - \tfrac{1}{2} \sum_{j,k=1}^d \lambda_j \lambda_k \langle M^j, M^k \rangle_t\right]$$

defines a martingale for any sequence of real numbers $\lambda_1, \ldots, \lambda_d$. We fix $t_0 \in \mathbb{R}_+$ and $A \in \mathcal{F}_{t_0}$, and consider the martingales

$$N_t^j := 1_A \times (M_t^j - M_{t \wedge t_0}^j) = \int_{t \wedge t_0}^t 1_A \, dM_s^j,$$

which obey $\langle N^j, N^k \rangle_t = 1_A \times (t - t \wedge t_0) \times 1_{\{j=k\}}$. By changing M to N and $\lambda := (\lambda_1, \ldots, \lambda_d)$ to $\sqrt{-1}\,\lambda$ in Z_t defined above, we obtain the martingale

$$Z_t = \exp\left[\sqrt{-1} \sum_{j=1}^d \lambda_j N_t^j + \tfrac{1}{2} |\lambda|^2 (t - t \wedge t_0) 1_A\right]$$

$$= 1_A \exp\left[\sqrt{-1} \sum_{j=1}^d \lambda_j (M_t^j - M_{t_0}^j) + \tfrac{1}{2} |\lambda|^2 (t - t_0)\right] + 1_{A^c}.$$

Now $\mathbb{E}(Z_t) = 1$ reads

$$\mathbb{E}\left[1_A \times \exp\left(\sqrt{-1} \sum_{j=1}^d \lambda_j (M_t^j - M_{t_0}^j)\right)\right] = \mathbb{P}(A) \times \exp\left(-\tfrac{1}{2} |\lambda|^2 (t - t_0)\right).$$

This shows the independence relations and the Gaussian character, which guarantee the claim. ◇

Remark 6.5.8 It is sometimes convenient to extend the Itô integral to an even larger space. We say that a continuous process (M_t) is a **local martingale** if and only if there exists a sequence of stopping times T_n increasing to infinity such that for all n, $(M_{t \wedge T_n})$ is a martingale of $\mathcal{M}_c^\infty$. This is in fact equivalent to producing a localizing sequence such that the stopped processes are martingales, since we can then build from it, and from hitting times of integers τ_n by M, another localizing sequence $T_n \wedge \tau_n$ satisfying the assumption. Of course, martingales are local martingales. Let Λ^0 be the space of previsible processes such that $\int_0^t X_s^2 \, ds < \infty$ for any $t > 0$ almost surely. The Itô integral extends to a linear map from Λ^0 into the space of local martingales. Itô's formula also extends to this framework, but it is often less directly useful, as the Itô integral in it is then only a local martingale.

We shall need the following Burholder–Davis–Gundy inequality.

Proposition 6.5.9 *Fix $p \geq 1$. There exists a positive constant C_p such that for any càdlàg adapted process $H \in \Lambda$ and for any $s, t \geq 0$, we have*

$$\mathbb{E}\left[\left|\int_s^t H \, dB\right|^{2p}\right] \leq C_p \, \mathbb{E}\left[\left|\int_s^t H^2\right|^p\right].$$

Proof Up to modifying H, we can take $s = 0$ for simplicity. Consider the continuous martingale $M_t := \int_0^t H \, dB$. Note that by using a non-decreasing sequence of stopping times going almost surely to infinity, we can assume that H and M are bounded. Applying Itô's formula (6.4) to M, we get

$$|M_t|^{2p} = 2p \int_0^t |M_s|^{2p-1} \operatorname{sign}(M_s) \, H_s \, dB_s + p(2p-1) \int_0^t |M_s|^{2(p-1)} H_s^2 \, ds$$

and hence, taking the expected value and using Hölder's inequality,

$$\mathbb{E}\left[|M_t|^{2p}\right] = p(2p-1) \, \mathbb{E}\left[\int_0^t |M_s|^{2(p-1)} H_s^2 \, ds\right]$$

$$\leq p(2p-1) \, \mathbb{E}\left[\sup_{0 \leq s \leq t} |M_s|^{2(p-1)} \int_0^t H_s^2 \, ds\right]$$

$$\leq p(2p-1) \, \mathbb{E}\left[\sup_{0 \leq s \leq t} |M_s|^{2p}\right]^{(p-1)/p} \times \mathbb{E}\left[\left|\int_0^t H^2\right|^p\right]^{1/p}.$$

Using Doob's inequality (recall Theorem 6.3.2), we obtain

$$\mathbb{E}\left[|M_t|^{2p}\right] \leq p(2p-1) \, \mathbb{E}\left[\left(\frac{2p}{2p-1}\right)^{2p} |M_t|^{2p}\right]^{(p-1)/p} \times \mathbb{E}\left[\left|\int_0^t H^2\right|^p\right]^{1/p}$$

and hence, finally,

$$\mathbb{E}\left[|M_t|^{2p}\right] \leq \left(\frac{(2p)^{2p-1}}{2(2p-1)^{2p-3}}\right)^p \times \mathbb{E}\left[\left|\int_0^t H^2\right|^p\right]. \qquad \diamond$$

Proposition 6.5.10 *For any $d \geq 2$, almost surely a d-dimensional Brownian motion (B_s) never vanishes at any positive time s.*

Proof It is obviously enough to consider a two-dimensional Brownian motion $(B_s) = (B_s^1, B_s^2)$. Suppose first that it starts at $B_0 \neq 0$, and denote by τ its hitting time for 0. Applying Itô's formula (6.6) to $\log |B_t| = \frac{1}{2} \log\left(|B_t^1|^2 + |B_t^2|^2\right)$ and recalling that the log function is harmonic in $\mathbb{R}^2$, we obtain for $0 \leq t < \tau$

$$\log |B_t| = \int_0^t \frac{B_s^1 \, dB_s^1 + B_s^2 \, dB_s^2}{|B_s^1|^2 + |B_s^2|^2}.$$

For any $n \in \mathbb{Z}$, let τ_n denote the hitting time for the sphere of radius e^{-n} (centred at 0). If $\mathbb{P}(\tau < \infty) > 0$, choose an $N \in \mathbb{N}^*$ such that $\mathbb{P}(\tau_{-N} > \tau) > 0$. choose also $n_0 > -\log|B_0|$. For any $n \geq n_0$, $(\log|B_{t \wedge \tau_{-N} \wedge \tau_n}|)$ is a bounded continuous martingale and, by Proposition 6.4.5, $(\log|B_{\tau_{-N} \wedge \tau_n}|)_{n \geq n_0}$ is a discrete martingale. Now, since $\log|B_{\tau_{-N} \wedge \tau_n}| = N 1_{\{\tau_{-N} < \tau_n\}} - n 1_{\{\tau_{-N} > \tau_n\}}$, we have $\mathbb{E}[\log|B_{\tau_{-N} \wedge \tau_{n_0}}|] = N - (N+n) \mathbb{P}(\tau_{-N} > \tau_n) \longrightarrow -\infty$ as $n \to \infty$, a contradiction, which proves that $\tau = \infty$ almost surely.

Suppose now that $B_0 = 0$. By the strong Markov property (Corollary 6.3.5) and the above argument applied to every $(B_{\tau_n + t} - B_{\tau_n})$, we see that almost surely (B_s) never vanishes after the time τ_n. This yields the claim, since by the continuity of (B_s) at 0 we have $\lim_{n \to \infty} \tau_n = 0$ almost surely. $\diamond$

6.6   The Stratonovich integral

This modification of the Itô integral proves to be convenient for expressing the solutions of linear stochastic differential equations (and for stochastic calculations on manifolds; see [IW], Section VI.6).

Definition 6.6.1   *Let* $B^1, \ldots, B^d$ *be d independent standard real Brownian motions, and let* X, Y *belong to* $\mathcal{S}$. *The* **Stratonovich integral** *of* X *with respect to* Y *is defined almost surely , for all* $t \in \mathbb{R}_+$, *by*

$$\int_0^t X \circ dY := \int_0^t X \; dY + \frac{1}{2} \langle X, Y \rangle_t. \tag{6.8}$$

Proposition 6.6.2   *Let* $B^1, \ldots, B^d$ *be d independent standard real Brownian motions, let* $X^1, \ldots, X^n$ *belong to* $\mathcal{S}_b$, *let* $X := (X^1, \ldots, X^n)$ *and let* F *be a function of class* C^3 *on* $\mathbb{R}^n$, *having bounded derivatives. Then, almost surely for all* $t \in \mathbb{R}_+$, *we have*

$$F(X_t) = F(X_0) + \sum_{j=1}^n \int_0^t \frac{\partial F}{\partial x^j}(X_s) \circ dX_s^j. \tag{6.9}$$

Proof   We apply Definition (6.8) to $\int_0^t (\partial F / \partial x^j)(X_s) \circ dX_s^j$, and Itô's formula (6.5) to $(\partial F / \partial x^j)(X_s)$ and to $F(X_s)$, to obtain

$$\int_0^t \frac{\partial F}{\partial x^j}(X_s) \circ dX_s^j = \int_0^t \frac{\partial F}{\partial x^j}(X_s) \; dX_s^j + \frac{1}{2} \int_0^t \frac{\partial^2 F}{\partial x^j \, \partial x^k}(X_s) \; d\langle X^j, X^k \rangle_s$$

$$= F(X_s) - F(X_0).   \qquad \diamond$$

Remark 6.6.3   In an analogous way, the so-called **backward integral** is given by $2 \int_0^t X \circ dY - \int_0^t X \; dY$. This happens to be the limit in probability of the stochastic Riemannian sums $\sum_{j=0}^{N-1} X_{t_{j+1}}(Y_{t_{j+1}} - Y_{t_j})$, in the spirit of Lemma 6.5.3. Equivalently, $\int_0^t X \circ dY$ is the limit in probability of the stochastic Riemannian sums $\sum_{j=0}^{N-1} \frac{1}{2}(X_{t_j} + X_{t_{j+1}})(Y_{t_{j+1}} - Y_{t_j})$.

6.7   Notes and comments

The theory of martingales was developed by Doob, in connection with potential theory [Do].

Brownian motion was studied by Einstein and Langevin, in relation to heat transport and the kinetic theory of gases. The construction of the associated measure on continuous paths is due to Wiener; see also [Lév].

There is an extensive literature on Brownian motion. Properties of sample paths such as Hölderianity, the existence of a quadratic variation, the structure of the set of zeros and the existence of local times have been extensively studied and are described in many references; see for example [Lév], [IMK], [Bi], [RY], [KS] and [Dt]. In addition, Brownian motion allows a reinterpretation of potential-theoretical notions such as capacity, balayage and potentials; see [PSt] and [Do].

Stochastic calculus was introduced by Bernstein [Be], Itô [I] and Gihman ([Gi1], [Gi2]); see the first pages of [IW] for a historical account. A short classical exposition was given by McKean [MK]. Short introductions are given in [Bi] and [Dt].

The theory can be extended to Lévy processes (with Poisson integrals) and Markov processes [KW], and beyond [Me]. Many treatises and lectures on this theory have been published, some of them including a general study of stochastic processes; see in particular [GS], [J], [IW], [KS], [RY], [RW2], [Ka], [O] and [Pr].

Chapter Seven
Brownian motions on groups of matrices

The main application of Itô's calculus is to provide solutions to stochastic differential equations driven by Wiener processes (and, more generally, semi-martingales). Existence and uniqueness are based, as in the deterministic case, on Picard's iteration method. The simplest equations of this type are linear ones, and in this chapter we construct left and right Brownian motions on groups of matrices as solutions to linear stochastic differential equations. We establish in particular that the solution of such an equation lives in the subgroup associated with the Lie subalgebra generated by the coefficients of the equation. We also consider reversed processes, Hilbert–Schmidt estimates, approximation by stochastic exponentials, Lyapunov exponents and diffusion processes.

We then concentrate on some important examples: the Heisenberg group, PSL(2), SO(d), PSO(1, d), the affine group $\mathbb{A}^d$ and the Poincaré group $\mathcal{P}^{d+1}$. By a projection, we obtain the spherical and hyperbolic Brownian motions, and relativistic diffusion in Minkowski space.

Relativistic diffusion is a continuous process on a relativistic phase space. Its law is invariant under any Lorentz transformation. It can be lifted to the Poincaré group, where it describes the motion of a solid body subjected to infinitely many infinitely small boosts which are stationary, uniformly distributed and independent, in the proper frame and proper time of the body. (Note that this diffusion should not be viewed as a model for physical Brownian motion in a relativistic setting, in so far as physical Brownian motion should be produced by boosts coming from interaction with a medium. They should not occur stationarily in the proper time of the solid.)

7.1 Stochastic differential equations

We consider stochastic differential equations (SDEs for short) of the homogeneous type

$$X_s^x = x + \int_0^s \sigma(X_t^x)\, dB_t + \int_0^s b(X_t^x)\, dt,$$

where (B_t) is a Brownian motion of $\mathbb{R}^{d'}$, σ is a function from $\mathbb{R}^d$ into $\mathcal{L}(\mathbb{R}^{d'}, \mathbb{R}^d)$, b is a function from $\mathbb{R}^d$ into $\mathbb{R}^d$, $x \in \mathbb{R}^d$, s runs over $\mathbb{R}_+$ and the unknown is the process $(X_s^x)_{s \geq 0}$. This is also denoted by

$$dX_t = \sigma(X_t) \, dB_t + b(X_t) \, dt.$$

Theorem 7.1.1 *Take two functions* σ, b *on* $\mathbb{R}^d$, *where* σ *is* $\mathcal{L}(\mathbb{R}^{d'}, \mathbb{R}^d)$-*valued, and* b *is* $\mathbb{R}^d$-*valued, which are globally Lipschitz: there exists a constant* C *such that*

$$\|\sigma(x) - \sigma(x')\| + |b(x) - b(x')| \leq C \, |x - x'|, \quad \text{for all } x, x' \in \mathbb{R}^d.$$

Let $x \in \mathbb{R}^d$, *and let* (B_t) *be a Brownian motion of* $\mathbb{R}^{d'}$, *with filtration* $\big(\mathcal{F}_s := \sigma\{B_t \,|\, 0 \leq t \leq s\}\big)$. *Then there exists a unique continuous* $(\mathcal{F}_s)$-*adapted process* (X_s^x) *such that*

$$X_s^x = x + \int_0^s \sigma(X_t^x) \, dB_t + \int_0^s b(X_t^x) \, dt, \quad \text{for any } s \geq 0.$$

Proof We drop the index x, and choose the Euclidean norm as the norm on $\mathbb{R}^d, \mathbb{R}^{d'}$ and $\mathcal{L}(\mathbb{R}^{d'}, \mathbb{R}^d) \equiv \mathbb{R}^{dd'}$. We set $X_s^0 \equiv x$, and for $n \in \mathbb{N}$,

$$X_s^{n+1} := x + \int_0^s \sigma(X_t^n) \, dB_t + \int_0^s b(X_t^n) \, dt$$

and

$$E_s^n := \mathbb{E}\left[\sup_{0 \leq t \leq s} \left|X_t^{n+1} - X_t^n\right|^2 \right].$$

In particular, we have

$$E_s^0 = \mathbb{E}\left[\sup_{0 \leq t \leq s} \left|\sigma(x)B_t + b(x)t\right|^2 \right] \leq 2 \, \|\sigma(x)\|^2 \sum_{j=1}^{d'} \mathbb{E}\left[\sup_{0 \leq t \leq s} |B_t^j|^2 \right] + 2s^2 \, |b(x)|^2$$

$$\leq 2\big(\|\sigma(0)\| + C \, |x|\big)^2 (d' \times 4s) + 2s^2 \big(|b(0)| + C \, |x|\big)^2$$

$$= \mathcal{O}\big[(1 + |x|)^2 (4d' + s)s\big] < \infty.$$

We proceed by induction, taking any $n \in \mathbb{N}^*$ and supposing that for some constant $C' \geq 2C^2$,

$$E_u^{n-1} \leq \big(1 + |x|\big)^2 \times \big[(4d' + s)C'\big]^n \times u^n/n!, \quad \text{for } 0 \leq u \leq s.$$

Thus $\sup\limits_{0 \leq v \leq \cdot} \|\sigma(X_v^n) - \sigma(X_v^{n-1})\|$ is square integrable, so that $\int_0^{\cdot} [\sigma(X_v^n) - \sigma(X_v^{n-1})] \, dB_v$ is a square-integrable $\mathbb{R}^d$-valued martingale (recall Section 6.4; we mean that its components in the canonical basis of $\mathbb{R}^d$ are square-integrable martingales), to which we want to apply Doob's inequality (Theorem 6.3.2) and Itô's isometric identity (recall Remark 6.4.4). Theorem 6.3.2 was written for real square-integrable martingales, and Remark 6.4.4 dealt with real

Brownian motion, but a vector version is easily deduced as follows. Writing for short

$$\left[\sigma(X_v^n) - \sigma(X_v^{n-1})\right] =: \phi(v) = \left((\phi(v)_j^i)\right)_{1 \le i \le d, 1 \le j \le d'},$$

by Theorem 6.3.2 and Remark 6.4.4 we have

$$\mathbb{E}\left[\sup_{0 \le t \le u} \left| \int_0^t \left[\sigma(X_v^n) - \sigma(X_v^{n-1})\right] dB_v \right|^2 \right]$$

$$= \mathbb{E}\left[\sup_{0 \le t \le u} \sum_{i=1}^d \left| \sum_{j=1}^{d'} \int_0^t \phi(v)_j^i \, dB_v^j \right|^2 \right]$$

$$\le d' \sum_{i=1}^d \sum_{j=1}^{d'} \mathbb{E}\left[\sup_{0 \le t \le u} \left| \int_0^t \phi(v)_j^i \, dB_v^j \right|^2 \right] \le 4d' \sum_{i=1}^d \sum_{j=1}^{d'} \mathbb{E}\left[\int_0^u |\phi(v)_j^i|^2 \, dv\right]$$

$$= 4d' \, \mathbb{E}\left[\int_0^u \|\sigma(X_v^n) - \sigma(X_v^{n-1})\|^2 \, dv\right].$$

Using this and the Schwarz inequality, we obtain for $0 \le u \le s$

$$E_u^n = \mathbb{E}\left[\sup_{0 \le t \le u} \left| \int_0^t \left[\sigma(X_v^n) - \sigma(X_v^{n-1})\right] dB_v + \int_0^t \left[b(X_v^n) - b(X_v^{n-1})\right] dv \right|^2 \right]$$

$$\le 2\,\mathbb{E}\left[\sup_{0 \le t \le u} \left| \int_0^t \left[\sigma(X_v^n) - \sigma(X_v^{n-1})\right] dB_v \right|^2 \right] + 2\,\mathbb{E}\left[\sup_{0 \le t \le u} \left| \int_0^t \left[b(X_v^n) - b(X_v^{n-1})\right] dv \right|^2 \right]$$

$$\le 8d' \, \mathbb{E}\left[\int_0^u \|\sigma(X_v^n) - \sigma(X_v^{n-1})\|^2 \, dv\right] + 2\,\mathbb{E}\left[u \int_0^u |b(X_v^n) - b(X_v^{n-1})|^2 \, dv\right]$$

$$\le 2\,C^2 \times (4d' + u) \times \int_0^u \mathbb{E}[|X_v^n - X_v^{n-1}|^2] \, dv \le C'(4d' + s) \int_0^u E_v^{n-1} \, dv$$

$$\le (1 + |x|)^2 \times [(4d' + s)C']^{n+1} \times u^{n+1}/(n+1)!,$$

which proves by induction that for $0 \le u \le s$,

$$E_u^n \le (1 + |x|)^2 \left[(4d' + s)C'\right]^{n+1} \frac{u^{n+1}}{(n+1)!}.$$

Hence the series $\sum_n \sqrt{E_s^n}$ converges, the variable $\sum_n \sup_{0 \le t \le s} |X_t^{n+1} - X_t^n|$ is integrable and then the sequence X^n converges almost surely uniformly on compact subsets of $\mathbb{R}_+$ to a continuous process X, which is necessarily the wanted solution. In particular, it is clearly $(\mathcal{F}_s)$-adapted.

As to uniqueness, we see in the same way as above, if X, Y are two solutions, that

$$E_s := \mathbb{E}\left(\sup_{0 \le t \le s} |X_t - Y_t|^2 \right) \le (4d' + s)C' \int_0^s E_u \, du.$$

Hence, if $E_s \le N$ is finite, we obtain $E_u \le N\left[(4d' + s)C'\right]^n (u^n/n!)$ (for $0 \le u \le s$) by induction, whence $E_s = 0$ for any $s > 0$, which means that X and Y are indistinguishable.

Now we can reduce the proof to this case, by localization, using the stopping times $T_N := \inf\left\{ t \,|\, |X_t| + |Y_t| > \sqrt{N} \right\}$: in fact, we can apply the above to $(X_t^N, Y_t^N) := (X_{t \wedge T_N}, Y_{t \wedge T_N})$ and to

$$E_s(N) := \mathbb{E}\left(\sup_{0 \le t \le s} |X_t^N - Y_t^N|^2 \right) = \mathbb{E}\left(\sup_{0 \le t \le s \wedge T_N} |X_t - Y_t|^2 \right) \le N.$$

This shows that almost surely $X_t = Y_t$ for all $t \in [0, T_N]$, and then for all $t \ge 0$ since $\lim_{N \to \infty} T_N = \infty$, by continuity of X, Y.   ◇

We now prove a comparison theorem for real SDEs. Consider a real $(\mathcal{F}_t)$-Brownian motion (B_t), and Lipschitz real functions σ, b on $\mathbb{R}$, so that Theorem 7.1.1 applies, guaranteeing the existence and uniqueness of the real diffusion (X_s), which is the strong (i.e., $(\mathcal{F}_t)$-adapted) solution to the following stochastic differential equation:

$$X_s = X_0 + \int_0^s \sigma(X_t) \, dB_t + \int_0^s b(X_t) \, dt.$$

Consider moreover two adapted continuous processes $(\beta_1(t)), (\beta_1(t))$, and real $(\mathcal{F}_t)$-adapted processes $(X_s^1), (X_s^2)$, solving that the following stochastic differential equations (for $j = 1, 2$):

$$X_s^j = X_0^j + \int_0^s \sigma(X_t^j) \, dB_t + \int_0^s \beta_j(t) \, dt.$$

Theorem 7.1.2 *Suppose that the following comparison assumption almost surely holds:*

$$\beta_1(t) \le b(X_t^1) \quad and \quad b(X_t^2) \le \beta_2(t) \quad for \ all \ t \ge 0, \quad and \quad X_0^1 \le X_0^2.$$

Then, almost surely, $X_t^1 \le X_t^2$ for all $t \ge 0$.

Proof By the usual localization argument, we can suppose that σ and b are uniformly bounded and Lipschitz. For any $n \in \mathbb{N}^*$, we denote by φ_n a continuous function from $\mathbb{R}$ into $[0, 2n]$, null on $]-\infty, 0] \cup [1/n, \infty[$, and such that $\int_0^{1/n} \varphi_n = 1$. We now set $\phi_n(x) := \int_0^x dy \int_0^y \varphi_n$. Note that $(x - 1/n)^+ \le \phi_n(x) \le x^+$ for any $x \in \mathbb{R}$.

Applying Itô's formula, we have

$$\phi_n(X_s^1 - X_s^2) = \phi_n(X_s^1 - X_s^2) - \phi_n(X_0^1 - X_0^2)$$

$$= \int_0^s \phi_n'(X_t^1 - X_t^2) \left[\sigma(X_t^1) - \sigma(X_t^2)\right] dB_t$$

$$+ \int_0^s \phi_n'(X_t^1 - X_t^2)\left[\beta_1(t) - \beta_2(t)\right] dt + \frac{1}{2}\int_0^s \varphi_n(X_t^1 - X_t^2) \left|\sigma(X_t^1) - \sigma(X_t^2)\right|^2 dt.$$

The first integral on the right-hand side has mean 0, while the third is $\mathcal{O}(1/n)$, since

$$0 \le \int_0^s \varphi_n(X_t^1 - X_t^2) \left|\sigma(X_t^1) - \sigma(X_t^2)\right|^2 dt$$

$$\le \int_0^s (2n) \times \sup\left\{\left|\sigma(x) - \sigma(x')\right|^2 \,\Big|\, |x - x'| \le 1/n\right\}.$$

As to the second one, $J_s := \int_0^s \phi_n'(X_t^1 - X_t^2)\left[\beta_1(t) - \beta_2(t)\right] dt$, we have

$$J_s \le \int_0^s \phi_n'(X_t^1 - X_t^2)\left[b(X_t^1) - b(X_t^2)\right] dt \le C\int_0^s 1_{\{X_t^1 > X_t^2\}} |X_t^1 - X_t^2| \, dt$$

$$= C\int_0^s (X_t^1 - X_t^2)^+ \, dt.$$

Hence, letting $n \to \infty$, we obtain

$$\mathbb{E}\left[(X_s^1 - X_s^2)^+\right] \le C\int_0^s \mathbb{E}\left[(X_t^1 - X_t^2)^+\right] dt,$$

whence $\mathbb{E}\left[(X_s^1 - X_s^2)^+\right] = 0$ for all $s \ge 0$. This yields the result. ◇

7.2 Linear stochastic differential equations

We specialize Theorem 7.1.1 here to the case of main interest for the following discussion: linear equations. The main feature is that if we take convenient coefficients for such an equation in some Lie subalgebra $\mathcal{G}$, the resulting process (which solves the equation) will live in the associated subgroup (even if this is not a Lie subgroup), yielding a Brownian motion on this group.

Theorem 7.2.1 *Consider $A_0, A_1, \ldots, A_k \in M(d)$, $X_0 \in M(d)$ and an $\mathbb{R}^k$-valued standard Brownian motion $W_t = (W_t^1, \ldots, W_t^k)$. Then there exists a unique continuous $M(d)$-valued $(\sigma\{W_t \mid 0 \le t \le s\})$-adapted process (X_s), which is the solution of*

$$X_s = 1 + \sum_{j=1}^k \int_0^s X_t \, A_j \, dW_t^j + \int_0^s X_t \left(\frac{1}{2}\sum_{j=1}^k A_j^2 + A_0\right) dt. \tag{7.1}$$

Moreover, the right increments of (X_s) are independent and homogeneous: for any $t \geq 0$, the process $s \mapsto X_t^{-1} X_{s+t}$ has the same law as the process $s \mapsto X_s$, and is independent of the σ-field $\mathcal{F}_t$ generated by the Brownian motion (W_s) up to time t.

Furthermore this statement remains true if the constant time t is replaced by a stopping time T conditionally on the event $\{T < \infty\}$.

Proof The first part is merely a particular case of Theorem 7.1.1. Then, for any fixed $r > 0$, we have

$$X_{r+s} = X_r + \sum_{j=1}^{k} \int_r^{r+s} X_t \, A_j \, dW_t^j + \int_r^{r+s} X_t \left(\frac{1}{2} \sum_{j=1}^{k} A_j^2 + A_0 \right) dt,$$

or, equivalently,

$$X_r^{-1} X_{r+s} = \mathbf{1} +$$

$$\sum_{j=1}^{k} \int_0^s X_r^{-1} X_{r+t} \, A_j \, d(W_{r+t}^j - W_r^j) + \int_0^s X_r^{-1} X_{r+t} \left(\frac{1}{2} \sum_{j=1}^{k} A_j^2 + A_0 \right) dt.$$

Hence $X_r^{-1} X_{r+t}$ satisfies the same equation (7.1) as X_s, up to shifting the Brownian motion (W_t) by the time r. This proves the result, since this shifted Brownian motion is Brownian as well, and is independent of $\mathcal{F}_r$. Finally, this proof is also valid with a stopping time T instead of r, restricted to the event $\{T < \infty\}$. ◇

Note The process $\left(W_t^A := \sum_{j=1}^k A_j W_t^j \right)$ is a Brownian motion on the Lie subalgebra $\mathcal{G}$ generated by $A_0, \ldots, A_k$. More precisely, this is a continuous process with independent and homogeneous increments, and we have, for any $u \in \mathcal{G}^*$ and any $t \geq 0$: $\mathbb{E}\left[e^{\sqrt{-1} \, u(W_t^A)} \right] = e^{-t^2/2 \, \alpha(u,u)}$, where $\alpha := \sum_{j=1}^k A_j \otimes A_j$ is a non-negative bilinear form on $\mathcal{G}^*$, which determines the law of (W_t^A).

In fact, by considering a basis $(E_1, \ldots, E_\ell)$ of $\mathcal{G}$, and its dual basis $(E_1^*, \ldots, E_\ell^*)$, we can obtain the claimed formula by writing $A_j = \sum_n \lambda_j^n E_n$, $u = \sum_m \mu^m E_m^*$ and

$$u(W_t^A) = \sum_{j,n} \mu^n \lambda_j^n \, W_t^j, \quad \alpha(u,u) = \sum_j u(A_j)^2 = \sum_j \left[\sum_n \mu^n \lambda_j^n \right]^2.$$

The matrix A_0 of eqn (7.1) is known as the **drift** component of the process (X_s).

By applying Itô's formula, we get the following theorem (in which the right Lie derivatives $\mathcal{L}_A$ are defined on $\mathcal{M}(d)$ as in Section 1.1.4).

Theorem 7.2.2 *Consider the solution (X_s) to the linear SDE (7.1), and a function ϕ of class C^2 on $\mathcal{M}(d)$. We then have the following. (i) The process*

(X_s), *the unique solution of eqn (7.1), is almost surely $GL(d)$-valued. (ii) We have*

$$\phi(X_s) = \phi(1) + \sum_{j=1}^{k} \int_0^s \mathcal{L}_{A_j} \phi(X_t) \, dW_t^j + \int_0^s \left[\frac{1}{2} \sum_{j=1}^{k} (\mathcal{L}_{A_j})^2 + \mathcal{L}_{A_0} \right] \phi(X_t) \, dt$$

(7.2)

or, equivalently, in Stratonovich form (recall Section 6.6),

$$\phi(X_s) = \phi(1) + \sum_{j=1}^{k} \int_0^s \mathcal{L}_{A_j} \phi(X_t) \circ dW_t^j + \int_0^s \mathcal{L}_{A_0} \phi(X_t) \, dt. \qquad (7.3)$$

The second-order operator $\mathcal{A} := \frac{1}{2} \sum_{j=1}^{k} (\mathcal{L}_{A_j})^2 + \mathcal{L}_{A_0}$ *is called the* **infinitesimal generator** *(or* **generator** *for short) of the process* (X_s).

Proof We apply Itô's formula (6.5). We have

$$\phi(X_s) - \phi(1) = \sum_{1 \le a,b \le d} \int_0^s \frac{\partial \phi}{\partial X^{ab}}(X_t) \, dX_t^{ab}$$

$$+ \frac{1}{2} \sum_{1 \le a,b,\alpha,\beta \le d} \int_0^s \frac{\partial^2 \phi}{\partial X^{ab} \partial X^{\alpha\beta}}(X_t) \, d\langle X^{ab}, X^{\alpha\beta} \rangle_t$$

$$= \sum_{1 \le a,b \le d} \int_0^s \frac{\partial \phi}{\partial X^{ab}}(X_t) \times \left[\sum_{j=1}^{k} (X_t A_j)^{ab} \, dW_t^j + \left(X_t \left(\frac{1}{2} \sum_{j=1}^{k} A_j^2 + A_0 \right) \right)^{ab} dt \right]$$

$$+ \frac{1}{2} \sum_{1 \le a,b,\alpha,\beta \le d} \sum_{j=1}^{k} \int_0^s \frac{\partial^2 \phi}{\partial X^{ab} \partial X^{\alpha\beta}}(X_t) \, (X_t A_j)^{ab} (X_t A_j)^{\alpha\beta} \, dt$$

$$= \sum_{j=1}^{k} \int_0^s \mathcal{L}_{A_j} \phi(X_t) \, dW_t^j + \int_0^s \left(\frac{1}{2} \sum_{j=1}^{k} \mathcal{L}_{A_j^2} + \mathcal{L}_{A_0} \right) \phi(X_t) \, dt$$

$$+ \frac{1}{2} \sum_{1 \le a,b,\alpha,\beta \le d} \sum_{j=1}^{k} \int_0^s \left[(\mathcal{L}_{A_j})^2 \phi(X_t) - \mathcal{L}_{A_j^2} \phi(X_t) \right] dt,$$

since

$$\mathcal{L}_A \phi(X) = \frac{d_0}{d\varepsilon} \phi(X e^{\varepsilon A}) = \frac{d_0}{d\varepsilon} \phi(X + \varepsilon X A) = \sum_{1 \le a,b \le d} \frac{\partial \phi}{\partial X^{ab}}(X) \, (X A)^{ab}$$

and

$$(\mathcal{L}_A)^2 \phi(X) - \mathcal{L}_{A^2}\phi(X)$$

$$= \sum_{a,b} \frac{d_0}{d\varepsilon}\left[\frac{\partial\phi}{\partial X^{ab}}(X + \varepsilon X A)\big((X + \varepsilon X A)A\big)^{ab}\right] - \sum_{a,b} \frac{\partial\phi}{\partial X^{ab}}(X)(XA^2)^{ab}$$

$$= \sum_{1\leq a,b,\alpha,\beta\leq d} \frac{\partial^2\phi}{\partial X^{ab}\,\partial X^{\alpha\beta}}(X_t)\,(XA)^{ab}(XA)^{\alpha\beta}.$$

This proves eqn (7.2). Equation (7.3) follows at once from eqn (7.2) (applied to ϕ and to $\mathcal{L}_{A_j}\phi$) and from eqn (6.8) (of Definition 6.6.1).

To obtain the first assertion of the statement, let us apply eqn (7.2) to the particular function $\phi := [X \mapsto \det(X)]$. We clearly have

$$\mathcal{L}_A \det(X) = \frac{d_0}{d\varepsilon}\det(\mathbf{1} + \varepsilon A)\det(X) = \operatorname{Tr}(A)\det(X),$$

so that

$$\det(X_s) = 1 +$$

$$\sum_{j=1}^{k}\int_0^s \det(X_t)\,\operatorname{Tr}(A_j)\,dW_t^j + \int_0^s\left[\frac{1}{2}\sum_{j=1}^{k}(\operatorname{Tr}(A_j))^2 + \operatorname{Tr}(A_0)\right]\det(X_t)\,dt,$$

and then

$$\det(X_s) = \exp\left[\sum_{j=1}^{k}\operatorname{Tr}(A_j)\,dW_s^j + \operatorname{Tr}(A_0)\,s\right]$$

almost surely, since by Itô's formula the right-hand side satisfies the above equation for $\det(X_s)$, and since by Theorem 7.1.1 the solution to this equation is unique. This shows that, almost surely, $\det(X_s)$ does not vanish. $\diamond$

Note that the process (X_s) solving the linear SDE (7.1) also possesses the following trivial left invariance property: for any $g \in \mathcal{M}(d)$, $(g\,X_s)$ is the solution to eqn (7.1) where the starting matrix $\mathbf{1} = X_0$ is replaced by $g = X_0$.

Theorem 7.2.3 *Consider the solution (X_s) to the linear SDE (7.1), with coefficients $A_0, A_1, \ldots, A_k$ belonging to some Lie subalgebra $\mathcal{G}$. The process (X_s) takes, almost surely, its values in the group associated with $\mathcal{G}$ (recall Section 1.1.4).*

*This process is called a **left Brownian motion** on G (with drift A_0).*

Proof Let $\mathcal{V}_0 \subset\subset \mathcal{V} \subset\subset \mathcal{V}'$ be compact neighbourhoods of 0 in $\mathcal{M}(d)$ (where $\mathcal{V}_0 \subset\subset \mathcal{V}$ means that $\mathcal{V}_0$ is included in the interior of $\mathcal{V}$), such that the restriction of the exponential map to a neighbourhood of $\mathcal{V}'$ is a diffeomorphism. For $0 \leq t < \tau_{\mathcal{V}'} := \inf\{t > 0 \,|\, X_t \notin \exp(\mathcal{V}')\}$, we have $M_t := \exp^{-1}(X_t) \in \mathcal{V}'$, and d $\exp_{M_t}$ is an isomorphism from $\mathcal{M}(d)$ onto $T_{\exp(M_t)}GL(d) \equiv \exp(M_t) \times \mathcal{M}(d)$, which is given (recall Proposition 1.1.3.2) by

$$d \exp_{M_t}(B) = \exp(M_t) \times \sum_{k \in \mathbb{N}} \frac{\mathrm{ad}(-M_t)^k}{(k+1)!}(B) =: \exp(M_t) \times \alpha_{M_t}(B).$$

Moreover, the linear map α_{M_t} depends analytically on M_t, so that its inverse $\alpha_{M_t}^{-1}$ is well defined for $0 \le t < \tau_{\mathcal{V}'}$ and depends analytically on M_t as well. For such a t, we have

$$\mathcal{L}_{A_j} \exp^{-1}(X_t) = \frac{d_o}{d\varepsilon} \exp^{-1}(X_t + \varepsilon X_t A_j) = d \exp_{X_t}^{-1}(X_t A_j)$$

$$= (d \exp_{M_t})^{-1}(X_t A_j) = \alpha_{M_t}^{-1}(A_j).$$

Hence, applying eqn (7.3) of Theorem 7.2.2 with $\phi = \exp^{-1}$, we obtain, for $0 \le s < \tau_{\mathcal{V}'}$,

$$M_s = \sum_{j=1}^k \int_0^s \alpha_{M_t}^{-1}(A_j) \circ dW_t^j + \int_0^s \alpha_{M_t}^{-1}(A_0) \, dt.$$

Consider now a smooth function ψ from $\mathcal{M}(d)$ into $[0,1]$, equal to 1 on $\mathcal{V}_0$ and vanishing outside $\mathcal{V}$, and then the stochastic differential Stratonovich equation

$$M_s' = \sum_{j=1}^k \int_0^s \left[\alpha_{M_t'}^{-1}(A_j) \times \psi(M_t') \right] \circ dW_t^j + \int_0^s \alpha_{M_t'}^{-1}(A_0) \, \psi(M_t') \, dt, \qquad (7.4)$$

where $M_t' =: x_t^1 V_1 + \cdots + x_t^\ell V_\ell$ belongs to $\mathcal{G}$, $(V_1, \ldots, V_\ell)$ being some fixed basis of $\mathcal{G}$. Note that the matrices $\alpha_{M_t'}^{-1}(A_j) \times \psi(M_t')$ are well defined for any time t, by the choice of the function ψ, which causes them to vanish when M_t' does not belong to $\mathcal{V}$. Moreover, they belong to the Lie subalgebra $\mathcal{G}$, since the restriction of $\alpha_{M_t'}$ to $\mathcal{G}$ is an automorphism of $\mathcal{G}$. Furthermore, they depend smoothly on M_t', which allows us to compute the Itô correction in the above Stratonovich integral according to eqn (6.8), as follows:

$$\frac{1}{2} \sum_{j=1}^k \left\langle \left[\alpha_{M_t'}^{-1}(A_j) \times \psi(M_t') \right], W_t^j \right\rangle$$

$$= \frac{1}{2} \sum_{j=1}^k \left[\psi(M_t') \, d\left[\alpha^{-1}(A_j) \right]_{M_t'} + \alpha_{M_t'}^{-1}(A_j) \, d\psi_{M_t'} \right] \left(\langle M_t', W_t^j \rangle \right)$$

$$= \frac{1}{2} \psi(M_t') \sum_{j=1}^k \left[\psi(M_t') \, d\left[\alpha^{-1}(A_j) \right]_{M_t'} + \alpha_{M_t'}^{-1}(A_j) \, d\psi_{M_t'} \right] \left(\alpha_{M_t'}^{-1}(A_j) \right).$$

Hence eqn (7.4), seen as an equation relating to $(x_t^1, \ldots, x_t^\ell) \in \mathbb{R}^\ell$, satisfies the hypothesis of Theorem 7.1.1. Therefore, applying Theorem 7.1.1 provides a solution M_s' which lives in $\mathcal{G}$ for all times s (and is constant after the random time $\inf\{t > 0 \,|\, M_t' \notin \mathcal{V}\}$).

Now, by the above, M_t satisfies the same equation on the random time interval $[0, \tau_\mathcal{V}[$, with $\tau_\mathcal{V} = \inf\{s > 0 \,|\, M_s \notin \mathcal{V}\}$. Thus, by the uniqueness of the solution to this localized equation, viewing it as an equation relating to $\mathcal{M}(d) \equiv \mathbb{R}^{d^2}$ and applying Theorem 7.1.1, we must have $M_t = M_t'$, and then $M_t \in \mathcal{G}$ and $X_t \in G$, for $0 \le t < \tau_\mathcal{V}$. Here G denotes the group associated with $\mathcal{G}$.

In particular, we have $X_t \in G$ for $0 \le t \le \tau_{\mathcal{V}_0}$. Since the process $X_{\tau_{\mathcal{V}_0}}^{-1} X_{\tau_{\mathcal{V}_0}+t} =: X_t'$ satisfies the same equation as X_t by Theorem 7.2.1, we can apply the above to it, to obtain $X_t' \in G$ for $0 \le t \le \tau_{\mathcal{V}_0}$. This yields $X_t \in G$ for $0 \le t \le \tau_{\mathcal{V}_0^2}$ (where $\mathcal{V}_0^2 = \{MM' \,|\, M, M' \in \mathcal{V}_0\}$). By an obvious induction, we thus obtain $X_t \in G$ for $0 \le t \le \tau_{\mathcal{V}_0^n}$. The result follows, since the increasing sequence of neighbourhoods $(\mathcal{V}_0^n)$ exhausts $\mathcal{M}(d)$, so that the sequence $\tau_{\mathcal{V}_0^n}$ increases to infinity. $\diamond$

Remark 7.2.4 Theorem 7.2.2 $\big($and, in particular, Itô's formula (7.2)$\big)$ is valid on $C^2(G)$. Note that Theorem 7.2.3 is valid even if the group G is not closed.

Remark 7.2.5 The laws (ν_t) of the left Brownian motion (X_t) constitute a convolution semi-group on G, $\nu_{s+t} = \nu_s * \nu_t$, and we have $d/dt\, \nu_t(f) = \nu_t(\mathcal{A}f)$ for any test function f on G. The (infinitesimal) **generator** $\mathcal{A}$ (of Theorem 7.2.2) is thus associated with (ν_t).

Proof The second formula follows at once from eqn (7.2) and, for any non-negative s, t, using Theorem 7.2.1 we have

$$\int_G f \, d(\nu_s * \nu_t) = \int_{G^2} f(gh)\, \nu_s(dg)\, \nu_t(dh)$$

$$= \int_G \mathbb{E}\big[f(g\, X_s^{-1}\, X_{s+t})\big]\, \nu_s(dg) = \mathbb{E}\big[f(X_{s+t})\big] = \int_G f \, d\nu_{s+t}. \ \diamond$$

Definition 7.2.6 *Consider the left Brownian motion (X_s) solving the linear SDE (7.1), and its generator $\mathcal{A}$ (recall Theorem 7.2.2). The associated* **semi-group** *$(P_t)_{t \ge 0}$ is defined by $P_t f(g) = \mathbb{E}\big[f(gX_t)\big]$, for any $f \in C_b(G)$ and $g \in G$.*

Note that $P_t f(g)$ makes sense for $f \in C_b\big(\mathcal{M}(d)\big)$ and $g \in \mathcal{M}(d)$ as well. Owing to Remark 7.2.5, P_t acts by convolution with the law ν_t of X_t: $P_t f = f * \nu_t$.

Note From now on, we assume that the group G is closed in $\mathcal{M}(d)$ (and hence is a Lie group). Note that this implies that the functions in $C_b^2(G)$ admit a C_b^2-continuation on some neighbourhood of G.

Proposition 7.2.7

(i) (P_t) is a family of non-negative endomorphisms on $C_b(G)$, such that P_0 is the identity and $P_t 1 = 1$ for any $t \geq 0$.

(ii) (P_t) satisfies the **semi-group property**, $P_s P_t = P_{s+t}$.

(iii) The semi-group (P_t) is strongly continuous: $\lim\limits_{t \searrow 0} \|P_t f - f\| = 0$, for any $f \in C_b(\mathcal{M}(d))$.

(iv) For any $n \in \mathbb{N}^*$, $g \in G$, $f_0, \ldots, f_n \in C_b(G)$ and $0 \leq t_1 \leq \ldots \leq t_n$,

we have

$$\mathbb{E}\big[f_1(g\, X_{t_1}) \times \cdots \times f_n(g\, X_{t_n})\big] = P_{t_1}\Big[f_1\, P_{t_2-t_1}\big[f_2 \ldots P_{t_n-t_{n-1}} f_n\big]\Big](g).$$

Proof (i) is clear. (ii) follows from the independence and homogeneity of the right increments of (X_s): for any $s, t \geq 0$, we have

$$P_s P_t f(g) = \mathbb{E}\big[P_t f(gX_s)\big] = \mathbb{E}\big[f(gX_s\, X_s^{-1} X_{s+t})\big] = P_{s+t} f(g),$$

which is precisely the semi-group property $\big($which is equivalent to that of the convolution semi-group $(\nu_t)\big)$.

(iii) By eqn (7.2), (P_t) satisfies

$$P_s \phi = \phi + \int_0^s P_t\, \mathcal{A}\, \phi\, dt, \quad \text{for any } \phi \in C_b^2(\mathcal{M}(d)),\ s \geq 0.$$

This implies the strong continuity of (P_t) by the closedness of G, recall the note above.

(iv) Proceeding as for the semi-group property; we have

$$\mathbb{E}\big[f_1(g\, X_{t_1}) \ldots f_n(g\, X_{t_n})\big]$$

$$= \mathbb{E}\Big[f_1(g\, X_{t_1}) \ldots f_{n-2}(g\, X_{t_{n-2}}) \times \big(f_{n-1} \times (P_{t_n-t_{n-1}} f_n)\big)(g\, X_{t_{n-1}})\Big].$$

The claim follows, by an obvious induction on n. ◇

Proposition 7.2.8 *The semi-group* (P_t) *is* **Fellerian** *on* G, *which means that in addition to the properties described in Proposition 7.2.7, it maps* $C_b^2(G)$ *into* $C_b^2(G)$. *Moreover, we have*

$$\frac{d}{ds} P_s \phi = P_s\, \mathcal{A}\, \phi = \mathcal{A}\, P_s \phi \quad \text{for any } \phi \in C_b^2(G),\ s \geq 0.$$

This means in particular that $C_b^2(G)$ *is contained in the* **domain** *of the (infinitesimal) generator* $\mathcal{A}$, *i.e., the space of those* $f \in C_b(G)$ *for which there exists* $\mathcal{A}f \in C(G)$ *such that* $\lim\limits_{t \searrow 0} \big\|1/t(P_t f - f) - \mathcal{A}f\big\| = 0$ *(so that the graph of* $\mathcal{A}$ *is the closure of that of its restriction to* $C_b^2(G)$*).*

Proof The first claim is clear from the very definition of (P_s) in Definition 7.2.6, by dominated convergence. The first equality then follows at once from eqn (7.2), as expressed in the proof of part (iii) of Proposition 7.2.7. Finally, we deduce the second equality in the statement again from the definition of (P_s) by dominated convergence, using the expression for $\mathcal{A}$ (in Theorem 7.2.2) and noticing that the process (X_t) has moments of order 2; see Lemma 7.2.12. $\diamond$

Remark 7.2.9 Equation (7.1) in Theorem 7.2.1 was taken with increments on the right. We can consider similarly the analogous equation with increments on the left,

$$Y_s = 1 - \sum_{j=1}^{k} \int_0^s A_j \, Y_t \, dW_t^j + \int_0^s \left(\frac{1}{2} \sum_{j=1}^{k} A_j^2 - A_0 \right) Y_t \, dt. \tag{7.5}$$

Theorem 7.2.1 is valid in this case as well. Theorem 7.2.2 and eqn (7.2) hold with the left Lie derivatives $-\mathcal{L}'_A$ replacing the right Lie derivatives $\mathcal{L}_A$, so that the process (Y_s) has an infinitesimal generator

$$\frac{1}{2} \sum_{j=1}^{k} (\mathcal{L}'_{A_j})^2 - \mathcal{L}'_{A_0}.$$

Theorem 7.2.1 holds with the left increments $Y_{s+t} Y_t^{-1}$ considered instead of the right increments $X_t^{-1} X_{s+t}$ (and with right invariance instead of left invariance). Theorem 7.2.3 is valid as well: the process (Y_s) is almost surely G-valued if the coefficients A_j belong to $\mathcal{G}$.

The process (Y_s) is then called a **right Brownian motion** on G (with **drift** component $-A_0$).

Proposition 7.2.10 *The right Brownian motion (Y_t) of eqn (7.5) is almost surely the inverse of the left Brownian motion (X_t) of eqn (7.1): $X_t Y_t = 1$ almost surely for all $t \geq 0$.*

Proof Applying Itô's Formula (for integration by parts) to $X_t Y_t$ and using eqns (7.5) and (7.1) we obtain

$$X_t Y_t - 1 = \int_0^t X_s \, dY_s + \int_0^t dX_s \, Y_s + \langle X, Y \rangle_t$$

$$= \sum_{j=1}^{k} \int_0^t X_s \left[-A_j Y_s \, dW_s^j + \left(\frac{1}{2} \sum_{j=1}^{k} A_j^2 - A_0 \right) Y_s \, ds \right]$$

$$+ \sum_{j=1}^{k} \int_0^t \left[X_s A_j \, dW_s^j + X_s \left(\frac{1}{2} \sum_{j=1}^{k} A_j^2 + A_0 \right) ds \right] Y_s - \sum_{j=1}^{k} \int_0^t X_s A_j^2 Y_s \, ds = 0.$$

$\diamond$

Proposition 7.2.11 *Consider the solution* (X_s) *to the linear SDE (7.1), take* $r > 0$, *and set* $\widehat{X}^r_s := X^{-1}_{r-s} X_r$ *and* $\widehat{W}_s := W_{r-s} - W_r$, *for* $0 \le s \le r$. *Then the* **reversed process** $(\widehat{X}^r_s)$ *satisfies the following* **backward equation**: *for* $0 \le s \le r$,

$$\widehat{X}^r_s = 1 - \sum_{j=1}^{k} \int_0^s A_j \widehat{X}^r_t \, d\widehat{W}^j_t + \int_0^s \left(\frac{1}{2} \sum_{j=1}^{k} A_j^2 + A_0 \right) \widehat{X}^r_t \, dt. \tag{7.6}$$

As a consequence, if $A_0 = 0$, *for fixed* $s > 0$ *the variable* X^{-1}_s *has the same law as the variable* X_s. *More generally, for any* A_0 *and fixed* $s > 0$, *the variable* X^{-1}_s *has the same law as the solution* X^*_s *of*

$$X^*_s = 1 + \sum_{j=1}^{k} \int_0^s A_j X^*_t \, dW^j_t + \int_0^s \left(\frac{1}{2} \sum_{j=1}^{k} A_j^2 - A_0 \right) X^*_t \, dt.$$

Note The identity in law between X_s and X^{-1}_s can hold generally only for fixed s, since in a non-Abelian group they do not have their independent increments on the same side. A precise counterexample is given in the note in Section 7.6.3.

Proof Note that $(\widehat{W}_s)$ is also a Brownian motion, with filtration

$$\widehat{\mathcal{F}}_{[s,t]} := \sigma\{\widehat{W}_{s'} - \widehat{W}_s \,|\, s \le s' \le t\} = \sigma\{W_u \,|\, r - t \le u \le r - s\} = \mathcal{F}_{[r-t, r-s]},$$

so that by Remark 7.2.9, eqn (7.6) possesses a unique solution. Recall that we have

$$X_r = X_{r-s} + \sum_{j=1}^{k} \int_{r-s}^{r} X_t A_j \, dW^j_t + \int_{r-s}^{r} X_t \left(\frac{1}{2} \sum_{j=1}^{k} A_j^2 + A_0 \right) dt,$$

whence, setting $Z_s := (\widehat{X}^r)^{-1}_s = X^{-1}_r X_{r-s}$,

$$Z_s = 1 - \sum_{j=1}^{k} \int_{r-s}^{r} X^{-1}_r X_t A_j \, dW^j_t - \int_{r-s}^{r} X^{-1}_r X_t \left(\frac{1}{2} \sum_{j=1}^{k} A_j^2 + A_0 \right) dt.$$

We now consider a dyadic regular partition $r - s = s'_0 < \dots < s'_N = r$ of $[r-s, r]$, and set $s_n := r - s'_{N-n}$, so that $0 = s_0 < \dots < s_N = s$ is a partition of $[0, s]$. We approach the Itô stochastic integrals by Riemann sums (recall Lemma 6.5.3(ii)) as follows:

$$\int_{r-s}^{r} X_t \, dW^j_t = \lim_{N \to \infty} \sum_{n=0}^{N-1} X_{s'_n} \left(W^j_{s'_{n+1}} - W^j_{s'_n} \right)$$

$$= \lim_{N \to \infty} \sum_{n=1}^{N} X_{r - s_{N-n+1}} \left[\widehat{W}^j_{s_{N-n}} - \widehat{W}^j_{s_{N-n+1}} \right]$$

$$= -\lim_{N\to\infty} \sum_{n=0}^{N-1} X_{r-s_{n+1}} \left[\widehat{W}^j_{s_{n+1}} - \widehat{W}^j_{s_n} \right]$$

$$= -\lim_{N\to\infty} \sum_{n=0}^{N-1} X_{r-s_n} \left[\widehat{W}^j_{s_{n+1}} - \widehat{W}^j_{s_n} \right] - \lim_{N\to\infty} \sum_{n=0}^{N-1} (X_{r-s_{n+1}} - X_{r-s_n}) \left[\widehat{W}^j_{s_{n+1}} - \widehat{W}^j_{s_n} \right].$$

Hence

$$\int_{r-s}^{r} X_r^{-1} X_t \, dW_t^j$$

$$= -\lim_{N\to\infty} \sum_{n=0}^{N-1} Z_{s_n} \left[\widehat{W}^j_{s_{n+1}} - \widehat{W}^j_{s_n} \right] - \lim_{N\to\infty} \sum_{n=0}^{N-1} (Z_{s_{n+1}} - Z_{s_n}) \left[\widehat{W}^j_{s_{n+1}} - \widehat{W}^j_{s_n} \right].$$

Observe now that $\left[\widehat{W}^j_{s_{n+1}} - \widehat{W}^j_{s_n} \right] \in \mathcal{F}_{[r-s_{n+1}, r-s_n]}$, while $X_{r-s_n} \in \mathcal{F}_{r-s_n}$. Since, by Theorem 7.2.1, $Z_{s_n} = X_r^{-1} X_{r-s_n}$ is independent of $\mathcal{F}_{r-s_n}$ which contains $\mathcal{F}_{[r-s_{n+1}, r-s_n]}$, Lemma 6.5.3($ii$) yields

$$\lim_{N\to\infty} \sum_{n=0}^{N-1} Z_{s_n} \left[\widehat{W}^j_{s_{n+1}} - \widehat{W}^j_{s_n} \right] = \int_0^s Z_t \, d\widehat{W}_t^j.$$

Then, by polarization (using bilinearity), Lemma 6.5.3(i) yields

$$\lim_{N\to\infty} \sum_{n=0}^{N-1} (Z_{s_{n+1}} - Z_{s_n}) \left[\widehat{W}^j_{s_{n+1}} - \widehat{W}^j_{s_n} \right] = \int_0^s d\langle Z, \widehat{W}^j \rangle_t.$$

Thus we obtain

$$\int_{r-s}^{r} X_r^{-1} X_t \, dW_t^j = -\int_0^s Z_t \, d\widehat{W}_t^j - \int_0^s d\langle Z, \widehat{W}^j \rangle_t.$$

Notice this is just minus the backward integral (of Z.; see Remark 6.6.3).

This shows first that the martingale part of Z_s is $\sum_{j=1}^k \int_0^s Z_t A_j \, d\widehat{W}_t^j$, and then that

$$-\int_{r-s}^{r} X_r^{-1} X_t \, dW_t^j = \int_0^s Z_t \, d\widehat{W}_t^j + \int_0^s Z_t A_j \, dt.$$

Hence we have obtained

$$Z_s = \mathbf{1} + \sum_{j=1}^k \int_0^s Z_t A_j \, d\widehat{W}_t^j + \sum_{j=1}^k \int_0^s Z_t A_j^2 \, dt - \int_{r-s}^{r} Z_{r-t} \left(\frac{1}{2} \sum_{j=1}^k A_j^2 + A_0 \right) dt$$

$$= \mathbf{1} + \sum_{j=1}^k \int_0^s Z_t A_j \, d\widehat{W}_t^j + \int_0^s Z_{r-t} \left(\frac{1}{2} \sum_{j=1}^k A_j^2 - A_0 \right) dt.$$

Now, using Proposition 7.2.10, we see that $\widehat{X}_s^r = Z_s^{-1}$ solves eqn (7.5) with $(-A_0)$ instead of A_0, i.e., precisely eqn (7.6).

Assume, finally, that $A_0 = 0$, so that eqns (7.5) and (7.6) are the same, up to a Brownian change. Then $X_r = \widehat{X}_r^r$ has the same law as $Y_r = X_r^{-1}$ (recall Proposition 7.2.10). $\diamondsuit$

We shall need to control the L^2-norm of the solution to eqn (7.1) (and recall that we have already used this for proving Proposition 7.2.8). Note that the Hilbert–Schmidt norm that we use here on matrices, and denote by $\|\cdot\|_{HS}$ in the following lemma, is merely the Euclidean norm on $\mathcal{M}(d)$ seen as $\mathbb{R}^{d^2}$: $\|X\|_{HS} := \sqrt{\mathrm{Tr}({}^t X\, X)} = \|{}^t X\|_{HS}$.

Lemma 7.2.12 *For any $x \in \mathbb{R}^d$ and $A, X \in \mathcal{M}(d)$, we have $|Ax| \le \|A\|_{HS}\, |x|$, i.e., $\|A\|_{HS}$ dominates the Euclidean operator norm of A; and, moreover,*

$$\|AX\|_{HS} \le \|A\|_{HS} \|X\|_{HS}, \quad |\mathcal{L}_A| \|X\|_{HS}| \le \|A\|_{HS} \|X\|_{HS}$$
$$and \quad \left|(\mathcal{L}_A)^2\left[\|X\|_{HS}^2\right]\right| \le 4 \|A\|_{HS}^2 \|X\|_{HS}^2.$$

As a consequence, the process X_t solving eqn (7.1) in Section 7.2 satisfies the following L^2-norm control: for any $t \ge 0$,

$$\mathbb{E}\left[\|X_t\|_{HS}^2\right] \le d \times \exp\left[2\left(\|A_0\|_{HS} + \sum_{j=1}^{k} \|A_j\|_{HS}^2\right)t\right].$$

Proof A direct application of Schwarz's inequality in $\mathbb{R}^d$ yields $|Ax| \le \|A\|_{HS}\,|x|$ and $\mathrm{Tr}(XY) \le \|X\|_{HS}\,\|Y\|_{HS}$. Note that $\|{}^t X\, X\|_{HS} \le \|X\|_{HS}^2$, as is easily seen by writing ${}^t X\, X = P\, D\, {}^t P$, with $P\, {}^t P = \mathbf{1}$ and D diagonal and non-negative. This implies that

$$\|XY\|_{HS}^2 = \mathrm{Tr}\left({}^t Y\, {}^t X\, XY\right)$$
$$= \mathrm{Tr}\left(Y\, {}^t Y\, {}^t X\, X\right) \le \|Y\, {}^t Y\|_{HS}\, \|{}^t X\, X\|_{HS} \le \|X\|_{HS}^2\, \|Y\|_{HS}^2.$$

Otherwise, $\mathcal{L}_A\left[\|X\|_{HS}^2\right] = \mathrm{Tr}\left[{}^t A\, {}^t X X + {}^t X\, X A\right] = \mathrm{Tr}\left[({}^t A + A)\, {}^t X X\right]$ and

$$\mathcal{L}_A^2\left[\|X\|_{HS}^2\right] = \mathrm{Tr}\left[{}^t A^2\, {}^t X X + 2\, {}^t A\, {}^t X X A + {}^t X X\, A^2\right] = \mathrm{Tr}\left[({}^t A^2 + 2 A\, {}^t A + A^2)\, {}^t X X\right],$$

so that by Schwarz's inequality

$$\left|\mathcal{L}_A\left[\|X\|_{HS}^2\right]\right| \le \|{}^t A + A\|_{HS} \times \|X\|_{HS}^2 \le 2 \|A\|_{HS} \times \|X\|_{HS}^2,$$

and

$$\left| \mathcal{L}_A^2 \, \|X\|_{HS}^2 \right| \leq \|{}^t A^2 + 2A \, {}^t A + A^2 \|_{HS} \, \|X\|_{HS}^2$$

$$\leq \left[\|{}^t A^2\|_{HS} + 2 \, \|A \, {}^t A\|_{HS} + \|A^2\|_{HS} \right] \|X\|_{HS}^2 \leq 4 \, \|A\|_{HS}^2 \, \|X\|_{HS}^2.$$

Set $K := \|A_0\|_{HS} + \sum_{j=1}^{k} \|A_j\|_{HS}^2$, and for any $N \in \mathbb{N}$, $T_N := \inf \left\{ t \geq 0 \, | \right.$ $\left. \|X_t\|_{HS}^2 \geq N \right\}$.

Then, by applying Itô's formula (7.2) (with $\phi(X) = \|X\|_{HS}^2$ and $s = t \wedge T_N$), taking the expectation (note that the martingale part of $\|X_{t \wedge T_N}\|_{HS}^2$ has its L^2-norm bounded by $2N\sqrt{Kt}$) and using the above estimates, we obtain the following, for any $N \in \mathbb{N}$, $t \geq 0$:

$$\mathbb{E}\left[\|X_{t \wedge T_N}\|_{HS}^2 \right] \leq \|\mathbf{1}\|_{HS}^2 + \mathbb{E}\left[\int_0^{t \wedge T_N} 2K \, \|X_s\|_{HS}^2 \, ds \right] \leq d$$

$$+ 2K \int_0^t \mathbb{E}\left[\|X_{s \wedge T_N}\|_{HS}^2 \right] ds,$$

Hence, by (Gronwall) classical iteration, $\mathbb{E}\left[\|X_{t \wedge T_N}\|_{HS}^2 \right] \leq d \, e^{2Kt}$, and then, by $N \to \infty$ and Fatou's lemma, $\mathbb{E}\left[\|X_t\|_{HS}^2 \right] \leq d \, e^{2Kt}$. $\diamond$

The following simple observation will be useful in Section 7.3 and Section B.2.

Remark 7.2.13 For any $\mathcal{M}(d)$-valued continuous process $(\mathcal{A}_t)$, any real Brownian motion (W_t) and any $0 \leq u < v$, we have

$$\mathbb{E}\left[\left\| \int_u^v \mathcal{A}_t \, dW_t \right\|_{HS}^2 \right] = \int_u^v \mathbb{E}\left[\|\mathcal{A}_t\|_{HS}^2 \right] dt.$$

Proof By the very definition of $\| \cdot \|_{HS}$ and by Itô's formula and the linearity of the trace, we have

$$\mathbb{E}\left[\left\| \int_u^v \mathcal{A}_t \, dW_t \right\|_{HS}^2 \right] = \mathbb{E}\left[\mathrm{Tr}\left(\int_u^v {}^t \mathcal{A}_t \, dW_t \times \int_u^v \mathcal{A}_t \, dW_t \right) \right]$$

$$= \mathbb{E}\left[\int_u^v \mathrm{Tr}\left(\int_u^t {}^t \mathcal{A} \, dW \times \mathcal{A}_t \right) dW_t + \int_u^v \mathrm{Tr}\left({}^t \mathcal{A}_t \times \int_u^t \mathcal{A} \, dW \right) dW_t \right.$$

$$\left. + \int_u^v \mathrm{Tr}({}^t \mathcal{A}_t \, \mathcal{A}_t) \, dt \right] = \int_u^v \mathbb{E}\left[\|\mathcal{A}_t\|_{HS}^2 \right] dt. \qquad \diamond$$

7.3 Approximation of a left Brownian motion by exponentials

We consider here the left Brownian motion (X_t), which is the solution to eqn (7.1) with coefficients $A_0, A_1, \ldots, A_k$ in a given Lie subalgebra $\mathcal{G}$. According to Theorem 7.2.1 and Theorem 7.2.3, this is almost surely a G-valued process, with homogeneous and independent right increments $X_t^{-1} X_{t+s}$, and generator $\mathcal{A} = \frac{1}{2} \sum_{j=1}^{k} (\mathcal{L}_{A_j})^2 - \mathcal{L}_{A_0}$.

Of course, we could equally well have written this section for a right Brownian motion (Y_t) (solving eqn (7.5)) instead. The changes are straightforward, just as it was was to deduce Remark 7.2.9 from the preceding discussion in Section 7.2.

In the present section, we obtain the left Brownian motion (X_t) as a limit of products of exponentials. In particular, this provides alternative proofs for Theorem 7.2.3, stating that $(X_t), (Y_t)$ belong to the Lie subgroup G (the closed subgroup generated by $\exp[\mathcal{G}]$), and for the fact that for any fixed time t, X_t and X_t^{-1} have the same law (Proposition 7.2.11).

Taking the matrices A_j and the $\mathbb{R}^k$-valued Brownian motion $W = (W^1, \ldots, W^k)$ as fixed, for any times $t < t'$, we denote the exponential we need by

$$\mathcal{E}_A^W(t, t') := \exp\left[\sum_{j=1}^{k} A_j (W_{t'}^j - W_t^j) + A_0(t' - t)\right]. \tag{7.7}$$

Lemma 7.3.1 *With the same norm $\|\cdot\|_{HS}$ as in Lemma 7.2.12, we have*

$$\mathbb{E}\left[\left\|\mathcal{E}_A^W(t, t')\right\|_{HS}^2\right] \leq 2 \, \exp\left[2\left(\|A_0\|_{HS} + \sum_{j=1}^{k} \|A_j\|_{HS}^2\right)(t' - t)\right].$$

Proof By Lemma 7.2.12, we have for any $A \in \mathcal{M}(d)$

$$\|\exp(A)\|_{HS} \leq \sum_n \left\|\frac{A^n}{n!}\right\|_{HS} \leq \sum_n \frac{\|A\|_{HS}^n}{n!} = \exp[\|A\|_{HS}],$$

so that $\left(\text{using } \mathbb{E}\left[e^{2aW_{t'-t}}\right] = e^{2a^2(t'-t)}\right)$

$$\mathbb{E}\left[\left\|\mathcal{E}_A^W(t, t')\right\|_{HS}^2\right] \leq \mathbb{E}\left[\exp\left[2\left\|\sum_{j=1}^{k} A_j(W_{t'}^j - W_t^j) + A_0(t' - t)\right\|_{HS}\right]\right]$$

$$\leq \mathbb{E}\left[\exp\left[2\sum_{j=1}^{k} \|A_j\|_{HS} |W_{t'}^j - W_t^j| + 2\|A_0\|_{HS}(t' - t)\right]\right]$$

$$\leq 2 \, \exp\left[2\left(\|A_0\|_{HS} + \sum_{j=1}^{k} \|A_j\|_{HS}^2\right)(t' - t)\right]. \qquad \diamond$$

We can expand the exponential of eqn (7.7) up to second order, as follows.

Lemma 7.3.2 *With the same norm $\|\cdot\|_{HS}$ as in Lemma 7.2.12, we have for any $t, \varepsilon > 0$*

$$\mathcal{E}_A^W(t, t + \varepsilon) = \mathbf{1} + \sum_{j=1}^{k} A_j W_\varepsilon^{j,t} + \varepsilon A_0 + \frac{\varepsilon}{2} \sum_{j=1}^{k} A_j^2$$

$$+ \frac{1}{2} \sum_{i,j=1}^{k} (A_i A_j + A_j A_i) \int_0^\varepsilon W_s^{i,t} \, dW_s^{j,t} + R_t^\varepsilon,$$

where $W_s^{j,t} := (W_{t+s}^j - W_t^j)$ and $\mathbb{E}\left[\left\|R_t^\varepsilon\right\|_{HS}^2\right] = \mathcal{O}(\varepsilon^3)$ (uniformly with respect to t).
In particular, we have $\mathbb{E}[\|\mathcal{E}_A^W(t, t+\varepsilon) - \mathbf{1}\|_{HS}^2] = \mathcal{O}(\varepsilon^2)$ (uniformly with respect to t).

Proof By the definition of $\mathcal{E}_A^W$ in Definition 7.7, we have

$$\mathcal{E}_A^W(t, t+\varepsilon) = \mathbf{1} + \sum_{j=1}^k A_j W_\varepsilon^{j,t} + A_0\,\varepsilon + \frac{1}{2}\sum_{i,j=1}^k A_i A_j\,W_\varepsilon^{i,t}\,W_\varepsilon^{j,t} + R_t^\varepsilon,$$

with

$$R_t^\varepsilon := \frac{\varepsilon}{2}\sum_{j=1}^k (A_0 A_j + A_j A_0)W_\varepsilon^{j,t} + \frac{\varepsilon^2}{2}A_0^2 + \sum_{n=3}^\infty \frac{1}{n!}\left[\sum_{j=1}^k A_j\,W_\varepsilon^{j,t} + A_0\,\varepsilon\right]^n.$$

By Itô's formula, we have

$$W_\varepsilon^{i,t}\,W_\varepsilon^{j,t} = \int_0^\varepsilon W_s^{i,t}\,dW_s^{j,t} + \int_0^\varepsilon W_s^{j,t}\,dW_s^{i,t} + \varepsilon\,\mathbf{1}_{\{i=j\}},$$

so that

$$\sum_{i,j=1}^k A_i A_j\,W_\varepsilon^{i,t}\,W_\varepsilon^{j,t} = \sum_{i,j=1}^k (A_i A_j + A_j A_i)\int_0^\varepsilon W_s^{i,t}\,dW_s^{j,t} + \varepsilon\sum_{j=1}^k A_j^2.$$

Hence we have to control the terms of the remainder R_t^ε. By Lemma 7.2.12, we have first

$$\mathbb{E}\left[\left\|\varepsilon\sum_{j=1}^k (A_0 A_j + A_j A_0)W_\varepsilon^{j,t}\right\|_{HS}^2\right] \le 2k\,\varepsilon^2\,\|A_0\|_{HS}^2 \sum_{j=1}^k \|A_j\|_{HS}^2\,\mathbb{E}\left[|W_\varepsilon^{j,t}|^2\right] = \mathcal{O}(\varepsilon^3).$$

Then,

$$\mathbb{E}\left[\left\|\left[\sum_{j=1}^k A_j\,W_\varepsilon^{j,t} + A_0\,\varepsilon\right]^n\right\|_{HS}^2\right]^{1/2} \le \mathbb{E}\left[\left\|\left[\sum_{j=1}^k A_j\,W_\varepsilon^{j,t} + A_0\,\varepsilon\right]^{2n}\right\|_{HS}\right]^{1/2}$$

$$\le \mathbb{E}\left[\left(\sum_{j=1}^k \|A_j\|_{HS}\,|W_\varepsilon^{j,t}| + \varepsilon\,\|A_0\|_{HS}\right)^{2n}\right]^{1/2}$$

$$\le \mathbb{E}\left[\left[(k+1)^{n-1}\left(\sum_{j=1}^k \|A_j\|_{HS}^n\,|W_\varepsilon^{j,t}|^n + \varepsilon^n\,\|A_0\|_{HS}^n\right)\right]^2\right]^{1/2}$$

$$\le (k+1)^{n-1}\left(\sum_{j=1}^k \|A_j\|_{HS}^n\,\sqrt{\mathbb{E}\left[|W_\varepsilon^{j,t}|^{2n}\right]} + \varepsilon^n\,\|A_0\|_{HS}^n\right)$$

$$\le \left[(k+1)\sqrt{\varepsilon}\right]^n\left(\sum_{j=1}^k 2^{n/2}\,\|A_j\|_{HS}^n\,\sqrt{n!} + \varepsilon^{n/2}\,\|A_0\|_{HS}^n\right),$$

since $\mathbb{E}\left[|W_{\varepsilon}^{j,t}|^{2n}\right] = \varepsilon^n \times \prod_{j=1}^{n}(2j-1) \le (2\varepsilon)^n \times n!$. Setting $K := \sqrt{2} \max_{0 \le j \le k} \|A_j\|_{HS}$ and using the fact that $\sqrt{\mathbb{E}[\| \cdot \|_{HS}^2]}$ is a norm, this yields finally

$$
\mathbb{E}\left[\left\| \sum_{n=3}^{\infty} \frac{1}{n!} \left[\sum_{j=1}^{k} A_j W_{\varepsilon}^{j,t} + A_0 \varepsilon \right]^n \right\|_{HS}^2 \right]^{1/2} \le \sum_{n=3}^{\infty} \left[(k+1)K\sqrt{\varepsilon}\right]^n \frac{k\sqrt{n!} + \varepsilon^{n/2}}{n!}
$$

$$
= \mathcal{O}(\varepsilon^{3/2}).
$$

The last assertion follows at once.   ◇

Let $\tau_n := \{0 = t_0 < t_1 < \ldots < t_n < t_{n+1} = s\}$ be a subdivision of a given time interval $[0, s]$, and consider the $GL(d)$-valued process (Z_t^n) inductively defined by $Z_0^n = \mathbf{1}$ and

$$
Z_t^n = Z_{t_q}^n \times \mathcal{E}_A^W(t_q, t) \quad \text{for } 0 \le q \le n \text{ and } t_q \le t \le t_{q+1}.
$$

Lemma 7.3.3 Set $B := \frac{1}{2} \sum_{j=1}^{k} A_j^2 + A_0$. We have in the L^2-norm, for any $u \in [0, s]$,

$$
Z_u^n - \mathbf{1} - \sum_{j=1}^{k} \int_0^u Z_t^n A_j \, dW_t^j - \int_0^u Z_t^n B \, dt = M_u^n + e^{Cs} \mathcal{O}\left(u\sqrt{|\tau_n|}\right),
$$

where $|\tau_n| := \max\left\{|t_{q+1} - t_q| \,\big|\, 0 \le q \le n\right\}$ denotes the mesh of the subdivision τ_n,

$$
M_u^n := \frac{1}{2} \sum_{q=0}^{n} Z_{t_q}^n \times \sum_{i,j=1}^{k} (A_i A_j + A_j A_i) \int_{u \wedge t_q}^{u \wedge t_{q+1}} (W_t^i - W_{t_q}^i) \, dW_t^j
$$

is an $\mathcal{M}(d)$-valued centred continuous martingale and $C := \|A_0\|_{HS} + \sum_{j=1}^{k} \|A_j\|_{HS}^2$.

Proof (After [MK]) We take $\ell \in \{0, \ldots, n\}$ such that $t_\ell \le u \le t_{\ell+1}$, and set

$$
N_u^n := \sum_{q=0}^{n} Z_{t_q}^n \times \left[\mathcal{E}_A^W(u \wedge t_q, u \wedge t_{q+1}) - \mathbf{1} - \sum_{j=1}^{k} A_j (W_{u \wedge t_{q+1}}^j - W_{u \wedge t_q}^j) \right.
$$

$$
\left. - B(u \wedge t_{q+1} - u \wedge t_q) \right].
$$

We have

$$
\Lambda_u^n := Z_u^n - \mathbf{1} - \sum_{j=1}^{k} \int_0^u Z_t^n A_j \, dW_t^j - \int_0^u Z_t^n B \, dt - N_u^n
$$

$$
= Z_{t_\ell}^n - \mathbf{1} - \sum_{j=1}^{k} \int_0^{t_\ell} Z_t^n A_j \, dW_t^j - \int_0^{t_\ell} Z_t^n B \, dt - N_{t_\ell}^n
$$

$$- Z_{t_\ell}^n \left[\sum_{j=1}^k \int_{t_\ell}^u (\mathcal{E}_A^W(t_\ell, t) - \mathbf{1}) A_j \, dW_t^j - \int_{t_\ell}^u (\mathcal{E}_A^W(t_\ell, t) - \mathbf{1}) B \, dt \right]$$

$$= \sum_{q=0}^{\ell-1} Z_{t_q}^n \left[- \sum_{j=1}^k \int_{t_q}^{t_{q+1}} (\mathcal{E}_A^W(t_q, t) - \mathbf{1}) A_j \, dW_t^j - \int_{t_q}^{t_{q+1}} (\mathcal{E}_A^W(t_q, t) - \mathbf{1}) B \, dt \right]$$

$$+ Z_{t_\ell}^n \left[- \sum_{j=1}^k \int_{t_\ell}^u (\mathcal{E}_A^W(t_\ell, t) - \mathbf{1}) A_j \, dW_t^j - \int_{t_\ell}^u (\mathcal{E}_A^W(t_\ell, t) - \mathbf{1}) B \, dt \right]$$

(by induction)

$$= - \sum_{q=0}^\ell Z_{t_q}^n \left[\sum_{j=1}^k \int_{t_q}^{u \wedge t_{q+1}} (\mathcal{E}_A^W(t_q, t) - \mathbf{1}) A_j \, dW_t^j + \int_{t_q}^{u \wedge t_{q+1}} (\mathcal{E}_A^W(t_q, t) - \mathbf{1}) B \, dt \right].$$

Hence, setting $u_q := u \wedge t_q$ (for $0 \le q \le n+1$) and

$$Q_q^u := \sum_{j=1}^k \int_{u_q}^{u_{q+1}} (\mathcal{E}_A^W(t_q, t) - \mathbf{1}) A_j \, dW_t^j + \int_{u_q}^{u_{q+1}} (\mathcal{E}_A^W(t_q, t) - \mathbf{1}) B \, dt,$$

and using Lemmas 7.2.12 and 7.3.1, we deduce (by the independence of $\mathcal{F}_{t_q}$ and Q_q^u)

$$\sqrt{\mathbb{E}\left[\left\|\Lambda_u^n\right\|_{HS}^2\right]} \le \sum_{q=0}^n \sqrt{\mathbb{E}\left[\left\|Z_{t_q}^n\right\|_{HS}^2 \left\|Q_q^u\right\|_{HS}^2\right]} = \sum_{q=0}^n \sqrt{\mathbb{E}\left[\left\|Z_{t_q}^n\right\|_{HS}^2\right]} \times \sqrt{\mathbb{E}\left[\left\|Q_q^u\right\|_{HS}^2\right]}$$

$$\le \sqrt{2} \, e^{Cs} \times \sum_{q=0}^n \sqrt{\mathbb{E}\left[\left\|Q_q^u\right\|_{HS}^2\right]}, \qquad \text{with} \quad C = \left\|A_0\right\|_{HS} + \sum_{j=1}^k \left\|A_j\right\|_{HS}^2.$$

Now, by Remark 7.2.13 we have

$$\sqrt{\mathbb{E}\left[\left\|Q_q^u\right\|_{HS}^2\right]}$$

$$\le \sum_{j=1}^k \sqrt{\mathbb{E}\left[\left\|\int_{u_q}^{u_{q+1}} (\mathcal{E}_A^W(t_q, t) - \mathbf{1}) A_j \, dW_t^j\right\|_{HS}^2\right]} + \sqrt{\mathbb{E}\left[\left\|\int_{u_q}^{u_{q+1}} (\mathcal{E}_A^W(t_q, t) - \mathbf{1}) B \, dt\right\|_{HS}^2\right]}$$

$$\le \sum_{j=1}^k \sqrt{\mathbb{E}\left[\int_{u_q}^{u_{q+1}} \left\|(\mathcal{E}_A^W(u_q, t) - \mathbf{1}) A_j\right\|_{HS}^2 \, dt\right]} + \sqrt{\mathbb{E}\left[\int_{u_q}^{u_{q+1}} \left\|(\mathcal{E}_A^W(u_q, t) - \mathbf{1}) B\right\|_{HS}^2 \, dt\right]}$$

$$\le \sum_{j=1}^k \sqrt{\mathbb{E}\left[\int_{u_q}^{u_{q+1}} \left\|\mathcal{E}_A^W(u_q, t) - \mathbf{1}\right\|_{HS}^2 \left\|A_j\right\|_{HS}^2 \, dt\right]}$$

$$+ \sqrt{\mathbb{E}\left[\int_{u_q}^{u_{q+1}} \left\|\mathcal{E}_A^W(u_q, t) - \mathbf{1}\right\|_{HS}^2 \left\|B\right\|_{HS}^2 \, dt\right]}$$

$$\leq \left(\sum_{j=1}^{k} \|A_j\|_{HS} + \|B\|_{HS} \right) \times \sqrt{\mathbb{E}\left[\int_{u_q}^{u_{q+1}} \left\| \mathcal{E}_A^W(u_q, t) - \mathbf{1} \right\|_{HS}^2 dt \right]}$$

$$= \left[\int_{u_q}^{u_{q+1}} \mathcal{O}(t - u_q)^2 \, dt \right]^{1/2} = \mathcal{O}\left((u_{q+1} - u_q)^{3/2} \right),$$

by Lemma 7.3.2. Hence, so far we have obtained

$$\sqrt{\mathbb{E}\left[\|\Lambda_u^n\|_{HS}^2 \right]} \leq \sqrt{2} \, e^{Cs} \sum_{q=0}^{n} \mathcal{O}\left((u_{q+1} - u_q)^{3/2} \right) = e^{Cs} \times u \times \mathcal{O}\left(\sqrt{|\tau_n|} \right).$$

Finally, by Lemma 7.3.2 again, in the same way we have

$$N_u^n - M_u^n = \sum_{q=0}^{n} Z_{t_q}^n \, R_{u_q}^{u_{q+1} - u_q} = \sum_{q=0}^{n} Z_{t_q}^n \times \mathcal{O}\left((u_{q+1} - u_q)^{3/2} \right)$$

$$= \sqrt{2} \, e^{Cs} \sum_{q=0}^{n} \mathcal{O}\left((u_{q+1} - u_q)^{3/2} \right) = e^{Cs} \, \mathcal{O}\left(u \sqrt{|\tau_n|} \right),$$

and the martingale property for (M_u^n) follows at once from the independence between $Z_{t_q}^n$ and the centred martingale

$$[t_q, t_{q+1}] \ni u \longmapsto \sum_{i,j=1}^{k} (A_i A_j + A_j A_i) \int_{u \wedge t_q}^{u \wedge t_{q+1}} W_t^i \, dW_t^j. \qquad \diamond$$

Corollary 7.3.4 *With the same notation as in Lemma 7.3.3, in the L^2-norm, uniformly with respect to $u \in [0, s]$, we have*

$$Z_u^n - \mathbf{1} - \sum_{j=1}^{k} \int_0^u Z_t^n \, A_j \, dW_t^j - \int_0^u Z_t^n \, B \, dt = e^{Cs} \times \mathcal{O}\left(\sqrt{u(1+u)\,|\tau_n|} \right).$$

Proof (after [MK]) By Lemma 7.3.3, we have only to control the centred martingale (M_u^n). Since $(\|M_u^n\|_{HS})$ is a non-negative submartingale, we apply Doob's inequality (Theorem 6.3.2), to obtain the following: for any $s' \in [0, s]$,

$$\mathbb{E}\left[\max_{0 \leq u \leq s'} \|M_u^n\|_{HS}^2 \right] \leq 4 \, \mathbb{E}\left[\|M_{s'}^n\|_{HS}^2 \right] = \sum_{0 \leq q, q' \leq n} \mathbb{E}\left[\mathrm{Tr}\left(Z_{t_q}^n \, \mathcal{A}_{s'}^q \, {}^t\mathcal{A}_{s'}^{q'} \times {}^t Z_{t_{q'}}^n \right) \right]$$

where

$$\mathcal{A}_u^q := \sum_{i,j=1}^{k} (A_i A_j + A_j A_i) \int_{u \wedge t_q}^{u \wedge t_{q+1}} (W_t^i - W_{t_q}^i) \, dW_t^j.$$

By the independence of $\mathcal{A}_{s'}^{q'}$ from $\mathcal{A}_{s'}^q$, $Z_{t_q}^n$, $Z_{t_{q'}}^n$ when $q' > q$, we then have

$$\mathbb{E}\Big[\max_{0\le u\le s'}\big\|M^n_u\big\|^2_{HS}\Big]\le\sum_{q=0}^{n}\mathbb{E}\Big[\big\|Z^n_{t_q}\,\mathcal{A}^q_{s'}\big\|^2_{HS}\Big]\le\sum_{q=0}^{n}\mathbb{E}\Big[\big\|Z^n_{t_q}\big\|^2_{HS}\,\big\|\mathcal{A}^q_{s'}\big\|^2_{HS}\Big]$$

$$=\sum_{q=0}^{n}\mathbb{E}\Big[\big\|Z^n_{t_q}\big\|^2_{HS}\Big]\mathbb{E}\Big[\big\|\mathcal{A}^q_{s'}\big\|^2_{HS}\Big]\le 2\,e^{2Cs}\sum_{q=0}^{n}k^2\sum_{i,j=1}^{k}\big\|A_iA_j$$

$$+A_jA_i\big\|^2_{HS}\int_{s'\wedge t_q}^{s'\wedge t_{q+1}}\mathbb{E}\big[|W^i_{t-t_q}|^2\big]\,dt\le k^2\,e^{2Cs}\sum_{i,j=1}^{k}\big\|A_iA_j+A_jA_i\big\|^2_{HS}\times s'|\tau_n|$$

by Lemma 7.3.1 and Remark 7.2.13. This yields the claim. $\diamond$

Theorem 7.3.5 *The solution (X_t) to the stochastic linear differential equation (7.1) is approached in the L^2-norm by exponentials as follows. For any time $S\ge 0$, we have*

$$X_s=\lim_{|\tau_n|\to 0}\prod_{q=0}^{n}\exp\Bigg[\sum_{j=1}^{k}A_j(W^j_{t_{q+1}}-W^j_{t_q})+A_0(t_{q+1}-t_q)\Bigg],\tag{7.8}$$

where $|\tau_n|:=\max\big\{|t_{q+1}-t_q|\,\big|\,0\le q\le n\big\}$ denotes the mesh of a generic subdivision $\tau_n=\{0=t_0<t_1<\ldots<t_n<t_{n+1}=s\}$ of $[0,s]$, uniformly with respect to $s\in[0,S]$.

Proof Using Corollary 7.3.4 and eqn (7.1), for $n\in\mathbb{N}^*$ and $s\in[0,S]$ we have

$$Z^n_s-X_s=\sum_{j=1}^{k}\int_0^s(Z^n_t-X_t)\,A_j\,dW^j_t+\int_0^s(Z^n_t-X_t)B\,dt+\mathcal{O}\Big(\sqrt{s\,|\tau_n|}\Big).$$

Hence, setting $\alpha^n_s:=\mathbb{E}\big[\sup_{0\le u\le s}\big\|Z^n_u-X_u\big\|^2_{HS}\big]$, we have

$$\sqrt{\alpha^n_s}\le\sum_{j=1}^{k}\sqrt{\mathbb{E}\Bigg[\sup_{0\le u\le s}\bigg\|\int_0^u(Z^n_t-X_t)A_j\,dW^j_t\bigg\|^2_{HS}\Bigg]}$$

$$+\sqrt{\mathbb{E}\Bigg[\sup_{0\le u\le s}\bigg\|\int_0^u(Z^n_t-X_t)B\,dt\bigg\|^2_{HS}\Bigg]}+\mathcal{O}\Big[\sqrt{s|\tau_n|}\Big]$$

$$\le 2\sum_{j=1}^{k}\sqrt{\mathbb{E}\Bigg[\bigg\|\int_0^s(Z^n_t-X_t)A_j\,dW^j_t\bigg\|^2_{HS}\Bigg]}$$

$$+\sqrt{\mathbb{E}\Bigg[\bigg[\int_0^s\|(Z^n_t-X_t)B\|_{HS}\,dt\bigg]^2\Bigg]}+\mathcal{O}\Big[\sqrt{s|\tau_n|}\Big]$$

by Doob's inequality (Theorem 6.3.2) and since, by Lemma 7.2.12 (for $0\le u\le s$),

$$\bigg\|\int_0^u Z_tB\,dt\bigg\|^2_{HS}=\mathrm{Tr}\bigg[\int_0^u\int_0^u{}^t(Z_tB)\,(Z_{t'}B)\,dt\,dt'\bigg]=\int_0^u\int_0^u\mathrm{Tr}\big[{}^t(Z_tB)\,(Z_{t'}B)\big]\,dt\,dt'$$

$$\le\int_0^u\int_0^u\|Z_tB\|_{HS}\,\|Z_{t'}B\|_{HS}\,dt\,dt'=\bigg[\int_0^u\|Z_tB\|_{HS}\,dt\bigg]^2\le\bigg[\int_0^s\|Z_tB\|_{HS}\,dt\bigg]^2.$$

Hence, using Remark 7.2.13 we obtain

$$\sqrt{\alpha_s^n} \leq 2 \sum_{j=1}^{k} \sqrt{\mathbb{E}\left[\int_0^s \left\|(Z_t^n - X_t)A_j\right\|_{HS}^2 \, dt\right]} + \sqrt{\mathbb{E}\left[s \int_0^s \left\|(Z_t^n - X_t)B\right\|_{HS}^2 \, dt\right]}$$

$$+ \mathcal{O}\left[\sqrt{s|\tau_n|}\right] \leq 2 \sum_{j=1}^{k} \left[\int_0^s \alpha_t^n \left\|A_j\right\|_{HS}^2 \, dt\right]^{1/2} + \left[s \int_0^s \alpha_t^n \left\|B\right\|_{HS}^2 \, dt\right]^{1/2} + \mathcal{O}\left[\sqrt{s\,|\tau_n|}\right]$$

$$\leq \left[2 \sum_{j=1}^{k} \left\|A_j\right\|_{HS} + \sqrt{S} \left\|B\right\|_{HS}\right] \times \left[\int_0^s \alpha_t^n \, dt\right]^{1/2} + \mathcal{O}\left[\sqrt{s\,|\tau_n|}\right].$$

Setting

$$C' := 2\left[2 \sum_{j=1}^{k} \left\|A_j\right\|_{HS} + \sqrt{S} \left\|B\right\|_{HS}\right]^2,$$

we therefore have the following: for any $s \in [0, S]$,

$$\alpha_s^n \leq C' \int_0^s \alpha_t^n \, dt + b_n \, s, \quad \text{with } b_n = \mathcal{O}(|\tau_n|).$$

Hence $f_n(s) := e^{-C's} \int_0^s \alpha_t^n \, dt$ satisfies $f_n'(s) \leq b_n \, s \, e^{-C's}$, whence

$$f_n(s) \leq \int_0^s b_n \, t \, e^{-C't} \, dt = \frac{b_n\left(1 - e^{-C's} - C's\,e^{-C's}\right)}{C'^2},$$

and then

$$\mathbb{E}\left[\sup_{0 \leq s \leq S} \left\|Z_s^n - X_s\right\|_{HS}^2\right] = \alpha_S^n \leq C' e^{C'S} f_n(S) + b_n \, S \leq b_n(e^{C'S} - 1)/C' = \mathcal{O}(|\tau_n|).$$

This means, as wanted, that $\lim_{n \to \infty} Z_s^n = X_s$ in the L^2-norm, uniformly with respect to $s \in [0, S]$.   ◇

Remark 7.3.6 It is not much more difficult to show that the convergence of Theorem 7.3.5 also occurs almost surely (uniformly with respect to $s \in [0, S]$); see [MK].

7.4 Lyapunov exponents

Proposition 7.4.1 *With the solution (X_s) of the linear SDE (7.1) (with $X_0 \in GL(d)$ instead of 1) is associated a **Lyapunov exponent**, which is a deterministic real λ_1 such that*

$$\lim_{s \to \infty} s^{-1} \log \left\|X_s\right\| = \lambda_1,$$

almost surely and in the L^2-norm. Here $\|\cdot\|$ denotes any norm on $\mathcal{M}(d)$. Moreover, we have $|\lambda_1| \le (3/2) \sum_{j=1}^{k} \|A_j\|_{HS}^2 + \|A_0\|_{HS}$ (where the norm $\|\cdot\|_{HS}$ is as in Lemma 7.2.12).

Note

(1) Note that from the last formula giving $\det(X_s)$ in the proof of Theorem 7.2.2, we have at once $\lim_{s \to \infty} s^{-1} \log |\det(X_s)| = \operatorname{Tr}(A_0)$, almost surely and in the L^2-norm.

(2) For $\lambda_1 > 0$, the solution is said to be stable, and unstable for $\lambda_1 < 0$.

Proof We apply Itô's formula (7.2) to the function $\log \|\cdot\|_{HS}^2$. Note that by Theorem 7.2.2(i) and since $X_0 \in GL(d)$, $\|X_0^{-1}X_s\|_{HS}$ never vanishes. We obtain

$$\log \|X_0^{-1}X_s\|_{HS}^2 = \log d + 2 \sum_{j=1}^{k} \int_0^s \frac{\mathcal{L}_{A_j} \|X_0^{-1}X_t\|_{HS}}{\|X_0^{-1}X_t\|_{HS}} \, dW_t^j$$

$$+ \int_0^s (f_1 - f_2)(X_0^{-1}X_t) \, dt,$$

with

$$f_1 := \|\cdot\|_{HS}^{-2} \times \left[\frac{1}{2} \sum_{j=1}^{k} (\mathcal{L}_{A_j})^2 + \mathcal{L}_{A_0} \right] \left(\|\cdot\|_{HS}^2 \right)$$

and

$$f_2 := \sum_{j=1}^{k} \left(\frac{\mathcal{L}_{A_j} \|\cdot\|_{HS}}{\|\cdot\|_{HS}} \right)^2.$$

By Lemma 7.2.12, $f_1 - f_2$ is C^∞ bounded, since

$$|f_1| \le 2 \sum_{j=1}^{k} \|A_j\|_{HS}^2 + 2 \|A_0\|_{HS} \quad \text{and} \quad 0 \le f_2 \le \sum_{j=1}^{k} \|A_j\|_{HS}^2,$$

and the quadratic variation of the martingale part M_t of $\log \|X_0^{-1}X_s\|_{HS}^2$ is

$$\langle M, M \rangle_t = 4 \sum_{j=1}^{k} \int_0^s \left| \frac{\mathcal{L}_{A_j} \|X_0^{-1}X_t\|_{HS}}{\|X_0^{-1}X_t\|_{HS}} \right|^2 dt \le 4 \sum_{j=1}^{k} \|A_j\|_{HS}^2 \, s =: C \, s.$$

As a consequence, on the one hand, for any $\varepsilon \in]0, \frac{1}{2}[$ and for large s, we almost surely have

$$s^{-1} \log \|X_0^{-1}X_s\|_{HS} = \mathcal{O}(s^{\varepsilon - 1/2}) + s^{-1} \int_0^s (f_1 - f_2)(X_0^{-1}X_t) \, dt.$$

On the other hand, setting $T_\lambda := \inf \left\{ s \geq 1 \,\big|\, M_s^2 > \lambda\, s^2 \right\} \geq 1$ for any $\lambda \geq 1$ and using Doob's inequality (Theorem 6.3.2(ii) and Proposition 6.5.9, we obtain

$$\mathbb{P}\left[\sup_{s \geq 1} \left| \frac{M_s}{s} \right|^2 > \lambda \right] = \mathbb{P}[T_\lambda < \infty] \leq \lambda^{-2}\, \mathbb{E}\left[T_\lambda^{-4}\, M_{T_\lambda}^4\, \mathbb{1}_{\{T_\lambda < \infty\}} \right]$$

$$\leq \lambda^{-2} \sum_{n \geq 1} n^{-4}\, \mathbb{E}\left[M_{T_\lambda}^4\, \mathbb{1}_{\{n \leq T_\lambda < n+1\}} \right] \leq (\tfrac{4}{3})^4\, \lambda^{-2} \sum_{n \geq 1} n^{-4}\, \mathbb{E}\left[M_{n+1}^4 \right]$$

$$\leq (\tfrac{4}{3})^4\, \lambda^{-2}\, C_2 \sum_{n \geq 1} n^{-4}\, \mathbb{E}\left[\langle M, M \rangle_{n+1}^2 \right] \leq (\tfrac{4}{3})^4\, \lambda^{-2}\, C_2\, C^2 \sum_{n \geq 1} n^{-4} (n+1)^2 = C'\, \lambda^{-2}.$$

Hence

$$\left\| \sup_{s \geq 1} \left| \frac{M_s}{s} \right|^2 \right\|_2 = \int_0^\infty \mathbb{P}\left[\sup_{s \geq 1} \left| \frac{M_s}{s} \right|^2 > \lambda \right]\, d\lambda \leq 1 + \int_1^\infty C'\, \lambda^{-2}\, d\lambda = 1 + C' < \infty,$$

which proves that $\displaystyle\sup_{s \geq 1} \left| s^{-1} \log \left\| X_0^{-1} X_s \right\|_{HS} \right|$ is square integrable.

For $s > 0$, set $\varphi_s := \log \left\| X_0^{-1} X_s \right\|_{HS}^2$, and observe that $\varphi_{r+s} = \varphi_r + \varphi_s \circ \Theta_r - \log d$, whence the subadditivity property

$$\varphi_{r+s} \leq \varphi_r + \varphi_s \circ \Theta_r,$$

where Θ denotes the shift operator on Brownian trajectories.

As Θ_1 preserves the underlying probability and (by the 0–1 law) is ergodic, Kingman's subadditive ergodic theorem (see for example [N2] or [St]) ensures the almost sure convergence of φ_n / n, to the deterministic $\ell := \inf \left\{ \mathbb{E}(\varphi_n / n) \,\big|\, n \in \mathbb{N}^* \right\}$ as the integer n goes to infinity. By the above, this implies $|\ell| \leq \| f_1 - f_2 \|_\infty$ and the almost sure convergence of φ_s / s (using

$$\frac{\varphi_s}{s} = \frac{[s]}{s}\, \frac{\varphi_{[s]}}{[s]} + \frac{1}{s} \big[\varphi_{s-[s]} \circ \Theta_{[s]} - \log d \big]$$

and the fact that the random variables $\displaystyle\sup_{0 \leq h \leq 1} \varphi_h \circ \Theta_n$ are independent identically distributed and integrable), as the real s goes to infinity, and then it implies its L^2-convergence too, by dominated convergence.

The result follows, since we have $s^{-1} \log \| X_s \| = \mathcal{O}(s^{-1}) + \tfrac{1}{2}\, \varphi_s / s$. $\qquad \diamondsuit$

Remark 7.4.2 Using the exterior algebra, it can be proved that Proposition 7.4.1 admits the following generalization, which yields the Lyapunov exponents $\lambda_2, \dots, \lambda_d$ after λ_1 : $\lambda_1 + \cdots + \lambda_j := \lim_{s \to \infty} s^{-1} \log \| X_s^{\wedge j} \|$ exists in $\mathbb{R}$, almost surely and in the L^2-norm, and is deterministic. For $j = d$, $X_s^{\wedge d} = \det(X_s)$, and then, from the note following Proposition 7.4.1, we have $\lambda_1 + \cdots + \lambda_d = \mathrm{Tr}(A_0)$.

7.5   Diffusion processes

We introduce here an important notion about processes living in a space which does not need to be a group, and consider a particular case addressed in this book, of diffusion processes which can be derived from a group-valued Brownian motion by means of some projection.

Definition 7.5.1 (*i*) *A **Fellerian** (or **Markovian**) **semi-group** on a separable metric space E is a family $(P_t)_{t\geq 0}$ of non-negative continuous endomorphisms on $C_b(E)$, such that P_0 is the identity, $P_{s+t} = P_s\, P_t$, $P_t 1 = 1$ for any $s, t \geq 0$, and $\lim\limits_{t\searrow 0} \|P_t f - f\| = 0$ for any $f \in C_b(E)$. It extends automatically to the space of non-negative measurable functions on E, and the associated **kernel** is defined for any $x \in E$ and measurable subset A of E by $P_t(x, A) := P_t 1_A(x)$.*

(*ii*) *Given a separable metric space E endowed with a Borelian probability measure μ and a Fellerian semi-group (P_t), a continuous E-valued process (X_t) such that*

$$\mathbb{E}\big[f_0(X_0) \times f_1(X_{t_1}) \times \ldots \times f_n(X_{t_n})\big] = \int f_0\, P_{t_1}\Big[f_1\, P_{t_2-t_1}\big[f_2 \ldots P_{t_n-t_{n-1}} f_n\big]\Big]\, d\mu,$$

*for any $n \in \mathbb{N}^*$, $f_0, \ldots, f_n \in C_b(E)$ and $0 \leq t_1 \leq \ldots \leq t_n$, is called a **diffusion process** on E with semi-group (P_t) and initial law μ.*

Note that (taking a Dirac mass δ_x as μ, $n = 1$ and $f_0 \equiv 1$) we have in particular $P_t f(x) = \mathbb{E}_x\big[f(X_t)\big] = \mathbb{E}\big[f(X_t)\,\big|\,X_0 = x\big]$, where the index x in $\mathbb{E}_x$ specifies the initial value $X_0 = x \in E$.

By Proposition 7.2.7, any left Brownian motion is an example of a diffusion process $\big($on a subgroup G of $\mathcal{M}(d)\big)$. The other diffusion processes considered in this book are all constructed in the following way (up to a possible left quotient by some Kleinian group, as in Sections 8.1 and 8.2).

Proposition 7.5.2 *Given a left Brownian motion (g_t) on a group G and an independent random variable $\mathbf{g}$ on G with law ν, consider a continuous map $\mathrm{p} : G \to E$ such that for any $g \in G$, the law $P_t(x, \cdot)$ of $\mathrm{p}(gg_t)$ depends only on $x := \mathrm{p}(g)$ and t. Then (P_t) is a Fellerian semi-group and $\mathrm{p}(\mathbf{g}\, g_t)$ is a diffusion process on E with semi-group (P_t) and initial law $\nu \circ \mathrm{p}^{-1}$.*

We say that the diffusion process starts at $x \in E$ when its initial law is a Dirac mass at x.

Proof We have to prove the identity in Definition 7.5.1(*ii*), with $X_t = \mathrm{p}(\mathbf{g}\, g_t)$, $\mu = \nu \circ \mathrm{p}^{-1}$ here. Setting $h_k := f_k \circ \mathrm{p}$, we know that $\mathbb{E}\big[h_k(g\, g_t)\big] = (P_t f_k) \circ \mathrm{p}(g)$. Now, for any $n \geq 2$ and $g \in G$, by the independence and homogeneity of the right increments $g'_{t_n - t_{n-1}} := (g_{t_{n-1}})^{-1} g_{t_n}$ we have

$$\mathbb{E}\big[h_1(g\, g_{t_1}) \ldots h_n(g\, g_{t_n})\big] = \mathbb{E}\big[h_1(g\, g_{t_1}) \ldots h_{n-1}(g\, g_{t_{n-1}})\, h_n(g\, g_{t_{n-1}}\, g'_{t_n - t_{n-1}})\big]$$

$$= \mathbb{E}\big[h_1(g\, g_{t_1}) \ldots h_{n-1}(g\, g_{t_{n-1}}) \times (P_{t_n - t_{n-1}} f_n) \circ \mathrm{p}(g\, g_{t_{n-1}})\big]$$

$$= \mathbb{E}\Big[h_1(g\, g_{t_1}) \ldots h_{n-2}(g\, g_{t_{n-2}}) \times \big(f_{n-1} \times (P_{t_n - t_{n-1}} f_n)\big) \circ \mathrm{p}(g\, g_{t_{n-1}})\Big].$$

By an obvious induction on n, this yields

$$\mathbb{E}\big[h_1(g\,g_{t_1})\ldots h_n(g\,g_{t_n})\big] = P_{t_1}\Big[f_1\,P_{t_2-t_1}\big[f_2\ldots P_{t_n-t_{n-1}}f_n\big]\Big]\circ \mathsf{p}(g).$$

Therefore, by the independence of g we obtain

$$\mathbb{E}\big[h_0(\mathsf{g})\,h_1(\mathsf{g}\,g_{t_1})\ldots h_n(\mathsf{g}\,g_{t_n})\big] = \int h_0\, P_{t_1}\Big[f_1\,P_{t_2-t_1}\big[f_2\ldots P_{t_n-t_{n-1}}f_n\big]\Big]\circ \mathsf{p}\; d\nu$$

$$= \int f_0\, P_{t_1}\Big[f_1\,P_{t_2-t_1}\big[f_2\ldots P_{t_n-t_{n-1}}f_n\big]\Big]\; d\mu,$$

which is the wanted identity in Definition 7.5.1(ii). Note that the semi-group property of (P_t) follows directly, merely by taking $n=2$, $f_0 = f_1 \equiv 1$, and a Dirac mass as μ. $\diamondsuit$

Remark 7.5.3 Note that by Proposition 7.5.2, if (Q_t) denotes the semi-group of (g_t) (recall Definition 7.2.6), then for any $t \geq 0$ and $f \in C_b(E)$ we have $(P_t f) \circ \mathsf{p} = Q_t(f \circ \mathsf{p})$.

7.6 Examples of group-valued Brownian motions

Here, we specialize the contents of Section 7.2 to some classical examples of Lie groups, for which simple expressions for some left or right Brownian motions of interest can be given.

7.6.1 Exponential semimartingale Take any commuting $A_0, A_1, \ldots,$ $A_k \in \mathcal{M}(d)$. Then the associated left Brownian motion $\big($i.e., the corresponding solution (X_s) to the linear SDE (7.1)$\big)$ can be written $X_s = \exp(A_1\, W_s^1 + \cdots + A_k\, W_s^k + A_0\, s)$, as is straightforward from Itô's formula (6.4).

7.6.2 Left Brownian motion on the Heisenberg group $\mathcal{H}^3$ Take $k = 2$, $A_0 = 0$, and

$$A_1 := \begin{pmatrix} 0 & 1 & 0 \\ 0 & 0 & 0 \\ 0 & 0 & 0 \end{pmatrix}, \quad A_2 := \begin{pmatrix} 0 & 0 & 0 \\ 0 & 0 & 1 \\ 0 & 0 & 0 \end{pmatrix},$$

so that

$$[A_1, A_2] = \begin{pmatrix} 0 & 0 & 1 \\ 0 & 0 & 0 \\ 0 & 0 & 0 \end{pmatrix}$$

commutes with A_1, A_2, and the Lie algebra $\mathcal{G}$ generated by A_1, A_2 has basis $\big(A_1, A_2, [A_1, A_2]\big)$. Then $G = \exp(\mathcal{G})$ is the group $\mathcal{H}^3$ of upper triangular matrices that have the diagonal $(1, 1, 1)$. More precisely, we have

$$[a_1, a_2, a_3] := \exp\left(a_1 A_1 + a_2 A_2 + a_3 [A_1, A_2]\right) = \begin{pmatrix} 1 & a_1 & a_3 + a_1 a_2/2 \\ 0 & 1 & a_2 \\ 0 & 0 & 1 \end{pmatrix},$$

and then the law of $\mathcal{H}^3$ is given by

$$[a_1, a_2, a_3] * [a'_1, a'_2, a'_3] = \left[a_1 + a'_1, a_2 + a'_2, a_3 + a'_3 + \tfrac{1}{2}(a_1 a'_2 - a'_1 a_2)\right],$$

which is equivalent to the (Campbell–Hausdorff) formula

$$\exp(A)\, \exp(A') = \exp\left(A + A' + \tfrac{1}{2}[A, A']\right),$$

which holds for any $A, A' \in \mathcal{G}$.

The associated left Brownian motion (i.e., the solution (X_s) to the corresponding linear SDE (7.1)) can be written

$$X_s = \left[W_s^1, W_s^2, \frac{1}{2}\int_0^s \left(W_t^1\, dW_t^2 - W_t^2\, dW_t^1\right)\right],$$

as can be directly verified, since $dX_s = A_1\, dW_s^1 + A_2\, dW_s^2 + [A_1, A_2]\, W_s^1\, dW_s^2$ implies eqn (7.1). Moreover, setting

$$X_{s,t} = \left[W_t^1 - W_s^1, W_t^2 - W_s^2, \frac{1}{2}\int_s^t \left((W_u^1 - W_s^1)\, dW_u^2 - (W_u^2 - W_s^1)\, dW_u^1\right)\right]$$

for any $0 \le s \le t$, we have $X_s = X_{0,s}$, $(X_{s,t})_{t \ge s} \overset{\text{law}}{=} (X_{t-s})_{t \ge s}$ for any $s \ge 0$, and then the easy Chen formula $X_s * X_{s,t} = X_t$.

The third component of X_s is the **Lévy area** generated by the planar Brownian motion $(W_s^1, W_s^2) \equiv (W_s^1 + \sqrt{-1}\, W_s^2)$.

Theorem 7.6.2.1 *Consider a planar Brownian motion $B = W + \sqrt{-1}\, W'$, starting at 0. The Fourier transform of its Lévy area $\mathcal{A}_t := \frac{1}{2}\int_0^t \left[W_s\, dW'_s - W'_s\, dW_s\right]$ is given by*

$$\mathbb{E}\left[e^{\sqrt{-1}\, x\, \mathcal{A}_t}\right] = \left(\mathbb{E}\left[e^{-(x^2/8)\int_0^t (W_s)^2 ds}\right]\right)^2 = \frac{1}{\mathrm{ch}\,(x\,t/2)}.$$

Proof Consider the planar Brownian motion $\beta + \sqrt{-1}\, \beta'$, given by

$$\beta_t := \int_0^t \frac{W_s\, dW_s + W'_s\, dW'_s}{|B_s|}, \qquad \beta'_t := \int_0^t \frac{W_s\, dW'_s - W'_s\, dW_s}{|B_s|}.$$

We have

$$d|B_t| = d\,\beta_t + \frac{dt}{2|B_t|}, \qquad d\,\arg(B_t) = \frac{d\beta'_t}{2|B_t|} \quad \text{and} \quad \mathcal{A}_t := \frac{1}{2}\int_0^t |B_s|\, d\beta'_s.$$

By Theorem 7.1.1 and since the planar Brownian motion almost surely does not vanish at any positive time (recall Proposition 6.5.10), the first equation can be solved between times t' and t, when $0 < t' < t$, and then by continuity, between times 0 and t, for any $t > 0$. This shows that $(|B_t|)$ is adapted to the filtration generated by (β_s), and hence is independent of the Brownian motion β'. As a consequence, we obtain at once

$$\mathbb{E}\left[e^{\sqrt{-1}\, x \mathcal{A}_t}\right] = \mathbb{E}\left[e^{-(x^2/8)\int_0^t |B_s|^2 ds}\right] = \mathbb{E}\left[e^{-(x^2/8)\int_0^t W_s^2 ds - (x^2/8)\int_0^t W_s'^2 ds}\right],$$

which gives the first equality of the statement, by the independence of W and W'.

Note that, by scaling, we clearly have $\mathbb{E}[e^{\sqrt{-1}\, x \mathcal{A}_t}] = \mathbb{E}[e^{\sqrt{-1}\, xt \mathcal{A}_1}]$.

Now, for any real x, consider the function $\mathbb{R}_+ \ni t \mapsto \phi(t) := \operatorname{ch}(xt) - (\operatorname{th} x)\operatorname{sh}(xt)$. We have $\phi'' = x^2\phi$, $\phi > 0$, $\phi(0) = 1$, $\phi(1) = 1/\operatorname{ch} x$ and $\phi'(1) = 0$. In particular, ϕ is convex and decreasing. Hence $\varphi := 1_{[0,1[}\phi + \phi(1)1_{[1,\infty[} = \max\{\phi,\phi(1)\}$ is convex, of class C^1 and non-increasing. Set $F := \varphi'/\varphi$. Note that $-x\operatorname{sh} x \le F \le 0$ and $F = 0$ on $[1,\infty[$. We then have

$$F(s)W_s^2 - \log\,\varphi(s) - 2\int_0^s F(t)W_t\, dW_t = F(s)W_s^2 - \int_0^s F(t)\, d(W_t^2)$$

$$= \int_0^s W_t^2\, F'(t)\, dt = \int_0^{\min\{s,1\}} W_t^2\, F'(t)\, dt = x^2\int_0^{\min\{s,1\}} W_t^2\, dt - \int_0^s W_t^2\, F^2(t)\, dt,$$

so that the bounded process (Z_s) defined by

$$Z_s := \exp\left[\frac{1}{2}\left(F(s)W_s^2 - \log\,\varphi(s) - x^2\int_0^{\min\{s,1\}} W_t^2\, dt\right)\right]$$

can also be written

$$Z_s = \exp\left[\int_0^s F(t)\, W_t\, dW_t - \frac{1}{2}\int_0^s F^2(t)\, W_t^2\, dt\right].$$

This shows that (Z_s) is a local martingale (recall Remark 6.5.8): in fact, if we set $T_n := \inf\{t \ge 0 \mid |W_t| = n\}$, then $(Z_{\min\{s,T_n\}})$ is a martingale (this is easily seen using Itô's formula), so that $\mathbb{E}(Z_{\min\{1,T_n\}}) = 1$. Now, by letting n go to infinity, since (Z_s) is bounded, we obtain by dominated convergence

$$1 = \mathbb{E}(Z_1) = \mathbb{E}\left(\exp\left[\frac{1}{2}\left(F(1)W_1^2 - \log\,\varphi(1) - x^2\int_0^1 W_t^2\, dt\right)\right]\right)$$

$$= \varphi(1)^{-1/2}\mathbb{E}\left(\exp\left[-\frac{x^2}{2}\int_0^1 W_t^2\, dt\right]\right),$$

i.e.,

$$\mathbb{E}\left(\exp\left[-\frac{x^2}{2}\int_0^1 W_t^2\, dt\right]\right) = \sqrt{\phi(1)} = \frac{1}{\sqrt{\operatorname{ch} x}},$$

which gives the claim directly.   ◇

The content of the present section can easily be extended to the case of the general Heisenberg group $\mathcal{H}^{2d+1}$; see for example [EFLJ4], Section 4.4.

7.6.3   Brownian motions in $\mathbf{SL(2)}$ Take $k = 2$, and

$$A_1 := \begin{pmatrix} \frac{1}{2} & 0 \\ 0 & -\frac{1}{2} \end{pmatrix}, \quad A_2 := \begin{pmatrix} 0 & 1 \\ 0 & 0 \end{pmatrix}$$

in sl(2), so that $[A_1, A_2] = A_2$, and the Lie algebra generated by A_1, A_2 has basis (A_1, A_2). Take $A_0 = a\,A_1$, for $a \in \mathbb{R}$. Then the associated left Brownian motion (i.e., the corresponding solution (X_s) to the linear SDE (7.1)) can be written

$$X_s = \begin{pmatrix} \sqrt{y_s} & x_s/\sqrt{y_s} \\ 0 & 1/\sqrt{y_s} \end{pmatrix},$$

with $y_s := \exp(W_s^1 + as)$ and $x_s := \int_0^s y_t\, dW_t^2$.

Similarly, the right Brownian motion solving the linear SDE:

$$Y_s = 1 + \int_0^s dW_t^A\, Y_t + \int_0^s \left(\frac{1}{2}(A_1^2 + A_2^2) + A_0 \right) Y_t\, dt$$

can be written

$$Y_s = \begin{pmatrix} \sqrt{y_s} & x_s'\sqrt{y_s} \\ 0 & 1/\sqrt{y_s} \end{pmatrix}, \qquad \text{with} \qquad x_s' := \int_0^s y_t^{-1}\, dW_t^2,$$

as is easily verified by applying Itô's formula. (X_s^{-1}) solves the same SDE as (Y_s), but with $-A_j$ instead of A_j.

Note   Observe that we have here a precise counterexample to the identity in law of (X_s) and (X_s^{-1}) jointly for different values of s (even for $a = 0$; recall the note following Proposition 7.2.11). In fact, if (for $0 < u < s$) (X_u, X_s) and (X_u^{-1}, X_s^{-1}) had the same law, then $\left((X_\cdot)^{1,1}(X_\cdot)^{1,0}\right)_{u,s} = (x_u, x_s)$ and $\left((X^{-1})^{1,1}(X^{-1})^{1,0}\right)_{u,s} = -(y_u^{-1}x_u,\, y_s^{-1}x_s)$ would have the same law too. Now this is false, since the former is a martingale while the latter is not, as we can compute easily:

$$\mathbb{E}\left[y_s^{-1}x_s \,\middle|\, \mathcal{F}_u\right] = \mathbb{E}\left[e^{-W_s^1} \int_0^s e^{W_t^1}\, dW_t^2 \,\middle|\, \mathcal{F}_u\right]$$

$$= \mathbb{E}\left[e^{-W_s^1 + W_u^1}\,\middle|\,\mathcal{F}_u\right] e^{-W_u^1} \int_0^u e^{W_t^1}\, dW_t^2 + \mathbb{E}\left[\int_u^s e^{W_t^1 - W_s^1}\, dW_t^2 \,\middle|\, \mathcal{F}_u\right]$$

$$= e^{(s-u)/2}\, y_u^{-1}\, x_u + 0 \neq y_u^{-1}\, x_u \qquad \text{almost surely}.$$

Note that both of the Brownian motions (X_s) and (Y_s) live in the group

$$\mathbb{T}_2' := \left\{ T'_{x+\sqrt{-1}\,y} := \exp(xA_2)\,\exp(\log y\, A_1) \,\middle|\, x \in \mathbb{R},\, y \in \mathbb{R}_+^* \right\},$$

which is isomorphic to the affine group $\mathbb{A}^2$ (recall Proposition 1.4.3).

Since we can never have $X_s = -X_{s'}$ or $Y_s = -Y_{s'}$, these two Brownian motions can be identified with their images under the canonical projection $\mathrm{SL}(2) \to \mathrm{PSL}(2)$, and thus seen as Brownian motions on $\mathrm{PSL}(2)$, or on $\mathrm{PSO}(1,2)$ according to Proposition 1.1.5.1.

7.6.4 Left Brownian motion on SO(d)

Recall from Sections 1.2 and 1.3 that we see $\mathrm{SO}(d)$ as the subgroup of elements in $\mathrm{PSO}(1,d)$ which fix e_0, and $\mathrm{SO}(d-1)$ as the subgroup of of elements in $\mathrm{SO}(d)$ which fix e_1.

The solution (X_t) of the linear SDE (for independent standard real Brownian motions $W^{k\ell}$)

$$X_t = \mathbf{1} + \int_0^t X_s \sum_{1 \le k < \ell \le d} E_{k\ell} \, dW_s^{k\ell} + \frac{1}{2} \int_0^t X_s \sum_{1 \le k < \ell \le d} E_{k\ell}^2 \, ds \qquad (7.9)$$

is a left Brownian motion on $\mathrm{SO}(d)$, with infinitesimal generator $\frac{1}{2} \Xi_0$, the half Casimir operator on $\mathrm{SO}(d)$. Note that $\sum_{1 \le k < \ell \le d} E_{k\ell}^2$ equals $(1-d)$ times the orthogonal projector from $\mathbb{R}^{1,d}$ to $\mathbb{R}^d$ (since $E_{k\ell}^2 e_j = \langle e_j, e_k \rangle e_k + \langle e_j, e_\ell \rangle e_\ell$ for $k < \ell$, and then $\sum_{1 \le k < \ell \le d} E_{k\ell}^2 \, e_j = (d-1) \sum_{1 \le k \le d} \langle e_j, e_k \rangle e_k$).

Proposition 7.6.4.1 *By identifying the above $\mathrm{SO}(d)$-valued left Brownian motion (X_t) with $(\tilde{\sigma}_t, \tilde{\varrho}_t) \in \mathbb{S}^{d-1} \times \mathrm{SO}(d-1)$ by means of the decomposition $X_t = \mathcal{R}_{\tilde{\sigma}_t} \tilde{\varrho}_t$ of Proposition 3.3.2, we obtain the following: on the one hand, another $\mathrm{SO}(d)$-valued left Brownian motion $(\mathcal{R}_{\tilde{\sigma}_t})$, solving*

$$\mathcal{R}_{\tilde{\sigma}_t} = \mathbf{1} + \int_0^t \mathcal{R}_{\tilde{\sigma}_s} \sum_{k=2}^d E_{1k} \, d\widetilde{W}_s^k + \frac{1}{2} \int_0^t \mathcal{R}_{\tilde{\sigma}_s} \sum_{k=2}^d E_{1k}^2 \, ds,$$

where $d\widetilde{W}_t^k := -\sum_{\ell=2}^d \langle \tilde{\varrho}_t \, e_\ell, e_k \rangle \, dW_t^{1\ell}$ (for $2 \le k \le d$) defines a $(d-1)$-dimensional Brownian motion $\widetilde{W}$ (rotated by $\tilde{\varrho}_t^{-1}$ from $(W^{1\ell})$); and on the other hand, the $\mathrm{SO}(d-1)$-valued left Brownian motion $(\tilde{\varrho}_t)$, solving

$$\tilde{\varrho}_t = \mathbf{1} + \int_0^t \tilde{\varrho}_s \sum_{2 \le k < \ell \le d} E_{k\ell} \, dW_s^{k\ell} + \frac{1}{2} \int_0^t \tilde{\varrho}_s \sum_{2 \le k < \ell \le d} E_{k\ell}^2 \, ds.$$

As above, $\sum_{2 \le k < \ell \le d} E_{k\ell}^2$ is $(2-d)$ times the orthogonal projector on $\mathbb{R}^{d-1}$.

Note that for $d \ge 3$, the rotations ϱ mapping e_1 to $-e_1$ are polar. Almost surely, we have $X_t e_1 \ne -e_1$ for all $t > 0$; see Remark 7.7.2.5.

Proof

(1) From Equation (7.9), we obtain directly

$$\tilde{\sigma}_t = \mathcal{R}_{\tilde{\sigma}_t} e_1 = X_t\, e_1 = e_1 - \int_0^t X_s \sum_{\ell=2}^d e_\ell \; dW_s^{1\ell} + \frac{1}{2} \int_0^t X_s(d-1) e_1 \; ds$$

$$= e_1 - \int_0^t \mathcal{R}_{\tilde{\sigma}_s} \sum_{\ell=2}^d \tilde{\varrho}\, e_\ell \; dW_s^{1\ell} + \frac{d-1}{2} \int_0^t \tilde{\sigma}_s \; ds.$$

Hence

$$\mathcal{R}_{\tilde{\sigma}_t}(e_1) = \tilde{\sigma}_t = e_1 - \int_0^t \mathcal{R}_{\tilde{\sigma}_s} \sum_{k=2}^d e_k \; d\widetilde{W}_s^k + \frac{d-1}{2} \int_0^t \tilde{\sigma}_s \; ds$$

$$= \left(1 + \int_0^t \mathcal{R}_{\tilde{\sigma}_s} \sum_{k=2}^d E_{1k} \; d\widetilde{W}_s^k + \frac{1}{2} \int_0^t \mathcal{R}_{\tilde{\sigma}_s} \sum_{k=2}^d E_{1k}^2 \; ds\right) e_1.$$

As the planar rotation $\mathcal{R}_{\tilde{\sigma}_t}$ is prescribed by the image vector $\tilde{\sigma}_t = \mathcal{R}_{\tilde{\sigma}_t}(e_1)$, we obtain

$$\mathcal{R}_{\tilde{\sigma}_t} = 1 + \int_0^t \mathcal{R}_{\tilde{\sigma}_s} \sum_{k=2}^d E_{1k} \; d\widetilde{W}_s^k + \frac{1}{2} \int_0^t \mathcal{R}_{\tilde{\sigma}_s} \sum_{k=2}^d E_{1k}^2 \; ds.$$

(2) On the other hand, by Proposition 7.2.10, $(\mathcal{R}_{\tilde{\sigma}_t}^{-1})$ solves the following equation of type (7.5):

$$\mathcal{R}_{\tilde{\sigma}_t}^{-1} = 1 - \int_0^t \sum_{k=2}^d E_{1k}\, \mathcal{R}_{\tilde{\sigma}_s}^{-1} \; d\widetilde{W}_s^k + \frac{1}{2} \int_0^t \sum_{k=2}^d E_{1k}^2\, \mathcal{R}_{\tilde{\sigma}_s}^{-1} \; ds.$$

Thus, by Itô's integration-by-parts formula (6.7), we have

$$\tilde{\varrho}_t = \mathcal{R}_{\tilde{\sigma}_t}^{-1} X_t = 1 + \int_0^t d(\mathcal{R}_{\tilde{\sigma}_s}^{-1}) X_s + \int_0^t \mathcal{R}_{\tilde{\sigma}_s}^{-1} \; dX_s + \langle \mathcal{R}_{\tilde{\sigma}\cdot}^{-1}, X\cdot \rangle_t$$

$$= 1 - \int_0^t \sum_{k=2}^d E_{1k}\, \tilde{\varrho}_s \; d\widetilde{W}_s^k + \int_0^t \tilde{\varrho}_s \sum_{1\le k<\ell\le d} E_{k\ell} \; dW_s^{k\ell} + \frac{1}{2} \int_0^t \sum_{k=2}^d E_{1k}^2\, \tilde{\varrho}_s \; ds$$

$$+ \frac{1}{2} \int_0^t \tilde{\varrho}_s \sum_{1\le k<\ell\le d} E_{k\ell}^2 \; ds - \int_0^t \sum_{j=2}^d E_{1j}\, \tilde{\varrho}_s \sum_{1\le k<\ell\le d} E_{k\ell} \; d\langle \widetilde{W}^j, W^{k\ell} \rangle_s.$$

Now $\tilde{\varrho}_s E_{1\ell}\, \tilde{\varrho}_s^{-1} = -\sum_{j=2}^d E_{1j}\, \langle \tilde{\varrho}_s e_\ell, e_j \rangle$, and $\tilde{\varrho}_s$ commutes with $\sum_{\ell=2}^d E_{1\ell}$, so that

$$\tilde{\varrho}_t = 1 + \int_0^t \tilde{\varrho}_s \sum_{2\le k<\ell\le d} E_{k\ell} \; dW_s^{k\ell} + \int_0^t \tilde{\varrho}_s \sum_{k=2}^d E_{1k}^2 \; ds + \frac{1}{2} \int_0^t \tilde{\varrho}_s \sum_{2\le k<\ell\le d} E_{k\ell}^2 \; ds$$

$$- \int_0^t \tilde{\varrho}_s \sum_{i,j,m=2}^d E_{1i} \langle \tilde{\varrho}_s^{-1} e_j, e_i \rangle \sum_{1 \le k < \ell \le d} E_{k\ell} \langle \tilde{\varrho}_s e_m, e_j \rangle \, d\langle W^{1m}, W^{k\ell} \rangle_s$$

$$= 1 + \int_0^t \tilde{\varrho}_s \sum_{2 \le k < \ell \le d} E_{k\ell} \, dW_s^{k\ell} + \int_0^t \tilde{\varrho}_s \sum_{k=2}^d E_{1k}^2 \, ds + \frac{1}{2} \int_0^t \tilde{\varrho}_s \sum_{2 \le k < \ell \le d} E_{k\ell}^2 \, ds$$

$$- \int_0^t \tilde{\varrho}_s \sum_{i,j,\ell=2}^d E_{1i} \langle \tilde{\varrho}_s^{-1} e_j, e_i \rangle E_{1\ell} \langle e_\ell, \tilde{\varrho}_s^{-1} e_j \rangle \, ds$$

$$= 1 + \int_0^t \tilde{\varrho}_s \sum_{2 \le k < \ell \le d} E_{k\ell} \, dW_s^{k\ell} + \frac{1}{2} \int_0^t \tilde{\varrho}_s \sum_{2 \le k < \ell \le d} E_{k\ell}^2 \, ds. \qquad \diamondsuit$$

7.6.5   Left Brownian motions on PSO(1, d) and $\mathbb{A}^d$ The solution (Λ_s) of the linear SDE driven by $(W_s) \equiv (W_s^1, \ldots, W_s^d)$,

$$\Lambda_s = 1 + \int_0^s \Lambda_t \sum_{j=1}^d E_j \, dW_t^j + \frac{1}{2} \int_0^s \Lambda_t \sum_{j=1}^d E_j^2 \, dt, \qquad (7.10)$$

is a left Brownian motion on the Lorentz–Möbius Lie group $PSO(1, d)$, which (by Theorem 7.2.2) has generator $\frac{1}{2} \tilde{\Xi} := \frac{1}{2} \sum_{j=1}^d (\mathcal{L}_{E_j})^2$. Recall that $\sum_{j=1}^d E_j \, dW_t^j \equiv dW_t^E$ is a Brownian motion on the Lie algebra $so(1, d)$ (according to the note on page 151).

Consider also the solution (Z_s) of the linear SDE.

$$Z_s = 1 + \int_0^s Z_t \sum_{j=1}^d \tilde{E}_j \, dW_t^j + \frac{1}{2} \int_0^s Z_t \left(\sum_{j=1}^d \tilde{E}_j^2 - (d-1)E_1 \right) dt, \qquad (7.11)$$

which is a left Brownian motion on the affine group $\mathbb{A}^d$. Recall that for convenience, we set $\tilde{E}_1 = E_1$. According to eqn (3.3), the infinitesimal generator of (Z_s) and of its associated convolution semi-group ν_t is

$$\frac{1}{2} \left[\sum_{j=2}^d (\mathcal{L}_{\tilde{E}_j})^2 + (\mathcal{L}_{E_1})^2 + (1-d)\mathcal{L}_{E_1} \right] = \frac{1}{2} \mathcal{D}.$$

Let us apply the Iwasawa decomposition to the left Brownian motion (Λ_s) solving eqn (7.10): we have $\Lambda_s = T_{z_s} \varrho_s$, with $T_{z_s} = \mathrm{Iw}(\Lambda_s) \in \mathbb{A}^d$ and $\varrho_s = \mathrm{I}^w(\Lambda_s) \in SO(d)$.

Theorem 7.6.5.1 (*i*) *The first Iwasawa projection $T_{z_s} = \mathrm{Iw}(\Lambda_s)$ of the left Brownian motion (Λ_s) is a left Brownian motion on the affine group $\mathbb{A}^d$, which solves eqn (7.11) (but with the following change of the Brownian motion: the d-dimensional white noise dW_t is replaced by its image $d(\varrho^{-1}W)_t$ under the random* *rotation ϱ_t^{-1}). In particular, the $\mathbb{H}^d$-valued processes $(\Lambda_s e_0)$ and $(Z_s e_0)$ have the same law. Moreover, the $\mathbb{R}^{d-1} \times \mathbb{R}_+^*$-valued process $z_s = (x_s, y_s)$ is given by (for $2 \le j \le d$)*

$$y_s = \exp\left[(\varrho^{-1}W)_s^1 - \tfrac{d-1}{2}\, s\right] \quad and \quad x_s^j = \int_0^s y_t \; d(\varrho^{-1}W)_t^j. \tag{7.12}$$

(ii) The second Iwasawa projection $\varrho_s = \mathrm{I}^{\mathrm{w}}(\Lambda_s)$ of the left Brownian motion (Λ_s) is a right Brownian motion on $\mathrm{SO}(d)$, the inverse of which is a left Brownian motion that has the same law as the component $\mathcal{R}_{\tilde{\sigma}_t}$ of Proposition 7.6.4.1 (and then can be identified with the spherical Brownian motion of Proposition 7.6.6.2).

Proof *(i)* By eqn (7.10) and Lemma 1.4.2, we have

$$T_{z_s} e_0 = \Lambda_s e_0 = e_0 + \int_0^s T_{z_t} \sum_{j=1}^d \mathrm{Ad}(\varrho_t)\left[E_j \; dW_t^j + \frac{1}{2}\, E_j^2 \; dt\right] \varrho_t e_0$$

$$= e_0 + \int_0^s T_{z_t} \sum_{j=1}^d \left[E_{\varrho_t e_j} \; dW_t^j + \frac{1}{2}\,(E_{\varrho_t e_j})^2 \; dt\right] e_0.$$

Now, since $\varrho_t \in \mathrm{SO}(d)$, on the one hand we have

$$\sum_{j=1}^d (E_{\varrho_t(e_j)})^2 = \sum_{j=1}^d \sum_{i,k=1}^d \langle \varrho_t e_j, e_i\rangle\langle \varrho_t e_j, e_k\rangle\, E_i\, E_k = \sum_{k=1}^d E_k^2,$$

and on the other hand, setting $d(\varrho^{-1}W)_t^k := -\sum_{j=1}^d \langle \varrho_t e_j, e_k\rangle \; dW_t^j$, we find

$$\sum_{j=1}^d E_{\varrho_t e_j} \; dW_t^j = -\sum_{j,k=1}^d \langle \varrho_t e_j, e_k\rangle E_k \; dW_t^j = \sum_{k=1}^d E_k \; d(\varrho^{-1}W)_t^k.$$

Hence

$$T_{z_s} e_0 = e_0 + \int_0^s T_{z_t} \sum_{j=1}^d \left[E_j \; d(\varrho^{-1}W)_t^j + \frac{1}{2}\, E_j^2 \; dt\right] e_0.$$

Note that by Lévy's characterization (Theorem 6.5.7), $(\varrho^{-1}W)$ is another d-dimensional Brownian motion, since for $1 \le k, \ell \le d$ we have

$$\langle (\varrho^{-1}W)^k, (\varrho^{-1}W)^\ell\rangle_t = \int_0^t \sum_{j=1}^d \langle \varrho_s e_j, e_k\rangle\langle \varrho_s e_j, e_\ell\rangle \; ds$$

$$= -\int_0^t \langle \varrho_s^{-1} e_k, \varrho_s^{-1} e_\ell\rangle \; ds = 1_{\{k=\ell\}}\, t.$$

Now, using eqn (1.15), i.e., $E_j = \tilde{E}_j - E_{1j}$ (so that $E_j\, e_0 = \tilde{E}_j\, e_0$), we have

$$
T_{z_s}\, e_0 - e_0 = \int_0^s T_{z_t} \sum_{j=1}^d \left[\tilde{E}_j\, d(\varrho^{-1}W)_t^j + \frac{1}{2} E_j^2\, dt \right] e_0
$$

$$
= \int_0^s T_{z_t}\, d(\varrho^{-1}W)_t^{\tilde{E}}\, e_0 + \frac{1}{2} \int_0^s T_{z_t}\, dt \sum_{j=1}^d E_j^2\, e_0.
$$

Then, using the commutation relations (1.13) (and $E_{11} = 0$), we obtain

$$
\sum_{j=1}^d (E_j^2 - \tilde{E}_j^2)\, e_0 = \sum_{j=2}^d \left((E_j + \tilde{E}_j)(E_j - \tilde{E}_j) + [E_j, \tilde{E}_j] \right) e_0
$$

$$
= \sum_{j=2}^d [E_j, E_{1j}]\, e_0 = -\sum_{j=2}^d E_1\, e_0 = -(d-1)\, E_1\, e_0.
$$

So far, we have obtained

$$
T_{z_s}\, e_0 - e_0 = \int_0^s T_{z_t} \left[\sum_{j=1}^d \tilde{E}_j\, d(\varrho^{-1}W)_t^j + \frac{1}{2} \left(\sum_{j=1}^d \tilde{E}_j^2 - (d-1)\, E_1 \right) dt \right] e_0.
$$

As, by eqn (7.11) the same equation holds with (Z_s) instead of (T_{z_s}) (and W instead of $\varrho^{-1}W$), and since both processes are $\mathbb{A}^d$-valued, the first claim will follow directly from the uniqueness of the $\mathbb{R}^{d-1} \times \mathbb{R}_+^*$-valued process (z_s) solving such an equation.

To show this uniqueness (which is not a priori obvious, on account of the right action on e_0), we observe that (since $\tilde{E}_j e_0 = e_j$ and $E_j^2 e_0 = e_0$, and writing W instead of $\varrho^{-1}W$) this equation can also be written as

$$
T_{z_s}\, e_0 - e_0 = \int_0^s T_{z_t} \left[\sum_{j=1}^d e_j\, dW_t^j + \frac{d}{2} e_0\, dt \right].
$$

Then eqn (1.19) yields

$$
y_s = \langle T_{z_s} e_0, e_0 + e_1 \rangle^{-1} \quad \text{and} \quad x_s^j = -\langle T_{z_s} e_0, e_j \rangle\, y_s \quad \text{for } 2 \le j \le d,
$$

while, by Definition 1.4.1

$$
T_{z_s} e_j = x_s^j (e_0 + e_1) + e_j,
$$

and

$$
T_{z_s} e_1 = \left(y_s + \frac{|x_s|^2}{2\, y_s} \right)(e_0 + e_1) - \frac{1}{2 y_s}(e_0 - e_1) - \frac{x_s}{y_s}.
$$

Hence we obtain

$$dy_s^{-1} = -y_s^{-1} \, dW_s^1 + \frac{d}{2} y_s^{-1} \, ds \quad \text{and} \quad d\left(\frac{x_s^j}{y_s}\right) = -\frac{x_s^j}{y_s} \, dW_s^1 + dW_s^j + \frac{d}{2} \frac{x_s^j}{y_s} \, ds,$$

whence by Itô's formula,

$$dy_s = y_s \, dW_s^1 + \frac{2-d}{2} y_s \, ds \quad \text{and} \quad dx_s^j = y_s \, dW_s^j.$$

These are linear SDEs, which directly yield the wanted uniqueness and eqn (7.12) as well.

(ii) By Proposition 7.2.10, $(T_{z_s}^{-1})$ solves the following equation of type (7.5):

$$T_{z_s}^{-1} = 1 - \sum_{j=1}^{d} \int_0^s \tilde{E}_j \, T_{z_t}^{-1} \, d(\varrho^{-1}W)_t^j + \int_0^s \frac{1}{2}\left(\sum_{j=1}^{d} \tilde{E}_j^2 + (d-1)\, E_1\right) T_{z_t}^{-1} \, dt.$$

Thus, by Itô's integration-by-parts formula (6.7), we have

$$\varrho_s = T_{z_s}^{-1} \Lambda_s = 1 + \int_0^s d(T_{z_t}^{-1})\Lambda_t + \int_0^s T_{z_t}^{-1} \, d\Lambda_t + \langle T_{z.}^{-1}, \Lambda.\rangle_s$$

$$= 1 - \sum_{j=1}^{d} \int_0^s \tilde{E}_j \, \varrho_t \, d(\varrho^{-1}W)_t^j + \sum_{j=1}^{d} \int_0^s \varrho_t \, E_j \, dW_t^j + \frac{1}{2}\int_0^s \varrho_t \sum_{j=1}^{d} E_j^2 \, dt$$

$$+ \int_0^s \frac{1}{2}\left(\sum_{j=1}^{d} \tilde{E}_j^2 + (d-1)\, E_1\right) \varrho_t \, dt - \int_0^s \sum_{j,k=1}^{d} \tilde{E}_j \, \varrho_t \, E_k \, d\langle(\varrho^{-1}W)^j, W^k\rangle_t.$$

Now, as in (i) above, we have

$$\sum_{j=1}^{d} \int_0^s \varrho_t \, E_j \, dW_t^j = \sum_{j=1}^{d} \int_0^s E_{\varrho_t e_j} \, \varrho_t \, dW_t^j = \sum_{j=1}^{d} \int_0^s E_j \, \varrho_t \, d(\varrho^{-1}W)_t^j$$

and

$$\varrho_t \sum_{j=1}^{d} E_j^2 = \sum_{j=1}^{d} E_{\varrho_t e_j}^2 \, \varrho_t = \sum_{j=1}^{d} E_j^2 \, \varrho_t.$$

Moreover,

$$\sum_{j,k=1}^{d} \tilde{E}_j \, \varrho_t \, E_k \, \langle(\varrho^{-1}W)^j, W^k\rangle_t$$

$$= -\sum_{j,k=1}^{d} \tilde{E}_j \, E_{\varrho_t e_k} \, \varrho_t \, \langle \varrho_t e_k, e_j\rangle \, dt = \sum_{j=1}^{d} \tilde{E}_j \, E_j \, \varrho_t \, dt.$$

Therefore (again since $\sum_{j=1}^d (\tilde{E}_j^2 - E_j^2) = \sum_{j=2}^d (E_j + \tilde{E}_j)E_{1j} + (d-1)\,E_1$)

$$\varrho_s = 1 - \sum_{j=1}^d \int_0^s E_{1j}\,\varrho_t\,d(\varrho^{-1}W)_t^j$$

$$+ \int_0^s \frac{1}{2}\Big[\sum_{j=1}^d (\tilde{E}_j^2 + E_j^2 - 2\tilde{E}_j E_j) + (d-1)E_1\Big]\varrho_t\,dt$$

$$= 1 - \int_0^s \sum_{j=1}^d E_{1j}\,\varrho_t\,d(\varrho^{-1}W)_t^j \int_0^s \frac{1}{2}\sum_{j=1}^d E_{1j}^2\,\varrho_t\,dt.$$

This means that (ϱ_s) is a right Brownian motion on $\mathrm{SO}(d)$. Finally, by Proposition 7.2.10, its inverse (ϱ_s^{-1}) is a left Brownian motion, which has the same law as the component $\mathcal{R}_{\tilde{\sigma}_t}$ of Proposition 7.6.4.1.   ◇

7.6.6   Spherical Brownian motion   We present here our first example of a diffusion process (which does not live in a group) in the sense of Section 7.5. We consider the projection $\mathrm{p} : \varrho \mapsto \varrho\,e_1$ from $\mathrm{SO}(d)$ onto $\mathbb{S}^{d-1}$ and the left Brownian motion (X_t) on $\mathrm{SO}(d)$ (recall Section 7.6.4). The following shows that these satisfy the assumption of Proposition 7.5.2.

Lemma 7.6.6.1 *The law of the process $(\varrho\,X_t\,e_1)$ depends on $\varrho \in \mathrm{SO}(d)$ only through $\varrho\,e_1$.*

In fact, it is not much more difficult to prove that the law of the $\mathrm{SO}(d)$-valued left Brownian motion (X_t) solving eqn (7.9) is invariant under any conjugation by a fixed $\tilde{\varrho} \in \mathrm{SO}(d-1)$.

Proof We have to show that the processes $(\varrho\,X_t\,e_1)$ and $(\varrho\,\tilde{\varrho}\,X_t\,e_1)$ have the same law, for any $\tilde{\varrho} \in \mathrm{SO}(d-1)$. Using Proposition 7.6.4.1, we can replace (X_t) by $(\varrho_t := \mathcal{R}_{\tilde{\sigma}_t})$, which solves

$$\varrho_t = 1 + \int_0^t \varrho_s \sum_{k=2}^d E_{1k}\,d\widetilde{W}_s^k + \frac{1}{2}\int_0^t \varrho_s \sum_{k=2}^d E_{1k}^2\,ds,$$

and we then have to deal with $\tilde{\varrho}_t := \tilde{\varrho}\,\varrho_t\,\tilde{\varrho}^{-1}$, which solves

$$\tilde{\varrho}_t = 1 + \int_0^t \tilde{\varrho}_s \sum_{k=2}^d \mathrm{Ad}(\tilde{\varrho})(E_{1k})\,d\widetilde{W}_s^k + \frac{1}{2}\int_0^t \tilde{\varrho}_s \sum_{k=2}^d \mathrm{Ad}(\tilde{\varrho})(E_{1k})^2\,ds.$$

Now, on the one hand we have

$$\sum_{k=2}^d \mathrm{Ad}(\tilde{\varrho})(E_{1k})\,d\widetilde{W}_s^k = -\sum_{j,k=2}^d E_{1j}\,\langle \tilde{\varrho}\,e_k, e_j\rangle\,d\widetilde{W}_s^k = \sum_{j=2}^d E_{1j}\,d\big(\tilde{\varrho}^{-1}\widetilde{W}\big)_s^j,$$

where $\tilde{\varrho}^{-1}\widetilde{W}$ denotes another Brownian motion of $\mathbb{R}^{d-1}$, rotated from $\widetilde{W}$ under $\tilde{\varrho}^{-1}$, while on the other hand we have

$$\sum_{k=2}^{d} \mathrm{Ad}(\tilde{\varrho})(E_{1k})^2 = \sum_{i,j,k=2}^{d} E_{1i}\,E_{1j}\,\langle \tilde{\varrho}\,e_k, e_i\rangle\langle \tilde{\varrho}\,e_k, e_j\rangle = \sum_{k=2}^{d} E_{1k}^2.$$

Hence changing (ϱ_t) to $(\tilde{\varrho}_t)$ amounts merely to rotating its driving $(d-1)$-dimensional Brownian motion $\widetilde{W}$ by $\tilde{\varrho}^{-1}$, which of course does not change its law, thereby yielding the wanted irrelevance of $\tilde{\varrho}$.   ◇

Lemma 7.6.6.1 allows us to apply Proposition 7.5.2, yielding the spherical semi-group, say (S_t), which satisfies the identity

$$S_t f\big(\mathrm{p}(\varrho)\big) = Q_t(f \circ \mathrm{p})(\varrho) = \mathbb{E}\big[f(\varrho\,X_t\,e_1)\big],\tag{7.13}$$

for any $t \geq 0$, $\varrho \in \mathrm{SO}(d)$ and $f \in C(\mathbb{S}^{d-1})$. As in Remark 7.5.3 (apart from the notation), (Q_t) denotes the semi-group of (X_t) here.

The associated diffusion process is called **spherical Brownian motion**. When it starts at e_1, this is the process $(\tilde{\sigma}_t)$ of Proposition 7.6.4.1. Moreover, we have the following analogue of Proposition 7.2.8.

Proposition 7.6.6.2 *The spherical Brownian semi-group (S_t) maps $C^2(\mathbb{S}^{d-1})$ into $C^2(\mathbb{S}^{d-1})$, and for any $f \in C^2(\mathbb{S}^{d-1})$ we have*

$$\frac{d}{dt}\,S_t f = \frac{1}{2}\,\Delta^{\mathbb{S}^{d-1}} S_t f = \frac{1}{2}\,S_t\,\Delta^{\mathbb{S}^{d-1}} f.$$

Moreover, (S_t) is self-adjoint with respect to the volume measure of $\mathbb{S}^{d-1}$, and covariant with $\mathrm{SO}(d)$: we have $S_t(f \circ \varrho) = (S_t f) \circ \varrho$ for any $t \geq 0$, $\varrho \in \mathrm{SO}(d)$.

By analogy with Proposition 7.2.8, we say that the half spherical Laplacian $\frac{1}{2}\Delta^{\mathbb{S}^{d-1}}$ is the (infinitesimal) generator of spherical Brownian motion and of its semi-group (S_t).

Proof Note first that a function ϕ on $\mathbb{S}^{d-1}$ belongs to $C^2(\mathbb{S}^{d-1})$ if and only if $\phi \circ \mathrm{p} \in C^2(\mathrm{SO}(d))$ (since any directional derivative on $\mathbb{S}^{d-1}$ can be obtained by using some planar rotation). Since, by Proposition 7.2.8, Q_t maps $C^2(\mathrm{SO}(d))$ into $C^2(\mathrm{SO}(d))$, this shows the first claim.

Then, by Definition 3.4.1 and eqn (7.13), we have

$$\frac{d}{dt}(S_t f) \circ \mathrm{p} = \frac{d}{dt}\,Q_t(f \circ \mathrm{p}) = \frac{1}{2}\,\Xi_0\,Q_t(f \circ \mathrm{p}) = \frac{1}{2}\,\Xi_0\big((S_t f) \circ \mathrm{p}\big)$$

$$= \frac{1}{2}\,\Delta^{\mathbb{S}^{d-1}}(S_t f) \circ \mathrm{p}.$$

In the same way, we obtain

$$\frac{d}{dt}(S_t f) \circ \mathrm{p} = \tfrac{1}{2}\, Q_t\, \Xi_0(f \circ \mathrm{p}) = \frac{1}{2}\, Q_t\big(\Delta^{\mathbb{S}^{d-1}} f \circ \mathrm{p}\big) = \tfrac{1}{2}\,(S_t\, \Delta^{\mathbb{S}^{d-1}} f) \circ \mathrm{p}.$$

This proves the equalities of the statement. Then, by eqn (7.13), for any $f \in C\big(\mathrm{SO}(d)\big)$ and $\varrho\,,\varrho' \in \mathrm{SO}(d)$ we have

$$S_t(f \circ \varrho)\big(\mathrm{p}(\varrho')\big) = \mathbb{E}\big[f(\varrho\varrho' X_t\, e_1)\big] = S_t f\big(\mathrm{p}(\varrho\varrho')\big) = (S_t f) \circ \varrho\big(\mathrm{p}(\varrho')\big).$$

Finally the self-adjointness of (S_t) follows from that of $\Delta^{\mathbb{S}^{d-1}}$ (recall Proposition 3.6.2.3): if (S_t^*) denotes the adjoint semi-group, it is clear that

$$\frac{d}{dt}\, S_t^*\, f = \frac{1}{2}\, S_t^*\, \Delta^{\mathbb{S}^{d-1}}\, f$$

for any $f \in C^2(\mathbb{S}^{d-1})$. Hence, for any $0 \le t \le s$,

$$\frac{d}{dt}\, S_t^*\, S_{s-t} f = 0,$$

and then $S_s^* f = S_s f$. $\diamond$

7.6.7   Hyperbolic Brownian motion   We present here a second example of a diffusion process, which is essential to our purposes in this book. We proceed in a way very similar to that in Section 7.6.6 for spherical Brownian motion.

We consider now the projection $\mathrm{p} = \pi_0 \;:\; g \mapsto g\, e_0$ from $\mathrm{PSO}(1,d)$ onto $\mathbb{H}^d$ and the left Brownian motion (Λ_t) on $\mathrm{PSO}(1,d)$, of Section 7.6.5 and eqn (7.10). The following shows that they satisfy the assumption of Proposition 7.5.2.

Lemma 7.6.7.1   *The law of the processes* $(g\,\Lambda_t\, e_0)$ *and* $(g\, Z_t\, e_0)$ *depends on* $g \in \mathrm{PSO}(d)$ *only through* $g\, e_0$.

Proof   Theorem 7.6.5.1(i) allows us to substitute $(g\,\Lambda_t\, e_0)$ for $(g\, Z_t\, e_0)$. We have to show the irrelevance of $\varrho \in \mathrm{SO}(d)$ in the law of $\varrho\,\Lambda_t\, e_0 = \Lambda_t^\varrho\, e_0$, where $\Lambda_t^\varrho :=\varrho\,\Lambda_t\,\varrho^{-1}$. Now, by eqn (7.10) for (Λ_t), we have

$$\Lambda_t^\varrho = 1 + \int_0^t \Lambda_s^\varrho \sum_{j=1}^{d} \mathrm{Ad}(\varrho)(E_j)\, dW_s^j + \frac{1}{2}\int_0^t \Lambda_s^\varrho \sum_{j=1}^{d} \mathrm{Ad}(\varrho)(E_j)^2\, ds,$$

and (as in the proof of Proposition 7.6.6.2) on the one hand we have

$$\sum_{j=1}^{d} \mathrm{Ad}(\varrho)(E_j)\, dW_s^j = -\sum_{j,k=1}^{d} E_k\, \langle \varrho\, e_j, e_k\rangle\, dW_s^j = \sum_{k=1}^{d} E_k\, d\big(\varrho^{-1} W\big)_s^k,$$

where $\varrho^{-1}W$ denotes another Brownian motion of $\mathbb{R}^d$, rotated from W by ϱ^{-1}, while on the other hand we have

$$\sum_{j=1}^d \mathrm{Ad}(\varrho)(E_j)^2 = \sum_{i,j,k=1}^d E_i\, E_k\, \langle \varrho\, e_j, e_i \rangle \langle \varrho\, e_j, e_k \rangle = \sum_{k=1}^d E_k^2.$$

Hence changing Λ_t to Λ_t^ϱ amounts merely to rotating its driving d-dimensional Brownian motion W by ϱ^{-1}, which of course does not change its law, thereby yielding the wanted irrelevance of ϱ.   ◇

Lemma 7.6.7.1 allows us to apply Proposition 7.5.2, yielding the **hyperbolic heat semi-group**, say (Q_t), which satisfies the identities

$$Q_t f(z) = \mathbb{E}\big[f(g\,\Lambda_t\, e_0)\big] = \mathbb{E}\big[f(g\, Z_t\, e_0)\big], \tag{7.14}$$

for any $t \geq 0$, $f \in C_b(\mathbb{H}^d)$, $z \in \mathbb{H}^d$ and $g \in \mathrm{PSO}(1,d)$ such that $g\, e_0 = z$.

The associated diffusion process is called **hyperbolic Brownian motion**. We have the following analogue of Propositions 7.2.8 and 7.6.6.2.

Theorem 7.6.7.2 *The hyperbolic heat semi-group* (Q_t) *maps* $C_b^2(\mathbb{H}^d)$ *into* $C_b^2(\mathbb{H}^d)$, *and for any* $f \in C_b^2(\mathbb{H}^d)$ *we have*

$$\frac{d}{dt}\, Q_t f = \frac{1}{2}\, \Delta\, Q_t f = \frac{1}{2}\, Q_t\, \Delta f. \tag{7.15}$$

Moreover, Q_t *is self-adjoint with respect to the volume measure of* $\mathbb{H}^d$, *and covariant with* $\mathrm{PSO}(1,d)$: *we have* $Q_t(f \circ g) = (Q_t f) \circ g$ *for any* $t \geq 0$, $g \in \mathrm{PSO}(1,d)$.

By analogy with Proposition 7.2.8, we say that the half hyperbolic Laplacian $\frac{1}{2}\,\Delta$ is the (infinitesimal) generator of hyperbolic Brownian motion and of its (hyperbolic heat) semi-group (Q_t).

Proof Note first that a function ϕ on $\mathbb{H}^d$ belongs to $C_b^2(\mathbb{H}^d)$ if and only if $\phi \circ \mathrm{p} \in C_b^2(\mathrm{PSO}(1,d))$ (since any directional derivative on $\mathbb{H}^d$ can be obtained by using some planar hyperbolic rotation). As by Proposition 7.2.8, the semi-group P_t maps $C_b^2(\mathrm{PSO}(1,d))$ into $C_b^2(\mathrm{PSO}(1,d))$, this shows the first claim.

Then, by Theorem 3.5.2, Corollary 3.5.3 and eqn (7.14), we have

$$\frac{d}{dt}(Q_t f) \circ \mathrm{p} = \frac{d}{dt}\, P_t(f \circ \mathrm{p}) = \frac{1}{2}\, \Xi\, P_t(f \circ \mathrm{p}) = \frac{1}{2}\, \Xi\big((Q_t f) \circ \mathrm{p}\big)$$

$$= \frac{1}{2}\, \Delta(Q_t f) \circ \mathrm{p}.$$

In the same way, we obtain

$$\frac{d}{dt}(Q_t f) \circ \mathrm{p} = \frac{1}{2}\, P_t\, \Xi(f \circ \mathrm{p}) = \frac{1}{2}\, P_t\big(\Delta f \circ \mathrm{p}\big) = \frac{1}{2}\,(Q_t\, \Delta f) \circ \mathrm{p}.$$

This proves eqn (7.15). Then, by eqn (7.14), for any $f \in C_b(\mathrm{PSO}(1, d))$ and $g, g' \in \mathrm{PSO}(1, d)$, we have

$$Q_t(f \circ g)(\mathrm{p}(g')) = \mathbb{E}\big[f(g\, g' \Lambda_t\, e_0)\big] = Q_t f\big(\mathrm{p}(gg')\big) = (Q_t f) \circ g\big(\mathrm{p}(g')\big).$$

Finally, the self-adjointness follows from that of Δ (recall Proposition 3.6.2.3): if (Q_t^*) denotes the adjoint semi-group, it is clear that

$$\frac{d}{dt}\, Q_t^*\, f = \frac{1}{2}\, Q_t^*\, \Delta f$$

for any $f \in C_b^2(\mathbb{H}^d)$, and hence, for any $0 \le t \le s$,

$$\frac{d}{dt}\, Q_t^*\, Q_{s-t} f = 0$$

and then $Q_s^* f = Q_s f$ ◇

We now describe the main features of the asymptotic behaviour of hyperbolic Brownian motion. Recall that the Poincaré coordinates (x_s, y_s) of the hyperbolic Brownian motion $(Z_s\, e_0)$ are given by the integrated formulae in eqn (7.12).

Proposition 7.6.7.3 *For any starting point $p \in \mathbb{H}^d$, a hyperbolic Brownian motion (z_s) converges almost surely, as $s \to \infty$, to a boundary point $z_\infty \in \partial \mathbb{H}^d$, the law of which is the harmonic measure μ_p.*

Proof First, the covariance with respect to $g \in \mathrm{PSO}(1, d)$ allows us to consider only the case $p = e_0$. Then it is clear from eqn (7.12) (in which we may and do drop the random rotation ϱ_t^{-1}) that y_s goes almost surely to 0, and that we almost surely have

$$x_s^j \longrightarrow x_\infty^j := \int_0^\infty e^{W_t^1 - d - 1/2\, t}\, dW_t^j, \quad \text{for } 2 \le j \le d.$$

Let us justify this last claim in detail. We must beware that $x_\infty^j \notin L^2$ (for $d = 2$ or 3). Note, however, that Proposition 6.3.3 ensures the almost sure convergence of $\int_0^\infty e^{2 W_t^1 - (d-1)t}\, dt$. Otherwise, by the Itô isometric identity, for $0 \le s \le s'$ we almost surely have

$$\mathbb{E}\Big[\big|x_{s'}^j - x_s^j\big|^2 \,\big|\, \mathcal{F}_\infty^1\Big] = \int_s^{s'} e^{2 W_t^1 - (d-1)t}\, dt.$$

Therefore, conditionally on the σ-field $\mathcal{F}_\infty^1$, (x_s^j) is almost surely a continuous martingale $\big($with respect to the filtration $(\mathcal{F}_s^j)\big)$ which converges to x_∞^j in $L^2\big(\mathbb{P}(\cdot \,|\, \mathcal{F}_\infty^1)\big)$, and hence it converges almost surely as well, by Remark 6.3.8(*i*). This provides the wanted almost sure convergence.

The above yields a limit point $z_\infty = (x_\infty, 0) \in \partial \mathbb{H}^d$. It remains to compute its law. Now, by Theorem 7.6.7.2, for any $\varrho \in \mathrm{SO}(d)$, $(\varrho\, z_s)$ has the same law as

(z_s), so that the the law of z_∞ must be $SO(d)$-invariant. As stated in Remark 3.6.1.3, this yields in fact the definition of μ_{e_0} (Definition 3.6.1.1).   ◇

We postpone to Section B.3 a statement specifying how near to the limiting geodesic (containing $p = z_0$ and ending at z_∞) the Brownian path is asymptotically located.

7.7   Relativistic diffusion

We now present relativistic diffusion in Minkowski space, first constructed and studied by Dudley ([D1], [D2], [D3]). It will appear here as a projection of a left Brownian motion on the Poincaré group, in the same vein as in the case of hyperbolic Brownian motion in Section 7.6 (by letting it appear as a projection of a left Brownian motion on the Lorentz–Möbius group or, alternatively, on the affine group). We shall then derive its striking structure, of an integrated hyperbolic Brownian motion, and also the easy part of its asymptotic behaviour.

7.7.1   Left Brownian motion on the Poincaré group $\mathcal{P}^{d+1}$ Related to
$PSO(1, d)$ is the *Poincaré group* $\mathcal{P}^{d+1}$. This group is the analogue in the present Lorentz–Minkowski set-up of the classical group of rigid motions. It is the Lie subgroup of $GL(d + 2)$ made up of the matrices of the form

$$\begin{pmatrix} \Lambda & \xi \\ 0 & 1 \end{pmatrix},$$

with $\Lambda \in PSO(1, d)$, $\xi \in \mathbb{R}^{1,d}$ (written as a column) and $0 \in \mathbb{R}^{1+d}$ (written as a row). Its Lie algebra is the set of matrices

$$\begin{pmatrix} A & v \\ 0 & 0 \end{pmatrix} \in \mathcal{M}(d + 2),$$

with $A \in so(1, d)$ and $v \in \mathbb{R}^{1,d}$.
   We take $k = d$,

$$A_j = \begin{pmatrix} E_j & 0 \\ 0 & 0 \end{pmatrix} \text{ for } 1 \leq j \leq d,$$

where E_j is the infinitesimal boost of Section 1.3, and

$$A_0 = \begin{pmatrix} 0 & e_0 \\ 0 & 0 \end{pmatrix}.$$

The solution (X_s) to the linear SDE (7.1) is thus a left Brownian motion on the Poincaré group $\mathcal{P}^{d+1}$. Set

$$X_s = \begin{pmatrix} \Lambda_s & \xi_s \\ 0 & 1 \end{pmatrix}.$$

Equation (7.1) is then equivalent to

$$\Lambda_s = 1 + \int_0^s \Lambda_t \, dW_t^E + \frac{1}{2} \sum_{j=1}^d \int_0^s \Lambda_t \, E_j^2 \, dt \quad \text{and} \quad \xi_s = \int_0^s \Lambda_t \, e_0 \, dt. \quad (7.16)$$

Hence, on the one hand (Λ_s) is the solution of eqn (7.10), and on the other hand we have $\dot{\xi}_s = \Lambda_s \, e_0$.

Consider the projection p defined by

$$\mathrm{p}\begin{pmatrix} \Lambda & \xi \\ 0 & 1 \end{pmatrix} := (\xi, \Lambda e_0).$$

This satisfies the criterion of Proposition 7.5.2, since by eqn (7.16) it amounts to the irrelevance of $\varrho \in SO(d)$ in the law of $(\varrho\,\xi_t, \varrho\,\dot{\xi}_t = \varrho\,\Lambda_t\,e_0)$, and the latter is clear from Section 7.6.7, according to which $(\varrho\,\Lambda_t\,e_0)$ is a hyperbolic Brownian motion starting at $\varrho\,e_0 = e_0$.

Corollary 7.7.1.1 *The left Brownian motion*

$$X_s = \begin{pmatrix} \Lambda_s & \xi_s \\ 0 & 1 \end{pmatrix}$$

*induces (by projection) a diffusion process $(\xi_s, \dot{\xi}_s)$ on $\mathbb{R}^{1,d} \times \mathbb{H}^d$ (in the sense of Definition 7.5.1) such that $(\dot{\xi}_s)$ is a hyperbolic Brownian motion. This $T^1\mathbb{R}^{1,d}$-valued diffusion process is a **relativistic diffusion**. Its generator is the **relativistic operator***

$$\mathcal{L} := \dot{\xi}\,\frac{\partial}{\partial\xi} + \frac{1}{2}\,\Delta_{\dot{\xi}} \equiv \sum_{j=0}^d \dot{\xi}^j\,\frac{\partial}{\partial\xi^j} + \frac{1}{2}\,\Delta_{\dot{\xi}}.$$

Note that ξ_s is parametrized by its arc length. Mechanically, ξ_s describes the trajectory of a relativistic particle of small mass indexed by its proper time, subjected to a white-noise acceleration (in the proper time). In any fixed Lorentz frame, the time coordinate $t(s) = (\xi_s)_0$ is strictly increasing and the velocity $\left(v_s^j := (\dot{\xi}_s)_j/(\dot{\xi}_s)_0 \mid 1 \le j \le d\right)$ is bounded by the velocity of light (which equals 1 in this set-up). In particular, these relativistic trajectories cannot be closed: in the terminology of [HE], they satisfy the 'causality condition'.

7.7.2   Asymptotic behaviour of the relativistic diffusion   Let us parametrize $\mathbb{H}^d$ by the polar coordinates $(\rho, \phi) \in \mathbb{R}_+ \times \mathbb{S}^{d-1}$ of Proposition 3.5.4, and denote by (ρ_s, ϕ_s) the polar coordinates of the hyperbolic Brownian motion (p_s) arising in Corollary 7.7.1.1. Recall that this means $\dot{\xi}_s \equiv p_s = (\mathrm{ch}\,\rho_s, \phi_s\,\mathrm{sh}\,\rho_s) \in \mathbb{H}^d$.

By eqn (3.11), giving the expression for Δ in these coordinates, by Proposition 7.6.6.2, defining spherical Brownian motion, and by Itô's formula, we have:

$$dp_s = dw_s + \frac{d-1}{2} \coth \rho_s \, ds, \quad d\phi_s = \frac{d\Psi_s}{\operatorname{sh} \rho_s}, \qquad (7.17)$$

where we denote by (Ψ_s) a spherical Brownian motion on $\mathbb{S}^{d-1}$, and by (w_s) a standard real Brownian motion, independent of (Ψ_s).

We know from Proposition 7.6.7.3 that almost surely, as $s \to \infty$, the radial diffusion (ρ_s) goes to infinity and the angular coordinate (ϕ_s) goes to some random limit $\phi_\infty \in \mathbb{S}^{d-1}$. Let us be more precise now.

Lemma 7.7.2.1 *We have* $\rho_s = w_s + (d - 1/2) s + \eta_s$, *for some almost surely converging (as $s \to +\infty$) process (η_s). Hence the random variable $\int_0^\infty (\operatorname{sh} \rho_u)^{-2} \, du$ is almost surely finite.*

Proof By the comparison theorem of Theorem 7.1.2 (since $\coth \rho_s > 1$) and by Proposition 6.3.3, we have almost surely

$$\rho_s \geq \rho_0 + w_s + \frac{d-1}{2} s \longrightarrow +\infty.$$

This proves that we almost surely have $\rho_s \geq (d-1)s/3$ for large enough s (which is stronger than the mere $\rho_s \to \infty$ already known from Proposition 7.6.7.3). Hence, setting $\eta_s := \rho_s - w_s - (d-1)s/2$, we deduce that

$$\eta_s = \eta_{s_0} + (d-1) \int_{s_0}^s \frac{ds}{e^{2\rho_s} - 1}$$

converges almost surely. $\diamond$

Consider a fixed observer located at the spatial origin 0, having a relativistic trajectory $(\tau, 0) \in \mathbb{R}_+ \times \mathbb{R}^d \subset \mathbb{R}^{1,d}$, with proper time τ. The relativistic diffusion has a trajectory $\xi_s = ((\xi_s)_0, (\xi_s)_1, \ldots, (\xi_s)_d) =: (t(s), \vec{\xi}_s) \in \mathbb{R}_+ \times \mathbb{R}^d$. As we have already seen in Section 2.4, at time τ in the observer's frame, the position of the relativistic diffusion is $Z(\tau) := \vec{\xi}_{s(\tau)}$, with $s(\tau)$ determined by $\tau = t(s(\tau))$, and the velocity $(d/d\tau)Z(\tau) = \phi_s$ th ρ_s has a norm less than 1, i.e., less than the velocity of light. Moreover, we have the following, which states in particular that the magnitude of the velocity asymptotically becomes that of light.

Corollary 7.7.2.2 *The mean Euclidean velocity $Z(\tau)/\tau$ goes almost surely to $\phi_\infty \in \mathbb{S}^{d-1}$ as $\tau \to \infty$.*

Proof By Proposition 7.6.7.3 and Lemma 7.7.2.1, and since $\lim\limits_{\tau \nearrow \infty} s(\tau) = +\infty$ and $d/d\tau Z(\tau) = \phi_{s(\tau)}$ th $\rho_{s(\tau)}$, we immediately have

$$\lim_{\tau \to \infty} \operatorname{th} \rho_{s(\tau)} = \lim_{\tau \to \infty} \sqrt{1 - \left| p^0_{s(\tau)} \right|^{-2}} = 1,$$

and then

$$\lim_{\tau \to \infty} \frac{dZ(\tau)}{d\tau} = \phi_\infty,$$

almost surely. The result follows at once.   ◇

Corollary 7.7.2.2 (a corollary to Proposition 7.6.7.3) yields the limiting random direction $\phi_\infty \in \mathbb{S}^{d-1}$, an asymptotic variable produced by the relativistic diffusion $(\xi_s, \dot{\xi}_s)$. We now obtain another asymptotic variable. Its geometric meaning is that it specifies how almost every trajectory of the relativistic diffusion goes away to infinity in the random direction ϕ_∞: in fact, these trajectories do so by approaching asymptotically some random affine hyperplane of $\mathbb{R}^{1,d}$, parallel to the tangent hyperplane to the light cone which contains $e_0 + \phi_\infty$.

Proposition 7.7.2.3 *The random variable* $\langle \xi_s, e_0 + \phi_\infty \rangle$ *converges almost surely as* $s \to \infty$, *yielding a second (strictly positive) asymptotic random variable provided by the relativistic diffusion* $(\xi_s, \dot{\xi}_s)$, *in addition to* ϕ_∞.

Proof From the above, we straightforwardly deduce that

$$\langle \xi_s - \xi_0, e_0 + \phi_\infty \rangle = \int_0^s \left[\operatorname{ch} \rho_t - \phi_t \cdot \phi_\infty \times \operatorname{sh} \rho_t \right] dt$$

$$= \int_0^s e^{-\rho_t} \, dt + \int_0^s \left[(\phi_\infty - \phi_t) \cdot \phi_\infty \right] \operatorname{sh} \rho_t \, dt.$$

Here, $\phi_1 \cdot \phi_2$ denotes the Euclidean inner product in $\mathbb{R}^d$. It is clear from Lemma 7.7.2.1 that $\int_0^s e^{-\rho_t} \, dt$ converges almost surely. Then, by eqn (7.17) and Lemma 7.7.2.1, we can use an $\mathbb{S}^{d-1}$-valued Brownian motion (Φ_t), independent of (w_s) and therefore of the radial process (ρ_s), to write $\phi_\infty = \Phi_0$ and $\phi_t = \Phi \left[\int_t^\infty (\operatorname{sh} \rho_u)^{-2} \, du \right]$. Denoting by φ the angular distance on $\mathbb{S}^{d-1}$ from Φ_0 (taken as the north pole), we then have

$$(\phi_\infty - \phi_t) \cdot \phi_\infty = 1 - \cos \varphi \left(\Phi \left[\int_t^\infty (\operatorname{sh} \rho_u)^{-2} \, du \right] \right).$$

Now, recall from Proposition 3.4.2(*iii*) that the spherical Laplacian decomposes according to

$$\Delta^{\mathbb{S}^{d-1}} = \frac{\partial^2}{\partial \varphi^2} + (d-2) \operatorname{cotg} \varphi \, \frac{\partial}{\partial \varphi} + (\sin \varphi)^{-2} \Delta^{\mathbb{S}^{d-2}},$$

so that (by Itô's formula) the so-called Legendre process $(\varphi_t := \varphi \circ \Phi_t \in [0, \pi])$ solves the stochastic differential equation $d\varphi_t = d\tilde{w}_t + d - 2/2 \operatorname{cotg} \varphi_t \, dt$, for some standard real Brownian motion $(\tilde{w}_t)$. Hence we almost surely have

$$1 - \cos \varphi_t = \int_0^t \sin \varphi_s \, d\tilde{w}_s + \frac{d-1}{2} \int_0^t \cos \varphi_s \, ds = \int_0^t \sin \varphi_s \, d\tilde{w}_s + \mathcal{O}(t)$$

near 0.

To proceed further, we need the following lemma, relating to an abstract Brownian martingale.

Lemma 7.7.2.4 *Let* $M_t := \int_0^t H_s \, dB_s$, *for a real Brownian motion* (B_s) *and* $H \in \Lambda$ *such that* $|H_s| \le h_s$, *for some locally bounded deterministic* (h_s). *Then, almost surely, for any* $\varepsilon > 0$, *we have* $|M_t| = o\left(\int_0^{2t} h^2\right)^{\frac{1-\varepsilon}{2}}$ *as* $t \searrow 0$.

Proof By Doob's inequality (Theorem 6.3.2(*i*)), for any $n \in \mathbb{N}$ we have

$$
\mathbb{P}\left[\sup_{0 \le t \le 2^{-n}} |M_t| > \left(\int_0^{2^{-n}} h^2\right)^{1-\varepsilon/2}\right] \le \left(\int_0^{2^{-n}} h^2\right)^{\varepsilon-1/2} \mathbb{E}\left[\left|\int_0^{2^{-n}} H_s \, dB_s\right|\right]
$$

$$
\le \left(\int_0^{2^{-n}} h^2\right)^{\varepsilon-1/2} \mathbb{E}^{1/2}\left[\int_0^{2^{-n}} H_s^2 \, ds\right] \le \left(\int_0^{2^{-n}} h^2\right)^{\varepsilon/2} = \mathcal{O}\left(2^{-n\varepsilon/2}\right),
$$

by the Schwarz inequality, the isometric Itô identity (6.3) and the hypothesis. Hence, by the first Borel–Cantelli lemma, there exists almost surely an $n_0 \in \mathbb{N}$ such that for any $t \in \,]0, 2^{-n_0}]$, choosing $n \ge n_0$ conveniently, we have

$$
2^{-n-1} \le t \le 2^{-n} \Rightarrow |M_t| \le \left(\int_0^{2^{-n}} h^2\right)^{(1-\varepsilon/2)} \le \left(\int_0^{2t} h^2\right)^{(1-\varepsilon/2)}. \qquad \diamond
$$

End of the proof of Proposition 7.7.2.3 By applying Lemma 7.7.2.4 with $h_s \equiv 1$ to the continuous martingale $\int_0^t \sin \varphi_s \, d\tilde{w}_s$, we deduce that almost surely

$$
\frac{1}{2} \varphi_t^2 \sim 1 - \cos \varphi_t = o\left(t^{(1-\varepsilon)/2}\right) \quad \text{near } 0.
$$

This implies that we can apply Lemma 7.7.2.4 again, with $h_s = o\left(s^{(1-\varepsilon/4)}\right)$, which (for a convenient ε) yields the following: almost surely,

$$
1 - \cos \varphi_t = o\left(\int_0^{2t} s^{(1-\varepsilon/2)} \, ds\right)^{(1-\varepsilon/2)} = o\left(t^{(3-4\varepsilon+\varepsilon^2)/4}\right) = o\left(t^{(3/5)}\right) \quad \text{near } 0.
$$

Then, by this and by Lemma 7.7.2.1, we almost surely have

$$
\int_0^\infty \left[(\phi_\infty - \phi_t) \cdot \phi_\infty\right] \operatorname{sh} \rho_t \, dt = \mathcal{O}(1) \int_0^\infty \left[\int_t^\infty (\operatorname{sh} \rho_u)^{-2} \, du\right] \operatorname{sh} \rho_t \, dt
$$

$$
= \mathcal{O}(1) \int_0^\infty \left[\int_t^\infty \exp[-2w_u - (d-1)u] \, du\right]^{3/5} \times \exp\left[w_t + \frac{d-1}{2} t\right] dt
$$

$$
= \mathcal{O}(1) \int_0^\infty \left[\int_t^\infty \exp\left[-(d-1-\varepsilon)u\right] \, du\right]^{3/5} \exp\left[\frac{d-1+\varepsilon}{2} t\right] dt < \infty
$$

(for $0 < \varepsilon < 1/11$). Hence, the following limit exists almost surely:

$$\lim_{s \to \infty} \langle \xi_s - \xi_0, e_0 + \phi_\infty \rangle = \int_0^\infty e^{-\rho_t}\, dt + \int_0^\infty \left[(\phi_\infty - \phi_t) \cdot \phi_\infty \right] \operatorname{sh} \rho_t\, dt > 0. \, \Diamond$$

It is proved in [Bl] that $\left(\phi_\infty, \lim_{s \to \infty} \langle \xi_s, e_0 + \phi_\infty \rangle \right)$ actually exhausts the asymptotic σ-field of the relativistic diffusion.

Remark 7.7.2.5 For $d \geq 3$, the Legendre process $\left(\varphi_t = \varphi \circ \Phi_t \in [0, \pi] \right)$ arising in the preceding proof of Proposition 7.7.2.3 (where it arises as the angular distance of the spherical Brownian motion (Φ_t) from a pole $P \in \mathbb{S}^{d-1}$) almost surely never hits $\{0, \pi\}$. In other words, for $d \geq 3$, the rotations ϱ mapping e_1 to $\pm e_1$ are polar for the SO(d)-valued left Brownian motion (X_t) of Proposition 7.6.4.1: we almost surely have $X_t\, e_1 \neq \pm e_1$ for all $t > 0$.

Proof By its very definition, the Legendre process $\varphi_t = d(P, \Phi_t)$ hits $]0, \pi[$ at arbitrary small positive times. Hence it is enough to consider the case of $\varphi_0 \in]0, \pi[$. We have to show that almost surely, (φ_t) never hits $\{0, \pi\}$, for $d \geq 3$. Recall that (φ_t) satisfies the stochastic differential equation $d\varphi_t = d\tilde{w}_t + \frac{d-2}{2} \cotg \varphi_t\, dt$. Consider now the function defined for $0 < \varphi < \pi$ by $f(\varphi) := \int_{\pi/2}^\varphi (\sin \psi)^{2-d}\, d\psi$, and the real stochastic process $f(\varphi_t)$, almost surely defined as long as (φ_t) does not hit $\{0, \pi\}$. Note that $\lim_\pi f = \infty = -\lim_0 f$. Applying Itô's formula, we obtain

$$d\big(f(\varphi_t) \big) = f'(\varphi_t) + \tfrac{1}{2} f''(\varphi_t)\, d\langle \varphi, \varphi \rangle_t$$

$$= (\sin \varphi_t)^{2-d} \left(d\tilde{w}_t + \frac{d-2}{2} \cotg \varphi_t\, dt \right) + \frac{2-d}{2} (\sin \varphi_t)^{1-d}\, \cos \varphi_t\, dt$$

$$= (\sin \varphi_t)^{2-d}\, d\tilde{w}_t.$$

From this, considering the time change $\tau_t := \inf\{ s \mid \int_0^s (\sin \varphi_u)^{4-2d}\, du > t \}$, we see that $f(\varphi_{\tau_t})$ is a continuous martingale having quadratic variation t (Theorem 7) and hence is a real Brownian motion W_t (starting at $f(\varphi_0)$), according to Lévy's theorem 6.5.7 (and Remark 6.5.8). As a consequence, we almost surely have $f(\varphi_s) = W\big(\int_0^s (\sin \varphi_u)^{4-2d}\, du \big)$ for all s smaller than the hitting time T of $\{0, \pi\}$ by (φ_s). Now, if T were finite, then as $s \nearrow T$, $f(\varphi_s)$ would go to $+\infty$ if $\varphi_T = \pi$ and to $-\infty$ if $\varphi_T = 0$, whereas by Proposition 6.3.3, $W\big(\int_0^s (\sin \varphi_u)^{4-2d}\, du \big)$ almost surely cannot go to infinity while keeping a given sign. This shows that T is almost surely infinite, thereby ending the proof. $\Diamond$

7.8   Notes and comments

Linear stochastic differential equations appeared in [Ha] as deserving specific treatment. Developments followed in the direction of products of random matrices (see in particular [FgK] and [L1]), and of stochastic flows, (see for example [CE1], [LJW], [CE2], [Ba] and [Ku]).

Lyapunov exponents have also been used in the context of (stochastic or deterministic) dynamical systems; see [Os], [Ru] and [L1].

The first mathematically convincing treatment of relativistic diffusions was due to Dudley ([D1], [D2], [D3]). A generalization to Lorentz manifolds appeared in [FLJ1]. Some other references on this subject are [Bl], [F4], [E], [FLJ2] and [BF]. See [Deb] for a physical point of view.

Chapter Eight
The central limit theorem for geodesics

Les gens qui veulent fortement une chose sont presque toujours bien servis par le hasard.

(Honoré de Balzac)

In this chapter, we provide a proof of the Sinai central limit theorem, generalized to the case of a cofinite and geometrically finite Kleinian group. This theorem, which completes the mixing property established in Section 5.3, shows that, asymptotically, geodesics behave chaotically, and yields a quantitative expression for this phenomenon.

The method we use is to establish such a result first for Brownian trajectories, which is easier because of their strong independence properties. Then we compare geodesics with Brownian trajectories, by means of a change of contour and time reversal. This requires us in particular to consider diffusion paths on the stable foliation and to derive the existence of a key potential kernel, using the spectral gap presented in Section 5.4 and the commutation relation of Section 2.6 (which is related to the instability of the geodesic flow; recall Proposition 2.3.9).

We take a cofinite Kleinian group Γ admitting a spectral gap, for example a geometrically finite group (according to Theorem 5.4.2.5). We shall make much use of the Lie derivatives on $\mathbb{F}^d$ associated with the flows.

Notation We lighten the notation of Sections 1.1.4 and 1.4 somewhat by setting

$$\mathcal{L}_0 := \mathcal{L}_{E_1} \quad \text{and} \quad \mathcal{L}_j := \mathcal{L}_{\tilde{E}_j}, \quad \text{for } 2 \leq j \leq d;$$

that is, for any $f \in C^1(\mathbb{F}^d)$ and $\beta \in \mathbb{F}^d$ (and $2 \leq j \leq d$),

$$\mathcal{L}_0 f(\beta) := \frac{d_0}{ds} f(\beta \theta_s) \quad \text{and} \quad \mathcal{L}_j f(\beta) := \frac{d_0}{ds} f(\beta \theta^+_{se_j}). \tag{8.1}$$

We need also to control the Lie derivatives associated with $so(d)$. Recall from Section 1.3 that $so(d)$ is spanned by the infinitesimal rotations $\{E_{kl} = \langle e_k, \cdot \rangle e_l - \langle e_l, \cdot \rangle e_k\}$. Thus we set

$$\mathcal{L}_{kl} := \mathcal{L}_{E_{kl}}, \quad \text{for } 1 \leq k, l \leq d.$$

Recall from eqn (3.4) that (for $2 \leq j \leq d$) we have

$$\mathcal{L}_0 f(\beta T_{x,y}) = y \frac{\partial}{\partial y} f(\beta T_{x,y}) \text{ and } \mathcal{L}_j f(\beta T_{x,y}) = y \frac{\partial}{\partial x^j} f(\beta T_{x,y}). \tag{8.2}$$

We recall from Section 5.1 the notation introduced after Proposition 5.1.1: $\Gamma \backslash \mathbb{F}^d$ and $\Gamma \backslash \mathbb{H}^d$ denote the quotients of $\mathbb{F}^d$ and $\mathbb{H}^d$ under the left action of the Kleinian group Γ, so that we can identify Γ-invariant functions on $\mathbb{F}^d$ and $\mathbb{H}^d$ with functions on $\Gamma \backslash \mathbb{F}^d$ and $\Gamma \backslash \mathbb{H}^d$, respectively. Similarly, $\mathbb{F}^d / SO(d-1)$ denotes the quotient of $\mathbb{F}^d$ under the right action of the rotation group $SO(d-1)$, so that we can identify $SO(d-1)$-invariant functions on $\mathbb{F}^d$ with functions on $\mathbb{F}^d / SO(d-1) \equiv T^1 \mathbb{H}^d \equiv \mathbb{H}^d \times \partial \mathbb{H}^d$; recall Section 2.2.2.

In the same way, we can identify Γ-left invariant and $SO(d-1)$-right invariant functions on $\mathbb{F}^d$ with functions on the two-sided quotient $\Gamma \backslash \mathbb{F}^d / SO(d-1)$, which we shall denote henceforth by $\mathcal{M}$, so that we have $\Gamma \backslash \mathbb{H}^d \equiv \Gamma \backslash \mathbb{F}^d / SO(d) \equiv \pi_0(\mathcal{M})$.

8.1   Dual $\mathbb{A}^d$-valued left Brownian motions

We return now to the $\mathbb{A}^d$-valued left Brownian motion (Z_s) solving eqn (7.11) in Section 7.6.5. Recall that it has generator

$$\frac{1}{2} \mathcal{D} = \frac{1}{2} \left[\sum_{j=2}^{d} \mathcal{L}_j^2 + \mathcal{L}_0^2 + (1-d)\mathcal{L}_0 \right],$$

and that according to Section 7.6.7, it projects on $\mathbb{H}^d$ under π_0 to the hyperbolic Brownian motion; more precisely, for any $\beta \in \mathbb{F}^d$, $\pi_0(\tilde{\beta} Z_s)$ is a hyperbolic Brownian motion on $\mathbb{H}^d$, starting at $\pi_0(\beta) = \tilde{\beta}(e_0)$. Recall also that according to eqn (7.12), for any $t \geq 0$ and for some $\mathbb{R} \times \mathbb{R}^{d-1}$-valued Brownian motion (w, W), we have

$$Z_t = T_{z_t}, \quad \text{with} \quad y_t = e^{w_t - (d-1)t/2} \quad \text{and} \quad x_t = \int_0^t y_s \, dW_s. \tag{8.3}$$

We shall also use as a key ingredient (in Section 8.5) the dual diffusion process (Z_s^*), which is the $\mathbb{A}^d$-valued left Brownian motion solving the equation obtained from eqn (7.11) simply by changing E_1 to $-E_1$. Its generator is

$$\frac{1}{2} \mathcal{D}^* = \frac{1}{2} \left[\sum_{j=2}^{d} \mathcal{L}_j^2 + \mathcal{L}_0^2 + (d-1)\mathcal{L}_0 \right],$$

i.e., the adjoint of $\frac{1}{2}\mathcal{D}$ with respect to λ, according to Corollary 3.3.11. We denote by $P_s^* = \exp\left[(s/2)\mathcal{D}^* \right]$ its semi-group, adjoint to the semi-group (P_s) of (Z_s) with respect to λ.

Remark 8.1.1 *(i) For any s, Z_s^* and Z_s^{-1} have the same law.*

(ii) We have $\mathcal{D}^* F = y^{1-d} \mathcal{D}(y^{d-1} F)$ and then $P_s^* F = y^{1-d} P_s(y^{d-1} F)$, for any $F \in C_b^2(\mathbb{A}^d)$ and $s \geq 0$.

Proof (i) follows at once from Proposition 7.2.11. For (ii), note that the coordinate function y makes sense on $\mathbb{A}^d$, and then that by applying eqn (3.4) we obtain $y^{1-d} \mathcal{L}_0(y^{d-1} F) = \mathcal{L}_0 F + (d-1)F$ at once. From this, we easily obtain $\mathcal{D}^* F = y^{1-d} \mathcal{D}(y^{d-1} F)$ and then the last expression in part (ii). $\diamond$

As with (Z_t) in eqn (8.3), there exists a unique $\hat{z}_t = (\hat{x}_t, \hat{y}_t) \in \mathbb{R}^{d-1} \times \mathbb{R}_+^*$ such that for some independent Brownian motions w, W and for any $t \geq 0$,

$$Z_t^* = T_{\hat{z}_t}, \quad \text{with} \quad \hat{y}_t = e^{w_t + (d-1)t/2} \quad \text{and} \quad \hat{x}_t = \int_0^t \hat{y}_s \, dW_s. \tag{8.4}$$

We shall need at a crucial step (in Section 8.7, in the form of Remark 8.1.3) the following technical lemma, the main ingredient of which is the calculation of the Lie derivatives of the convolution of a smooth function by the laws of (Z_s) and (Z_s^*).

Lemma 8.1.2 *For any $n \in \mathbb{N}$, both of the semi-groups (P_s) and (P_s^*) act on C^n bounded functions on $\mathrm{PSO}(1, d)$ which have bounded Lie derivatives of order $\leq n$. Moreover, this result remains valid when $\mathcal{L}_j$-derivatives only are considered: (P_s) and (P_s^*) act on functions on $\mathbb{A}^d$ that have bounded $\mathcal{L}_0, \mathcal{L}_2, \dots, \mathcal{L}_d$-derivatives.*

Proof

(1) By Proposition 1.3.2 and eqn (1.15), we have to consider the Lie derivatives $\mathcal{L}_0$, $\mathcal{L}_j$ and $\mathcal{L}_{1j}$ for $2 \leq j \leq d$, and $\mathcal{L}_{jk}$ for $2 \leq j < k \leq d$. Take some f that has bounded derivatives. Up to replacing the function f by its left-translated version $f_\gamma := f(\gamma \cdot)$, for any given $\gamma \in \mathrm{PSO}(1, d)$, it is enough to consider the Lie derivatives at the unit $\mathbf{1}$ of $\mathrm{PSO}(1, d)$. Using eqn (8.3), for any bounded function f on $\mathbb{A}^d$ and any $t \geq 0$ we have

$$P_t f(\mathbf{1}) = \mathbb{E}[f(Z_t)] = \mathbb{E}\big[f(T_{(x_t, \log y_t)})\big] = \mathbb{E}\big[f(\theta_{x_t}^+ \theta_{\log y_t})\big]. \tag{8.5}$$

By eqn (8.4) we similarly have

$$P_t^* f(\mathbf{1}) = \mathbb{E}[f(Z_t^*)] = \mathbb{E}\big[f(T_{(\hat{x}_t, \log \hat{y}_t)})\big] = \mathbb{E}\big[f(\theta_{\hat{x}_t}^+ \theta_{\log \hat{y}_t})\big]. \tag{8.6}$$

We deduce from eqns (8.3), (8.1), (8.5) and (1.16) that

$$\mathcal{L}_0 P_t f(\mathbf{1}) = \frac{d_0}{ds} \mathbb{E}\big[f(\theta_s \theta_{x_t}^+ \theta_{\log y_t})\big] = \mathbb{E}\left[\frac{d_0}{ds} f(Z_t \theta_{(e^s-1)x_t/y_t}^+ \theta_s)\right]$$

$$= \mathbb{E}\left[\mathcal{L}_0 f(Z_t) + \sum_{j=2}^d \frac{x_t^j}{y_t} \mathcal{L}_j f(Z_t)\right] = P_t \mathcal{L}_0 f(\mathbf{1}) + \mathbb{E}\left[y_t^{-1} \sum_{j=2}^d x_t^j \mathcal{L}_j f(Z_t)\right].$$

Similarly, for $2 \leq j \leq d$ we have

$$\mathcal{L}_j P_t f(\mathbf{1}) = \frac{d_0}{ds} \mathbb{E}\big[f(\theta_{se_j}^+ \theta_{x_t}^+ \theta_{\log y_t})\big] = \mathbb{E}\left[\frac{d_0}{ds} f(Z_t \theta_{se_j/y_t}^+)\right]$$

$$= \mathbb{E}\big[y_t^{-1} \mathcal{L}_j f(Z_t)\big]$$

and

$$\mathcal{L}_{1j} P_t f(\mathbf{1}) = \frac{d_0}{ds} \mathbb{E}[f(e^{s E_{1j}} Z_t)] = \mathbb{E}\left[\frac{d_0}{ds} f\left(Z_t \, \mathrm{Ad}(Z_t^{-1})(e^{s E_{1j}})\right)\right]$$

$$= \mathbb{E}\left[\frac{d_o}{ds} f\left(Z_t \, \exp\left[s \, \mathrm{Ad}(Z_t^{-1})(E_{1j})\right]\right)\right].$$

(2) By Lemma 3.3.9, for any $t \geq 0$ and $2 \leq j \leq d$, on the one hand we have

$$\mathrm{Ad}(Z_t^{-1})(E_{1j}) = y_t \, E_{1j} + x_t^j \, E_1 + \sum_{k=2}^{d} x_t^k \, E_{kj}$$

$$- \frac{|x_t|^2 + y_t^2 - 1}{2 \, y_t} \tilde{E}_j + \frac{x_t^j}{y_t} \sum_{k=2}^{d} x_t^k \, \tilde{E}_k,$$

whence

$$\mathcal{L}_{1j} P_t f(\mathbf{1}) = \mathbb{E}\left[\left(y_t \, \mathcal{L}_{1j} + x_t^j \, \mathcal{L}_0 - \frac{|x_t|^2 + y_t^2 - 1}{2 \, y_t} \mathcal{L}_j + \frac{x_t^j}{y_t} \sum_{k=2}^{d} x_t^k \mathcal{L}_k \right.\right.$$

$$\left.\left. + \sum_{k=2}^{d} x_t^k \mathcal{L}_{kj}\right) f(Z_t)\right],$$

and on the other hand, for $2 \leq i < j \leq d$,

$$\mathrm{Ad}(Z_t^{-1})(E_{ij}) = E_{ij} - \frac{x_t^i}{y_t} \tilde{E}_j + \frac{x_t^j}{y_t} \tilde{E}_i,$$

whence

$$\mathcal{L}_{ij} P_t f(\mathbf{1}) = P_t[\mathcal{L}_{ij} f](\mathbf{1}) + \mathbb{E}\left[\frac{x_t^j}{y_t} \mathcal{L}_i f(Z_t) - \frac{x_t^i}{y_t} \mathcal{L}_j f(Z_t)\right].$$

(3) Of course, we have the same formulae for $\mathcal{L}_0 P_t^* f$, $\mathcal{L}_j P_t^* f$, $\mathcal{L}_{1j} P_t^* f$, $\mathcal{L}_{kj} P_t^* f$, by eqn (8.6), where we merely change (x_t, y_t) to $(\hat{x}_t, \hat{y}_t)$. Hence the first sentence of the lemma follows from the integrability of the random variables x_t^j / y_t, y_t^{-1}, x_t^j, $x_t^j x_t^k / y_t$ and, similarly, with $(\hat{x}_t, \hat{y}_t)$ instead of (x_t, y_t), from the integrability of the random variables y_t^{-2}, y_t^2, $(x_t^j x_t^k)^2$, $\hat{y}_t^{-2}$, $\hat{y}_t^2$, $(\hat{x}_t^j \hat{x}_t^k)^2$.

Now eqns (8.3) and (8.4) imply at once the integrability of y_t^n, $\hat{y}_t^n$, for all $n \in \mathbb{Z}$, and by Proposition 2.1.1 we have

$$\mathbb{E}\big[|x_t^j x_t^k|^2\big] \leq \frac{1}{2}\Big(\mathbb{E}\big[|x_t^j|^4\big] + \mathbb{E}\big[|x_t^k|^4\big]\Big) \leq C_2\,\mathbb{E}\bigg[\Big|\int_0^t y_s^2\,ds\Big|^2\bigg]$$

$$\leq C_2\,t\,\mathbb{E}\bigg[\int_0^t y_s^4\,ds\bigg] < \infty,$$

and similarly for $\mathbb{E}\big[|\hat{x}_t^j\,\hat{x}_t^k|^2\big]$. This establishes the first sentence of the lemma.

(4) To prove the second sentence of the lemma, we deal with the second derivatives in the same way. Thus, from the above, for $2 \leq j, k \leq d$, we obtain in turn

$$\mathcal{L}_0^2 P_t f(1) = P_t \mathcal{L}_0^2 f(1) + \mathbb{E}\bigg[y_t^{-1}\sum_{j=2}^{d} x_t^j (\mathcal{L}_j \mathcal{L}_0 + \mathcal{L}_0 \mathcal{L}_j) f(Z_t)\bigg]$$

$$+ \mathbb{E}\bigg[y_t^{-2}\sum_{j,k=2}^{d} x_t^j x_t^k\,\mathcal{L}_j \mathcal{L}_k f(Z_t)\bigg],$$

$$\mathcal{L}_0 \mathcal{L}_j P_t f(1) = \mathbb{E}\bigg[y_t^{-1}\,\mathcal{L}_0 \mathcal{L}_j f(Z_t)\bigg] + \mathbb{E}\bigg[y_t^{-2}\sum_{k=2}^{d} x_t^j x_t^k\,\mathcal{L}_k \mathcal{L}_j f(Z_t)\bigg],$$

$$\mathcal{L}_j \mathcal{L}_0 P_t f(1) = \mathbb{E}\bigg[y_t^{-1}\,\mathcal{L}_j \mathcal{L}_0 f(\beta \circ Z_t)\bigg] + \mathbb{E}\bigg[y_t^{-2}\sum_{k=2}^{d} x_t^k\,\mathcal{L}_j \mathcal{L}_k f(Z_t)\bigg],$$

$$\mathcal{L}_j \mathcal{L}_k P_t f(1) = \mathbb{E}\bigg[y_t^{-2}\,\mathcal{L}_j \mathcal{L}_k f(Z_t)\bigg],$$

and so on for the rotational derivatives. Thus we see that the claim about second-order derivatives follows from the integrability of the random variables $(x_t^j\,x_t^k)^4$, $(\hat{x}_t^j\,\hat{x}_t^k)^4$. Very similarly, the lemma follows from the integrability (for any $n \in \mathbb{N}$) of the random variables $(x_t^j)^{2n}$, $(\hat{x}_t^j)^{2n}$. Finally, by Proposition 6.5.9, this again amounts to the integrability of y_t^n, $\hat{y}_t^n$ already stated. $\diamond$

We still denote the semi-groups corresponding to the right action of $(Z_s), (Z_s^*)$ on $\Gamma\backslash\mathbb{F}^d$ by $(P_s), (P_s^*)$. Using the notation $f_\beta(\gamma) := f(\beta\gamma)$ (for any $\beta \in \mathbb{F}^d, \gamma \in PSO(1,d)$), we thus merely have

$$P_t f(\beta) = P_t f_\beta(1),$$

for any non-negative measurable function f on $\Gamma\backslash\mathbb{F}^d$ and any $\beta \in \mathbb{F}^d, t \geq 0$.

Remark 8.1.3 By Lemma 8.1.2, both of the semi-groups (P_s) and (P_s^*) act on C^n-bounded functions on $\Gamma\backslash\mathbb{F}^d$ which have bounded Lie derivatives of order $\leq n$. Moreover, this result remains valid when $\mathcal{L}_j$-derivatives only are considered: (P_s) and (P_s^*) act on functions on $\Gamma\backslash\mathbb{F}^d$ that have bounded $\mathcal{L}_1, \ldots, \mathcal{L}_d$-derivatives.

Lemma 8.1.4 (*i*) *The hyperbolic heat semi-group* (Q_s) *acts on* Γ-*left-invariant functions on* $\mathbb{H}^d$: *we have* $Q_s[F \circ \pi_0](\gamma\, p) = Q_s[F \circ \pi_0](p)$ *for any* $p \in \mathbb{H}^d$, *any* $\gamma \in \Gamma$ *and any test function* F *on* $\Gamma\backslash\mathbb{F}^d$.

(*ii*) *The relative volume measure* $d^\Gamma p$ (*recall Proposition 5.1.1(ii)*) *is invariant under the hyperbolic Brownian semi-group.*

Proof (*i*) For any $\beta \in \mathbb{F}^d$ such that $\beta_0 = p$, we have $P_s F(\gamma\beta) = \mathbb{E}[F(\gamma\beta Z_s)] = \mathbb{E}[F(\beta Z_s)] = P_s F(\beta)$, which, by projection under π_0, directly yields $Q_s[F \circ \pi_0](\gamma\, p) = Q_s[F \circ \pi_0](p)$.

(*ii*) Specializing Proposition 5.1.1(*iv*) to $g = Z_s$, where (Z_s) is the PSO$(1, d)$-valued left Brownian motion solving eqn (7.11), we have

$$\int f(\beta\, Z_s)\, \tilde{\lambda}^\Gamma(d\beta) = \int f\, d\tilde{\lambda}^\Gamma.$$

Taking the expectation, this yields

$$\int P_s f\, d\tilde{\lambda}^\Gamma = \int f\, d\tilde{\lambda}^\Gamma,$$

where (P_s) denotes the Markovian semi-group associated with (Z_s). Let (Q_s) denote the hyperbolic heat semi-group of Section 7.6.7. Using eqn (7.14), for any Γ-left-invariant test function h on $\mathbb{H}^d$ we obtain

$$\int Q_s h(p)\, d^\Gamma p = \int (Q_s h) \circ \pi_0\, d\tilde{\lambda}^\Gamma = \int P_s(h \circ \pi_0)\, d\tilde{\lambda}^\Gamma$$

$$= \int h \circ \pi_0\, d\tilde{\lambda}^\Gamma = \int h(p)\, d^\Gamma p. \qquad \diamond$$

We shall establish that (Q_s) is actually self-adjoint with respect to $d^\Gamma p$, see Theorem 8.2.4.

8.2 Two dual diffusions

We now use the right action of PSO$(1, d)$ on $\mathbb{F}^d$ and $\Gamma\backslash\mathbb{F}^d$, and Proposition 5.1.1.

Definition 8.2.1 *We set* $\mu := \tilde{\lambda}^\Gamma/\mathrm{covol}(\Gamma)$, *and choose a* $\Gamma\backslash\mathbb{F}^d$-*valued random variable* ξ_0, *independent of* (Z_s) *and* (Z_s^*) *and having law* μ. *We set*

$$\xi_s := \xi_0\, Z_s \qquad and \qquad \xi_s^* := \xi_0\, Z_s^*, \qquad for\ any\ s \geq 0.$$

Proposition 8.2.2 *Both of the diffusions* (ξ_s) *and* (ξ_s^*) *are stationary (with law* μ*) on* $\Gamma\backslash\mathbb{F}^d$, *and dual to each other: for any test functions* φ, ψ *on* $\Gamma\backslash\mathbb{F}^d$ *and any* $s \geq 0$, *we have*

$$\mathbb{E}[\varphi(\xi_s)\, \psi(\xi_0)] = \mathbb{E}[\varphi(\xi_0^*)\, \psi(\xi_s^*)].$$

Proof The stationarity is clear from the $PSO(1, d)$ right invariance of the Liouville measure (recall Proposition 5.1.1(iv)). Then, using Remark 8.1.1, Proposition 5.1.1(iv) and Definition 8.2.1, we have

$$\mathbb{E}\big[\varphi(\xi_0^*)\,\psi(\xi_s^*)\big] = \mathbb{E}\big[\varphi(\xi_0)\,\psi(\xi_s^*)\big] = \mathbb{E}\big[\varphi(\xi_0)\,\psi(\xi_0\,Z_s^{-1})\big]$$

$$= \mathbb{E}\bigg[\int_{\Gamma\backslash\mathbb{F}^d} \varphi(\xi)\,\psi(\xi\,Z_s^{-1})\,\mu(d\xi)\bigg]$$

$$= \mathbb{E}\bigg[\int_{\Gamma\backslash\mathbb{F}^d} \varphi(\xi\,Z_s)\,\psi(\xi)\,\mu(d\xi)\bigg] = \mathbb{E}\big[\varphi(\xi_s)\,\psi(\xi_0)\big] = \mathbb{E}\big[\varphi(\xi_s)\,\psi(\xi_0^*)\big]. \qquad \diamond$$

Remark 8.2.3 By Section 7.6.7 and Remark 7.5.3, we have $P_t(f \circ \pi_0) = (Q_t f) \circ \pi_0$ for any bounded f on $\mathbb{H}^d$ and any $t \geq 0$. In the same way, by Lemmas 8.1.3 and 8.1.4(i), for any bounded measurable h on $\Gamma\backslash\mathbb{H}^d$, we have $P_t(h \circ \pi_0) = (Q_t h) \circ \pi_0$, and then $d/dt\, Q_t h = \frac{1}{2}\Delta Q_t h = \frac{1}{2} Q_t \Delta h$, i.e., eqn (7.15) remains valid on the quotient $\Gamma\backslash\mathbb{H}^d$.

Moreover, again by Section 7.6.7 and eqn (7.14) in particular, for any $z \in \Gamma\backslash\mathbb{H}^d$ and $\beta \in \Gamma\backslash\mathbb{F}^d$ such that $\beta_0 = z$, $z_t := (\beta\, Z_t)_0$ defines a diffusion on $\Gamma\backslash\mathbb{H}^d$, such that $Q_t f(z) = \mathbb{E}\big[f(z_t)\big]$ for any test function f on $\Gamma\backslash\mathbb{H}^d$. Such a (z_t) is a **hyperbolic Brownian motion** (starting at z) on $\Gamma\backslash\mathbb{H}^d \equiv \pi_0(\mathcal{M})$. Thus the diffusion (ξ_s) projects under π_0 to a stationary hyperbolic Brownian motion on $\Gamma\backslash\mathbb{H}^d$.

We deduce the following important property from its analogue in $L^2(dp)$, which was established in Theorem 7.6.7.2.

Theorem 8.2.4 *The hyperbolic Brownian semi-group (Q_s) is self-adjoint in $L^2(d^\Gamma p)$.*

Proof We have to take the quotient (on the left-hand side) by Γ. Taking a fundamental domain D and using Remark 8.2.3, we have

$$\int_D Q_s f \times h\, dp = \int_{\mathbb{H}^d} f\, Q_s(h 1_D)\, dp = \int_{\mathbb{F}^d} f \circ \pi_0 \times Q_s(h 1_D) \circ \pi_0\, d\tilde{\lambda}$$

$$= \int_{\mathbb{F}^d} f \circ \pi_0\, P_s\Big(h \circ \pi_0\, 1_{\pi_0^{-1}(D)}\Big)\, d\tilde{\lambda}$$

$$= \mathbb{E}\bigg[\int_{\mathbb{F}^d} f \circ \pi_0(\beta)\, h \circ \pi_0(\beta Z_s)\, 1_{\pi_0^{-1}(D)Z_s^{-1}}(\beta)\, d\tilde{\lambda}(\beta)\bigg].$$

As $\pi_0^{-1}(D)Z_s^{-1}$ is almost surely another fundamental domain for the action of Γ on $\mathbb{F}^d$, using the Γ-invariance of f, h and $\tilde{\lambda}$ we obtain

$$\int Q_s f \times h \, d^\Gamma p = \int_D Q_s f \times h \, dp$$

$$= \mathbb{E}\left[\int_{\mathbb{F}^d} f \circ \pi_0(\beta) \, h \circ \pi_0(\beta Z_s) \, 1_{\pi_0^{-1}(D)}(\beta) \, d\tilde{\lambda}(\beta)\right]$$

$$= \int_{\pi_0^{-1}(D)} f \circ \pi_0 \, P_s(h \circ \pi_0) \, d\tilde{\lambda} = \int_D f \, Q_s h \, dp = \int f \, Q_s h \, d^\Gamma p.$$

$\diamond$

By Theorem 5.4.2.5, the hyperbolic Brownian semi-group (Q_s) satisfies a Poincaré inequality: for any $h \in L^2(d^\Gamma p)$ having zero mean and any $s > 0$, we have

$$\frac{d}{ds} \int |Q_s h|^2(p) \, d^\Gamma p = \int_D Q_s h \, \Delta Q_s h = -\int_D \gamma(Q_s h) \leq -C_D^{-1} \int |Q_s h|^2(p) \, d^\Gamma p,$$

since by Proposition 5.1.1, $\int Q_s h(p) \, d^\Gamma p = 0$. This entails the following corollary.

Corollary 8.2.5 *The hyperbolic Brownian semi-group (Q_s) admits a spectral gap (and hence is mixing) in $L^2(d^\Gamma p)$: for any $h \in L^2(d^\Gamma p)$ having zero mean and any $s \geq 0$, we have*

$$\int |Q_s h|^2(p) \, d^\Gamma p \leq e^{-C_D^{-1} s} \int h^2(p) \, d^\Gamma p.$$

8.3   Spectral gap along the foliation

In Section 5.4, we established the existence of a Poincaré inequality and then of a spectral gap, at the level of hyperbolic space; in fact, in a cofinite fundamental domain. We now want to extend this crucial spectral-gap property to the level of the tangent space and, more precisely, to the foliation defined by the geodesic and horocyclic flows. This will be done using appropriate norms. In contrast to the Brownian case, exponential decay does not hold for all L^2 functions, but it does so for rotationally Hölderian ones, owing to the fact that the diffusion evolves along a foliation. This immediately implies the existence of a potential kernel for the foliated Brownian semi-group (P_t).

We shall use here the Poincaré coordinate $u(\varrho)$ considered in Theorem 2.6.1, which is also the extremity of the half-geodesic $(e_0, \varrho'_{e_0} e_1) \theta_{\mathbb{R}_+} \subset \mathbb{H}^d$ determined by the tangent vector $d_o/ds \, \varrho(e_0 + s \, e_1)$ at its starting point $\varrho e_0 = e_0$; see Figure 8.1 and recall Proposition 2.6.2. By decomposing $\varrho = \mathcal{R}_{\alpha,\sigma} \tilde{\varrho}$ with $\alpha \in [0, \pi]$, $\sigma \in \mathbb{S}^{d-2}$ and $\tilde{\varrho} \in SO(d-1)$, we have $u(\varrho) = \mathrm{cotg}\,(\alpha/2) \, \sigma$, and the above vector is $\varrho e_1 = (\cos \alpha)e_1 + (\sin \alpha)\sigma$.

We need to consider the following projection of generic functions on $\mathbb{H}^d \times \partial\mathbb{H}^d$ onto functions on $\mathbb{H}^d$.

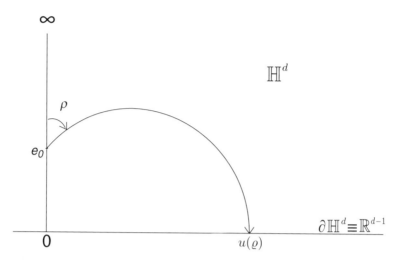

Fig. 8.1 The extremity $u(\rho)$ of the geodesic determined from e_0 by $\rho \in SO(d)$

Definition 8.3.1 *For any non-negative Borelian function F on $\mathbb{H}^d \times \partial\mathbb{H}^d$, and any $p \in \mathbb{H}^d$, we set*

$$\overline{F}(p) := \int_{\partial\mathbb{H}^d} F(p,\eta)\,\mu_p(d\eta),$$

where μ_p is the harmonic measure at p (recall Definition 3.6.1.1). For any non-negative Borelian function H on $\mathbb{F}^d$, any $(p,\eta) \in \mathbb{H}^d \times \partial\mathbb{H}^d$ and any $\beta \in \pi_1^{-1}(p,\eta)$, we set

$$\tilde{H}(p,\eta) := \int_{\pi_1^{-1}(p,\eta)} H\,d\varrho_{(p,\eta)} = \int_{SO(d-1)} H(\beta\,\varrho)\,d\varrho \quad and \quad \overline{H} := \overline{\tilde{H}},$$

where $d\varrho_{(p,\eta)} \equiv d\varrho$ denotes the normalized Haar measure on $\pi_1^{-1}(p,\eta) \equiv SO(d-1)$.

Note that by Propositions 3.6.2.4 and 3.6.2.2(ii) (for the same functions F, H as above), we have

$$\int_{\mathbb{H}^d} \overline{F}(p)\,dp = \int_{\mathbb{H}^d \times \partial\mathbb{H}^d} F\,d(\tilde{\lambda} \circ \pi_1^{-1}) \quad \text{and} \quad \int_{\mathbb{H}^d} \overline{H}(p)\,dp = \int_{\mathbb{F}^d} H\,d\tilde{\lambda}.$$

Note also that, thanks to the geometric property of harmonic measures (recall Remark 3.6.1.3), if F is Γ-invariant, then $\overline{F}$ is Γ-invariant too. Similarly, if H is Γ-invariant, then $\overline{H}$ is too. Then, for Γ-invariant functions F, H, recalling Proposition 5.1.1, we have

$$\int \overline{F}(p) \, d^{\Gamma} p = \int F \, d(\tilde{\lambda}^{\Gamma} \circ \pi_1^{-1}) \quad \text{and} \quad \int \overline{H}(p) \, d^{\Gamma} p = \int H \, d\tilde{\lambda}^{\Gamma}.$$

Definition 8.3.2 *Given $r > 0$, a Borelian function F on $\mathbb{H}^d \times \partial \mathbb{H}^d$ such that*

$$\|F\|_r := \sup_{\beta \in \mathbb{F}^d, \, \varrho \in SO(d) \smallsetminus SO(d-1)} \left| F \circ \pi_1(\beta \varrho) - F \circ \pi_1(\beta) \right| \times \ell(\varrho)^{-r} < \infty$$

*is said to be **rotationally r-Hölderian**. Here $\ell(\varrho)$ denotes the distance from ϱ to $SO(d-1)$ (which can be replaced by $|\sin(\alpha/2)|$ or $1/(|u(\varrho)|+1)$, recall Proposition 2.6.2).*

Remark 8.3.3 The semi-group (P_t) of (ξ_t) not only acts on $L^{\infty}(\Gamma \backslash \mathbb{F}^d, \tilde{\lambda}^{\Gamma})$ but also makes sense on $L^{\infty}(\mathcal{M}, \tilde{\lambda}^{\Gamma})$: (P_t) acts on $SO(d-1)$-invariant functions, that is, on functions on $\Gamma \backslash (\mathbb{H}^d \times \partial \mathbb{H}^d)$. Moreover, it is contracting in $L^2(\Gamma \backslash \mathbb{F}^d, \tilde{\lambda}^{\Gamma})$: $\|P_t F\|_2 \leq \|F\|_2$ if $t \geq 0$ and $F \in L^2(\Gamma \backslash \mathbb{F}^d, \tilde{\lambda}^{\Gamma})$.

Proof Since $\xi_s = \xi_0 Z_s$ and (Z_s) is $\mathbb{A}^d$-valued, the statement is clear from the commutativity between $SO(d-1)$ and $\mathbb{A}^d$: for any positive Borelian function H on $\mathbb{F}^d$ and any $\beta \in \mathbb{F}^d$, $\varrho \in SO(d-1)$, we have

$$P_t H \circ \pi_1(\beta \, \varrho) = \mathbb{E}\big[H \circ \pi_1(\beta \, \varrho \, Z_t)\big] = \mathbb{E}\big[H \circ \pi_1(\beta \, Z_t \, \varrho)\big]$$

$$= \mathbb{E}\big[H \circ \pi_1(\beta \, Z_t)\big] = P_t H \circ \pi_1(\beta).$$

And, by the Schwarz inequality and the right invariance of $\tilde{\lambda}$,

$$\|P_t F\|^2_{L^2(\Gamma \backslash \mathbb{F}^d, \tilde{\lambda}^{\Gamma})} = \int_{\Gamma \backslash \mathbb{F}^d} |P_t F|^2 \, d\tilde{\lambda}^{\Gamma} \leq \int_{\Gamma \backslash \mathbb{F}^d} P_t(F^2) \, d\tilde{\lambda}^{\Gamma} = \|F\|^2_{L^2(\Gamma \backslash \mathbb{F}^d, \tilde{\lambda}^{\Gamma})}. \quad \Diamond$$

Theorem 8.3.4 *For any $F \in L^{\infty}(\mathcal{M}, \tilde{\lambda}^{\Gamma})$ (and hence F is bounded, Γ-invariant and $SO(d-1)$-invariant), which is rotationally r-Hölderian and such that $\int F \, d\tilde{\lambda}^{\Gamma} = 0$, there exist $C = C(d)$ and $\delta = \delta(r, \Gamma) > 0$ such that*

$$\|P_t F\|_{L^2(\tilde{\lambda}^{\Gamma})} \leq C \left(\|F\|_{\infty} + \|F\|_r \right) e^{-\delta t} \quad \text{for all } t \geq 0.$$

Proof

(1) Using eqn (8.3), Definition 3.6.1.1 and Theorem 2.6.1, we have for any $p \in \mathbb{H}^d$, $\beta \in \pi_0^{-1}(p)$, and $t \geq 0$

$$\overline{P_t F}(p) = \int_{\partial \mathbb{H}^d} P_t F(p, \eta) \, \mu_p(d\eta) = \int_{SO(d)} \mathbb{E}\big[F \circ \pi_1(\beta \, \varrho \, T_{z_t})\big] \, d\varrho$$

$$= c_d \int_{[0,\pi] \times \mathbb{S}^{d-2}} \mathbb{E}\big[F \circ \pi_1(\beta \, \mathcal{R}_{\alpha,\sigma} \, T_{z_t})\big] (\sin \alpha)^{d-2} d\alpha \, d\sigma$$

$$\left(\text{by Proposition 3.3.2, with } c_d := \frac{\Gamma(d/2)}{\sqrt{\pi} \, \Gamma((d-1)/2)} \right)$$

$$= c_d \int_{[0,\pi] \times \mathbb{S}^{d-2}} \mathbb{E}\big[F \circ \pi_1(\beta \, T_{z'_t} \, \mathcal{R}_{\alpha'_t,\sigma'_t})\big] (\sin \alpha)^{d-2} \, d\alpha \, d\sigma$$

by the commutation relation of Theorem 2.6.1, with $z'_t = (x'_t, y'_t) \in \mathbb{R}^{d-1} \times \mathbb{R}^*_+$ such that $T_{z'_t} e_0 = \mathcal{R}_{\alpha,\sigma} \, T_{z_t} e_0$, and with $u(\mathcal{R}_{\alpha'_t,\sigma'_t}) = \big(u(\mathcal{R}_{\alpha,\sigma}) - x'_t\big)/y'_t$.

Then, Section 7.6.7 ensures that $T_{z'_t} e_0$ is a hyperbolic Brownian motion, just as $T_{z_t} e_0$ is. Hence, in the above integral, we can perform a change of Brownian motion $T_{z'_t} e_0 \mapsto T_{z_t} e_0$, which is equivalent to changing (z'_t) to (z_t). We thus obtain

$$\overline{P_t F}(p) = c_d \int_{[0,\pi] \times \mathbb{S}^{d-2}} \mathbb{E}\big[F \circ \pi_1(\beta \, T_{z_t} \, \mathcal{R}_{\alpha_t,\sigma_t})\big] (\sin \alpha)^{d-2} \, d\alpha \, d\sigma,$$

with $u(\mathcal{R}_{\alpha_t,\sigma_t}) = \big(u(\mathcal{R}_{\alpha,\sigma}) - x_t\big)/y_t$. By Proposition 2.6.2, the latter reads $\cotg (\alpha_t/2) \, \sigma_t = \big(\cotg (\alpha/2) \, \sigma - x_t\big)/y_t$. The above formula implies that for any $(p, \eta) \in \mathbb{H}^d \times \partial \mathbb{H}^d$ and any $\beta \in \pi_1^{-1}(p, \eta)$, we have

$$(P_t F - \overline{P_t F})(p, \eta) = c_d \int_{[0,\pi] \times \mathbb{S}^{d-2}} \mathbb{E}\big[F \circ \pi_1(\beta \, Z_t) - F \circ \pi_1(\beta \, Z_t \, \mathcal{R}_{\alpha_t,\sigma_t})\big]$$

$$\times (\sin \alpha)^{d-2} \, d\alpha \, d\sigma.$$

(2) We fix $0 < \varepsilon < 1/2$ and, for $t \geq 0$ and $\varrho \equiv \mathcal{R}_{\alpha,\sigma} \in \mathrm{SO}(d)$, set

$$E^1_t := 1_{\left\{ y_t > e^{-(d-1-\varepsilon) \, t/2} \right\}}, \quad E^2_t := (1 - E^1_t) \times 1_{\left\{ |u(\varrho) - x_t| \leq y_t \, e^{\varepsilon \, t/2} \right\}},$$

$$E^3_t := (1 - E^1_t - E^2_t)$$

and, for $1 \leq j \leq 3$,

$$A^j_t \equiv A^j_t(\beta)$$

$$:= c_d \int_{[0,\pi] \times \mathbb{S}^{d-2}} \mathbb{E}\Big[\big(F \circ \pi_1(\beta \, Z_t) - F \circ \pi_1(\beta \, Z_t \, \mathcal{R}_{\alpha_t,\sigma_t})\big) \times E^j_t\Big] (\sin \alpha)^{d-2} \, d\alpha \, d\sigma,$$

so that $(P_t F - \overline{P_t F})(p, \eta) = A^1_t + A^2_t + A^3_t$. Recall that (z_t) satisfies eqn (8.3). Then, on the one hand we have

$$\big\| A^j_t \big\|^2_{L^2(\tilde{\lambda}^\Gamma)} \leq 4 \, \|F\|^2_\infty \times \int_{\mathrm{SO}(d)} \mathbb{E}\big[E^j_t\big] d\varrho,$$

so that

$$\left\| A_t^1 \right\|_{L^2(\tilde{\lambda}^\Gamma)}^2 \leq 4 \left\| F \right\|_\infty^2 \times \mathbb{P}[w_t > \varepsilon \, t/2] = \left\| F \right\|_\infty^2 \times \mathcal{O}\!\left(e^{-\varepsilon^2 \, t/8} \right).$$

On the other hand, we have

$$\left\| A_t^2 \right\|_{L^2(\tilde{\lambda}^\Gamma)}^2 \leq 4 \left\| F \right\|_\infty^2 \int_{\mathrm{SO}(d)} \mathbb{P}\!\left[|u(\varrho) - x_t| \leq e^{-(d-1-2\varepsilon)\, t/2} \right] d\varrho$$

$$= 4 \left\| F \right\|_\infty^2 \, c_d \, \mathbb{E}\!\left[\int_{[0,\pi]\times\mathbb{S}^{d-2}} 1_{\left\{ |\cotg\left(\frac{\alpha}{2}\right)\sigma - x_t| \leq e^{-(d-1-2\varepsilon)\,t/2} \right\}} (\sin\alpha)^{d-2}\, d\alpha\, d\sigma \right]$$

$$= 4 \left\| F \right\|_\infty^2 \, c_d \, \mathbb{E}\!\left[\int_{\mathbb{R}^{d-1}} 1_{\left\{ |u - x_t| \leq e^{-(d-1-2\varepsilon)\,t/2} \right\}} \left(\frac{2}{1+|u|^2} \right)^{d-2} du \right]$$

$$\leq 2^d \left\| F \right\|_\infty^2 \, c_d \int_{\mathbb{R}^{d-1}} 1_{\left\{ |u| \leq e^{-(d-1-2\varepsilon)\,t/2} \right\}} du = \left\| F \right\|_\infty^2 \times \mathcal{O}\!\left(e^{-(d-1)(d-1-2\varepsilon)t/2} \right).$$

Finally we use Theorem 2.6.1, Proposition 2.6.2 and the Hölderian hypothesis that we have made about F to handle the third term as follows:

$$|A_t^3| \leq c_d \int_{[0,\pi]\times\mathbb{S}^{d-2}} \mathbb{E}\!\left[\left| F \circ \pi_1(\beta Z_t) - F \circ \pi_1(\beta Z_t \, \mathcal{R}_{\alpha_t, \sigma_t}) \right| \, 1_{\left\{ |u(\mathcal{R}_{\alpha_t, \sigma_t})| > e^{\varepsilon t/2} \right\}} \right]$$

$$\times \sin^{d-2}\alpha \, d\alpha \, d\sigma \leq \sup_{\beta \in \mathbb{F}^d, \, \varrho \in \mathrm{SO}(d),\, |u(\varrho)| > e^{\varepsilon\, t/2}} \left| F \circ \pi_1(\beta) - F \circ \pi_1(\beta\varrho) \right|$$

$$= \sup_{\beta \in \mathbb{F}^d, \, \varrho \in \mathrm{SO}(d),\, \ell(\varrho) = \mathcal{O}(e^{-\varepsilon\, t/2})} \left| F \circ \pi_1(\beta) - F \circ \pi_1(\beta\varrho) \right| = \left\| F \right\|_r \times \mathcal{O}\!\left(e^{-\varepsilon\, r\, t/2} \right).$$

(3) So far, we have shown the existence of $C = C(d), \delta = \delta(r) > 0$ such that for all $t \geq 0$,

$$\left\| P_t F - \overline{P_t F} \right\|_{L^2(\tilde{\lambda}^\Gamma)} \leq C(\|F\|_\infty + \|F\|_r)\, e^{-\delta t}.$$

Finally, on the one hand, by the above and by Remarks 8.2.3 and 8.3.3, we have

$$\left\| P_t F \right\|_{L^2(\tilde{\lambda}^\Gamma)} \leq \left\| P_{t/2}\!\left(P_{t/2} F - \overline{P_{t/2} F} \right) \right\|_{L^2(\tilde{\lambda}^\Gamma)} + \left\| P_{t/2}\!\left(\overline{P_{t/2} F} \right) \right\|_{L^2(\tilde{\lambda}^\Gamma)}$$

$$\leq \left\| P_{t/2} F - \overline{P_{t/2} F} \right\|_{L^2(\tilde{\lambda}^\Gamma)} + \left\| Q_{t/2}\!\left(\overline{P_{t/2} F} \right) \right\|_{L^2(d^\Gamma p)}$$

$$\leq C(\|F\|_\infty + \|F\|_r)\, e^{-\delta t/2} + \left\| Q_{t/2}\!\left(\overline{P_{t/2} F} \right) \right\|_{L^2(d^\Gamma p)},$$

and on the other hand, by Propositions 3.6.2.4 and 5.1.1, we have

$$\int \overline{P_{t/2}F}(p) \ d^{\Gamma}p = \int P_{t/2}F(p,\eta)\,\mu_p(d\eta)\,d^{\Gamma}p = \int P_{t/2}F\,d\tilde{\lambda}^{\Gamma} = \int F\,d\tilde{\lambda}^{\Gamma} = 0.$$

We conclude by using the spectral-gap property of the Brownian semi-group (Q_t) in $L^2(d^{\Gamma}p)$ (recall Corollary 8.2.5): there exists $\delta' = \delta'(\Gamma) > 0$ such that for any $t \geq 0$,

$$\|Q_{t/2}(\overline{P_{t/2}F})\|_{L^2(d^{\Gamma}p)} \leq e^{-\delta' t}\,\|\overline{P_{t/2}F}\|_{L^2(d^{\Gamma}p)}$$

$$= e^{-\delta' t}\,\|P_{t/2}F\|_{L^2(\tilde{\lambda}^{\Gamma})} \leq e^{-\delta' t}\,\|F\|_{L^2(\tilde{\lambda}^{\Gamma})}. \quad \diamondsuit$$

8.4   The resolvent kernel and conjugate functions

In this section, we introduce a resolvent (potential) kernel, which allows us to obtain conjugate functions to a given function f on $\mathbb{H}^d \times \partial\mathbb{H}^d$, provided it has some regularity. This construction will be crucial below for comparing geodesics with Brownian paths by means of a contour deformation.

We begin with a simplified theory of differential 1-forms on $\mathbb{F}^d$, containing just what is necessary to implement our contour deformation.

8.4.1   Differential 1-forms on $\mathbb{F}^d$

Definition 8.4.1.1 *We call any C^1 map from $\mathbb{F}^d$ into the dual Lie algebra $so(1,d)^*$ a **1-form**, and any C^1 map from $\mathbb{F}^d$ into the dual Lie subalgebra $\hat{\tau}_d{}^*$ a **longitudinal 1-form** (recall from Section 1.4 that $\hat{\tau}_d$ is the Lie algebra of $\mathbb{A}^d$).*

Using the canonical basis $(E_1, \tilde{E}_2, \ldots, \tilde{E}_d)$ of $\hat{\tau}_d$, any longitudinal 1-form ω can be identified with a system $(\omega^0, \omega^2, \ldots, \omega^d)$ of C^1 functions on $\mathbb{F}^d$, by writing $\omega = \omega^0\,E_1^* + \sum_{j=2}^d \omega^j\,\tilde{E}_j^*$.

Definition 8.4.1.2 *A longitudinal 1-form ω is **closed** if and only if (writing it as above) $(\mathcal{L}_i - \delta_{0i})\,\omega^j = \mathcal{L}_j\,\omega^i$ for $0 \leq i < j \leq d$.*

Proposition 8.4.1.3 *If a longitudinal 1-form ω is closed, then for any $\beta \in \mathbb{F}^d$, any $T_{x,y} \in \mathbb{A}^d$ and any C^1 path $\gamma \equiv (T_{x_s,y_s})_{0 \leq s \leq t}$ from $\mathbf{1}$ to $T_{x,y}$, the **line integral***

$$\int_{\beta\gamma[0,t]} \omega \; := \; \int_0^t \omega^0(\beta T_{x_s,y_s})\,y_s^{-1}\,dy_s + \sum_{j=2}^d \int_0^t \omega^j(\beta T_{x_s,y_s})\,y_s^{-1}\,dx_s^j \qquad (8.7)$$

does not depend on the choice of the path γ. Moreover, setting $z_s := (x_s, y_s)$, we have

$$\int_{\beta\gamma[0,t]} \omega = \int_0^t \omega(\beta T_{z_s})\big(T_{z_s}^{-1}\,T_{z_{s+ds}}\big).$$

Proof By Proposition 1.4.3, we have

$$T^{-1}_{x_s,y_s} T_{x_{s+ds},y_{s+ds}} = T_{(x_{s+ds}-x_s)/y_s,y_s+ds/y_s} = y_s^{-1}\Big(\sum_{j=2}^{d} \dot{x}_s^j \tilde{E}_j + \dot{y}_s E_1\Big) ds + o(ds),$$

whence

$$\int_{\beta\gamma[0,t]} \omega = \int_0^t \omega(\beta T_{x_s,y_s})\Big(T^{-1}_{x_s,y_s} T_{x_{s+ds},y_{s+ds}}\Big).$$

Now, by eqn (8.2), for $0 \leq i < j \leq d$ we have

$$\frac{\partial}{\partial y}\big[\omega^j(\beta T_{x,y})y^{-1}\big] = y^{-2}\big[\mathcal{L}_0\omega^j - \omega^j\big](\beta T_{x,y})$$

and

$$\frac{\partial}{\partial x^j}\big[\omega^i(\beta T_{x,y})y^{-1}\big] = y^{-2}\mathcal{L}_j\omega^i(\beta T_{x,y}),$$

so that the closedness of ω implies that of the 1-form $\tilde{\omega}_\beta$, defined on $\mathbb{R}^d$ by

$$\tilde{\omega}_\beta(x,y) := \omega^0(\beta T_{x,y})\,y^{-1}\,dy + \sum_{j=2}^{d}\omega^j(\beta T_{x,y})\,y^{-1}\,dx^j.$$

Hence, applying Stokes's theorem to $\tilde{\omega}_\beta$ yields the result.   ◇

8.4.2   Lift of f to the 1-form ω^f

Definition 8.4.2.1

(1) *We introduce the following resolvent kernel, for any $q \in \mathbb{N}^*$ and any bounded measurable function ϕ on $\mathbb{F}^d$:*

$$\mathcal{U}^q\phi(\beta) := \int_0^\infty e^{-qt}\,\phi(\beta\,\theta_t)\,dt. \tag{8.8}$$

(2) *For any Γ-invariant and $SO(d-1)$-invariant bounded Borelian function f on $\mathbb{F}^d$ such that $\int f\,d\tilde{\lambda}^\Gamma = 0$ and the derivatives $\mathcal{L}_j f, \mathcal{L}_j\mathcal{L}_k f$ (for $2 \leq j,k \leq d$) exist and are bounded, we set (for $2 \leq j \leq d$)*

$$f_j := -\mathcal{U}^1\mathcal{L}_j f \quad and \quad \omega^f := f\,E_1^* + \sum_{j=2}^{d} f_j\,\tilde{E}_j^*. \tag{8.9}$$

Lemma 8.4.2.2 *For f as in Definition 8.4.2.1, the 1-form ω^f is closed, with bounded coefficients f, f_j. Furthermore, for $2 \leq j,k \leq d$, the derivatives $\mathcal{L}_j f_k = \mathcal{U}^2\mathcal{L}_j\mathcal{L}_k f$ and $\mathcal{L}_0 f_k = \mathcal{L}_k f + f_k$ exist and are bounded on $\mathbb{F}^d$.*

Proof The commutation relation in eqn (1.16) implies, for $2 \leq j \leq d$,

$$\mathcal{L}_j \mathcal{U}^q \phi(\beta) = \frac{d_o}{ds} \int_0^\infty e^{-qt} \, \phi(\beta \, \theta_{se_j}^+ \theta_t) \, dt = \frac{d_o}{ds} \int_0^\infty e^{-qt} \, \phi\big(\beta \, \theta_t \, \theta_{se^{-t}e_j}^+\big) \, dt$$

$$= \mathcal{U}^{q+1} \mathcal{L}_j \phi(\beta),$$

provided that ϕ has bounded $\mathcal{L}_j$-derivatives. On the other hand, for bounded ϕ we have

$$\mathcal{L}_0 \mathcal{U}^q \phi(\beta) = \int_0^\infty e^{-qt} \frac{d_o}{dt} \, \phi(\beta \, \theta_t) \, dt = q \mathcal{U}^q \phi(\beta) - \phi(\beta).$$

By applying this to $\phi = -\mathcal{L}_k f$, we obtain the existence and boundedness of each $\mathcal{L}_j f_k$. Moreover, for $2 \leq j, k \leq d$, since the matrices $\theta_{se_j}^+$ and $\theta_{se_k}^+$ commute, on the one hand we have $\mathcal{L}_j f_k - \mathcal{L}_k f_j = \mathcal{U}^2 [\mathcal{L}_k, \mathcal{L}_j] f = 0$, and on the other hand $\mathcal{L}_0 f_k - \mathcal{L}_k f - f_k = 0$. This yields the result, according to Definition 8.4.1.2. ◇

8.5   Contour deformation

In this section, we take advantage of the closed form ω^f presented in Section 8.4, to change the integration path in $t^{-1/2} \int_0^t f(\beta \, \theta_s) \, ds$: we substitute the diffusion path $\xi[0, t'] := \{\xi_s \, |0 \leq s \leq t' := 2t/(d-1)\}$ for the geodesic path $\beta[0, t] := \{\beta \, \theta_s |0 \leq s \leq t\}$. In this contour deformation, two residual terms appear, which we prove to be asymptotically negligible. Recall that the diffusion (ξ_s) was introduced in Definition 8.2.1, and has a stationary law $\mu = \mathrm{covol}(\Gamma)^{-1} \tilde{\lambda}^\Gamma$, proportional to $\tilde{\lambda}^\Gamma$ (recall Proposition 8.2.2).

More precisely, the aim of this section is to establish the following theorem.

Theorem 8.5.1 *Let f be any Γ-invariant and $SO(d-1)$-invariant bounded Borelian function on $\mathbb{F}^d$ such that $\int f \, d\tilde{\lambda}^\Gamma = 0$ and its derivatives $\mathcal{L}_j f, \mathcal{L}_j^2 f$ (for $2 \leq j \leq d$) are bounded. Consider the associated 1-form ω^f as in Definition 8.4.2.1. Then, for any real a, we have*

$$\lim_{t \to \infty} \left\{ \int \exp\left(\frac{a \sqrt{-1}}{\sqrt{t}} \int_0^{(d-1)t/2} f(\beta \, \theta_s) \, ds \right) \mu(d\beta) \right.$$

$$\left. - \mathbb{E}\left[\exp\left(\frac{-a \sqrt{-1}}{\sqrt{t}} \int_{\xi[0, t]} \omega^f \right) \right] \right\} = 0.$$

Proof

(1) Using the geodesic invariance of the measure $\tilde{\lambda}^\Gamma$ (recall Proposition 5.1.1(iv)) and then that of the law μ (of Definition 8.2.1), for any $t > 0$, we have

$$\int \exp\left(\frac{a\sqrt{-1}}{\sqrt{t}} \int_0^{(d-1)t/2} f(\beta\,\theta_s)\,ds\right)\,d\mu(\beta)$$

$$= \int \exp\left(\frac{a\sqrt{-1}}{\sqrt{t}} \int_\beta^{\beta\,\theta_{(d-1)t/2}} \omega^f\right)\mu(d\beta).$$

We now write the change of contour, using Proposition 8.4.1.3 and Lemma 8.4.2.2. We have

$$\int_\beta^{\beta\,\theta_{(d-1)t/2}} \omega^f = \int_\beta^{\beta\,T_{\hat{x}_t,\hat{y}_t}} \omega^f - \int_{\beta\,T_{0,\hat{y}_t}}^{\beta\,T_{\hat{x}_t,\hat{y}_t}} \omega^f - \int_{\beta\,\theta_{(d-1)t/2}}^{\beta\,T_{0,\hat{y}_t}} \omega^f.$$

Here $\hat{z}_t = (\hat{x}_t, \hat{y}_t) \in \mathbb{R}^{d-1} \times \mathbb{R}_+^*$ is given by eqn (8.4), in which the expression for $\hat{y}_t$ explains why we use (Z_t^*) instead of (Z_t) (recall eqns (8.5) and (8.6)); see Figure 8.2.

(2) Since each f_j is bounded, by eqn (8.7), for any $t > 0$ (and for some positive constant C) we have

$$\int_{\beta\,T_{0,\hat{y}_t}}^{\beta\,T_{\hat{x}_t,\hat{y}_t}} \omega^f = \hat{y}_t^{-1}\sum_{j=2}^{d}\int_0^1 \hat{x}_t^j \times f_j\left(\beta\,\theta^+_{s\,\hat{x}_t^j\,e_j}\,\theta_{\log\,\hat{y}_t}\right)\,ds \le C\sum_{j=2}^{d}\frac{|\hat{x}_t^j|}{\hat{y}_t}.$$

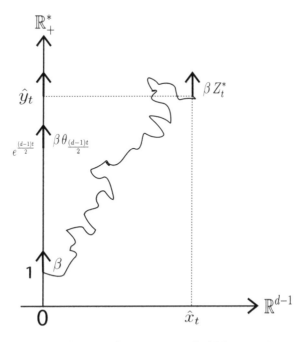

Fig. 8.2 Change of contour: proof of Theorem 8.5.1

Then, by eqn (8.4) and Corollary 6.2.3(4), we have

$$\frac{|\hat{x}_t^j|}{\hat{y}_t} = \int_0^t e^{w_s - w_t + ((d-1)/2)(s-t)} \, dW_s^j = \int_0^t e^{w_{t-s} - w_t - (d-1/2)s} \, dW_{t-s}^j$$

$$\stackrel{\text{LAW}}{\equiv} \int_0^t e^{-w_s - (d-1/2)s} \, dW_s^j \stackrel{\text{LAW}}{\equiv} x_t^j \longrightarrow x_\infty^j \in \mathbb{R},$$

the latter holding almost surely by the proof of Proposition 7.6.7.3. Hence, uniformly with respect to $\beta \in \mathbb{F}^d$, $t^{-1/2} \int_{\beta \, T_{0,\hat{y}_t}}^{\beta \, T_{\hat{x}_t, \hat{y}_t}} \omega^f$ goes to 0 in probability as $t \to \infty$. However, $x_\infty^j \notin L^2$ (for $d = 2$ or 3).

(3) Then, for any $t > 0$, we have

$$\int_{\beta \, \theta_{(d-1)t/2}}^{\beta \, T_{0,\hat{y}_t}} \omega^f = \int_{(d-1)t/2}^{w_t + (d-1)t/2} f(\beta \, \theta_s) \, ds = \int_0^{w_t} f\left(\beta \, \theta_{s+(d-1)t/2}\right) ds,$$

and then

$$\mathbb{E}\left[\int \exp\left[\frac{a\sqrt{-1}}{\sqrt{t}} \int_{\beta \, \theta_{(d-1)t/2}}^{\beta \, T_{0,\hat{y}_t}} \omega^f \right] \mu(d\beta) \right]$$

$$= \mathbb{E}\left[\int \exp\left[\frac{a\sqrt{-1}}{\sqrt{t}} \int_0^{w_t} f\left(\beta \, \theta_{s+(d-1)t/2}\right) ds \right] \mu(d\beta) \right]$$

$$= \mathbb{E}\left[\int \exp\left[\frac{a\sqrt{-1}}{\sqrt{t}} \int_0^{w_1\sqrt{t}} f(\beta \, \theta_s) \, ds \right] \mu(d\beta) \right]$$

$$\xrightarrow{t \to \infty} \mathbb{E}\left[\int \exp\left[a\sqrt{-1} \, w_1 \int f \, d\mu \right] \mu(d\beta) \right] = 1,$$

by ergodicity (recall Theorems 5.3.1 and 5.2.3) and since $\int f \, d\mu = 0$. This means that under the law $\mu \otimes \mathbb{P}$, $t^{-1/2} \int_{\beta \, \theta_{(d-1)t/2}}^{\beta \, T_{0,\hat{y}_t}} \omega^f$ goes to 0 in probability as $t \to \infty$.

(4) So far, and since $T_{\hat{x}_t, \hat{y}_t} = Z_t^*$, we have proved that

$$\int \exp\left(\frac{a\sqrt{-1}}{\sqrt{t}} \int_0^{(d-1)t/2} f(\beta \, \theta_s) \, ds \right) \mu(d\beta)$$

$$- \mathbb{E}\left[\int \exp\left(\frac{a\sqrt{-1}}{\sqrt{t}} \int_\beta^{\beta \, Z_t^*} \omega^f \right) \mu(d\beta) \right]$$

goes to 0 as $t \to \infty$. Finally, performing the time reversal allowed by Proposition 8.2.2, we have

$$\mathbb{E}\left[\int \exp\left(\frac{a\sqrt{-1}}{\sqrt{t}}\int_\beta^{\beta\, Z_t^*}\omega^f\right)\mu(d\beta)\right] = \mathbb{E}\left[\exp\left(\frac{a\sqrt{-1}}{\sqrt{t}}\int_{\xi_0^*}^{\xi_t^*}\omega^f\right)\right]$$

$$= \mathbb{E}\left[\exp\left(\frac{a\sqrt{-1}}{\sqrt{t}}\int_{\xi_t}^{\xi_0}\omega^f\right)\right] = \mathbb{E}\left[\exp\left(\frac{-a\sqrt{-1}}{\sqrt{t}}\int_{\xi[0,t]}\omega^f\right)\right].$$

$\diamond$

8.6 The divergence of ω^f

We take an $SO(d-1)$-invariant bounded function f on $\mathbb{F}^d$, with bounded Hölderian derivatives of order 1 and 2, and such that $\int f\, d\lambda^\Gamma = 0$.

Lemma 8.6.1 *The potentials $\mathcal{U}^1 f$ and $\mathcal{U}^2 f$ are bounded and Hölderian on $\mathbb{F}^d$.*

Proof The boundedness is obvious from eqn (8.8). We fix a Hölder exponent for f, say $0 < r < 1$. We consider a rotation $\varphi_s := \exp\left[s\sum_{1\le k<j\le d} a_{kj} E_{kj}\right]$, for $s > 0$ and fixed real a_{kj}. For $q \ge 1$ and $\beta \in \mathbb{F}^d$, we have

$$s^{-r}\left|\mathcal{U}^q f(\beta\varphi_s) - \mathcal{U}^q f(\beta)\right| \le \int_0^\infty e^{(r-q)t}(se^t)^{-r}\left|f(\beta\,\varphi_s\,\theta_t) - f(\beta\,\theta_t)\right|dt,$$

which is bounded with respect to (s,β) if $(se^t)^{-r}\,|\,f(\beta\varphi_s\theta_t) - f(\beta\,\theta_t)|$ or, equivalently, $s^{-r}\,|f\big(\beta\,\theta_{-t}\,\varphi_{se^{-t}}\,\theta_t\big) - f(\beta)|$ is bounded with respect to (s,β) and $t \ge 0$. This will be ensured if we can prove that $\ell(\theta_{-t}\,\varphi_{se^{-t}}\,\theta_t) = \mathcal{O}(s)$, uniformly with respect to t, where ℓ denotes the distance from $SO(d-1)$ (whose restriction to $SO(d)$ was introduced in Definition 8.3.2, where it could be replaced by $1/(|u(\varrho)| + 1)$, according to Proposition 2.6.2). Then, since E_{kj} commutes with θ_t for $k, j \ge 2$, it is sufficient to consider the case of $\varphi_s = \exp[s\, E_{1j}] = \mathcal{R}_{s,e_j}$ (for $2 \le j \le d$, recalling Proposition 3.3.2).

Now by Theorem 2.6.1, we have $\ell(\theta_{-t}\,\varphi_s\,\theta_t) = \ell(\theta_{-t}\,T_{x,y}\,\mathcal{R}_{\alpha,\sigma})$, with $T_{x,y}e_0 = \varphi_s\,\theta_t\,e_0$ and, by Proposition 2.6.2,

$$u(\mathcal{R}_{-\alpha,\sigma}) = u(\mathcal{R}_{\alpha,\sigma}^{-1}) = u(\varphi_s^{-1})\,e^{-t} = u(\mathcal{R}_{-s,e_j})\,e^{-t} = -e^{-t}\cot g\,(s/2)\,e_j.$$

By eqn (1.18),

$$T_{x,y}\,e_0 = \varphi_s\,\theta_t\,e_0 = (\text{ch}\,t)e_0 + (\text{sh}\,t\,\cos s)e_1 + (\text{sh}\,t\,\sin s)e_j$$

is equivalent to: $x = y\,\text{sh}\,t\,\sin s\, e_j$, $2y\,\text{sh}\,t\,\cos s = y^2 + (y\,\text{sh}\,t\,\sin s)^2 - 1$, and $2y\,\text{ch}\,t = y^2 + (y\,\text{sh}\,t\,\sin s)^2 + 1$, whence $y = (\text{ch}\,t - \text{sh}\,t\,\cos s)^{-1}$ and $x = (\text{sh}\,t\,\sin s/(\text{ch}\,t - \text{sh}\,t\,\cos s))\,e_j$. Hence we obtain

$$\ell(\theta_{-t}\,\varphi_{se^{-t}}\,\theta_t) = \ell(\theta_{-t}\,T_{x,y}\,\mathcal{R}_{\alpha',\sigma'}) = \ell(T_{e^{-t}x,e^{-t}y}\,\mathcal{R}_{\alpha',\sigma'})$$

$$\leq \operatorname{dist}\big((e^{-t}x, e^{-t}y); (0,1)\big) + \ell(\mathcal{R}_{\alpha',\sigma'})$$

$$= \mathcal{O}\left(\left|\frac{e^{-t}(\operatorname{sh}t)\,\sin(se^{-t})}{\operatorname{ch}t - \operatorname{sh}t\,\cos(se^{-t})}\right| + \left|\frac{e^{-t}}{\operatorname{ch}t - \operatorname{sh}t\,\cos(se^{-t})} - 1\right| + \left[1 + e^{-t}\,\big|\cotg\big(\tfrac{se^{-t}}{2}\big)\big|\right]^{-1}\right).$$

Since for $0 \leq s \leq 2\pi$ we have $\big(\operatorname{ch}t - \operatorname{sh}t\,\cos(se^{-t})\big)^{-1} = e^t\big(1 + \mathcal{O}(s^2)\big)$, this shows that $\mathcal{U}^q\phi$ is rotationally Hölderian on $\mathbb{F}^d$.

The Hölder property in the geodesic direction is clear since it does not need any commutation. Finally, we obtain the Hölder property in the horocyclic directions, using similarly

$$s^{-r}\left|\mathcal{U}^q f(\beta\theta^+_{se_j}) - \mathcal{U}^q f(\beta)\right| \leq \int_0^\infty e^{(r-q)t}\,\sup_{\beta,s}\left\{s^{-r}\,\big|\,f(\beta\,\theta_{-t}\,\theta^+_{se^{-t}e_j}\,\theta_t) - f(\beta)\big|\right\}dt,$$

and the simpler commutation formula of eqn (1.16):

$$\ell\big(\theta_{-t}\,\theta^+_{se^{-t}e_j}\,\theta_t\big) = \ell\big(\theta^+_{se^{-2t}e_j}\big) = \mathcal{O}(se^{-2t}) = \mathcal{O}(s), \quad \text{for } 2 \leq j \leq d. \qquad \diamond$$

We have the following general statement, derived from Itô's formula.

Proposition 8.6.2 *Consider a closed longitudinal 1-form Ω having continuous $\mathcal{L}_k$-derivatives (for $k \in \{0, 2, \ldots, d\}$). Then, for any $t \geq 0$ we almost surely have*

$$\int_{\xi[0,t]} \Omega = M_t^\Omega + \tfrac{1}{2}\int_0^t \operatorname{div}\Omega\,(\xi_s)\,ds,$$

with

$$\operatorname{div}\Omega := \mathcal{L}_0(\Omega(E_1)) + \sum_{j=2}^d \mathcal{L}_j\big(\Omega(\tilde{E}_j)\big) + (1-d)\,\Omega(E_1),$$

and a continuous martingale M_t^Ω having quadratic variation

$$\langle M^\Omega \rangle_t = \int_0^t \left(\Omega(E_1)^2 + \sum_{j=2}^d \Omega(\tilde{E}_j)^2\right)(\xi_s)ds.$$

Proof Set $e := (e_0, \ldots, e_d) \in \mathbb{F}^d$ and $F(\beta) := \int_e^\beta \Omega$. Then F is a C^2 function on $\mathbb{F}^d$, and Itô's formula (7.3) can be applied to the function $Z_s \mapsto F(\xi_s) = F(\xi_0\,Z_s)$ and to eqn (7.11), yielding

$$\int_{\xi[0,t]} \Omega = F(\xi_t) - F(\xi_0) = \int_0^t \sum_{j=0,2,\ldots,d} \mathcal{L}_j F(\xi_s)\,dW_s^j + \frac{1}{2}\int_0^t \mathcal{D}F(\xi_s)\,ds.$$

Now, writing $\Omega = \Omega^0 E_1^* + \sum_{j=2}^d \Omega^j \tilde{E}_j^*$ and using eqn (8.7), we have

$$\mathcal{L}_0 F(\xi) = \frac{d_0}{d\varepsilon} \int_\xi^{\xi\,\theta_\varepsilon} \Omega = \frac{d_0}{d\varepsilon} \int_0^\varepsilon \Omega^0(\xi\,\theta_s)\, ds = \Omega^0(\xi) = \Omega(E_1)(\xi)$$

for any $\xi \in \mathbb{F}^d$, and, similarly, $\mathcal{L}_j F = \Omega^j = \Omega(\tilde{E}_j)$ for $2 \le j \le d$. Hence we obtain the claimed expression for $\langle M^\Omega \rangle_t$, and for div $\Omega = DF$ as well, using the expression in eqn (3.3) for $\mathcal{D}$. $\diamond$

Let us apply Proposition 8.6.2 to ω^f (defined in eqn (8.9) in Definition 8.4.2.1). This is legitimate, by Lemmas 8.4.2.2 and 8.6.1. Noticing moreover that $\mathcal{L}_j f_j = -\mathcal{U}^2 \mathcal{L}_j^2 f$, we obtain by eqn (8.9) and Lemma 8.4.2.2

$$\int_{\xi[0,t]} \omega^f = M_t^f + \frac{1}{2} \int_0^t K f(\xi_s)\, ds \tag{8.10}$$

at once, where

$$K f := \mathcal{L}_0 f - \sum_{j=2}^d \mathcal{U}^2 \mathcal{L}_j^2 f + (1-d) f, \tag{8.11}$$

and M_t^f is a continuous martingale with a quadratic variation given by

$$\langle M^f \rangle_t = \int_0^t \left(f^2 + \sum_{j=2}^d f_j^2 \right)(\xi_s)\, ds. \tag{8.12}$$

Lemma 8.6.3 *The function Kf of eqn (8.11) is Γ-invariant, $SO(d-1)$-invariant, bounded (which we can summarize by $Kf \in L^\infty(\mathcal{M}, \mu)$) and Hölderian.*

Proof We have already observed in Lemma 8.4.2.2 that the f_j and the $\mathcal{L}_j f_j$ are bounded. $\mathcal{L}_0 f$ is $SO(d-1)$-invariant since f is and since the geodesic flow commutes with the $SO(d-1)$ action on $\mathbb{F}^d$. We must now show that $\sum_{j=2}^d \mathcal{U}^2 \mathcal{L}_j^2 f$ is $SO(d-1)$-invariant. The commutation of the geodesic flow with the $SO(d-1)$ action shows that it is sufficient to verify the $SO(d-1)$-invariance of $\sum_{j=2}^d \mathcal{L}_j^2 f$. Now, considering any $\varrho \in SO(d-1)$, and using the $SO(d-1)$-invariance of f and Lemma 1.4.2, we have

$$\sum_{j=2}^d \mathcal{L}_j^2 f (\beta\,\varrho) = \sum_{j=2}^d \frac{d_0^2}{dt^2} f(\beta\,\varrho\,\theta_{te_j}^+)$$

$$= \sum_{j=2}^d \frac{d_0^2}{dt^2} f\left(\beta\,\varrho\, e^{t\tilde{E}_j}\varrho^{-1}\right) = \sum_{j=2}^d \frac{d_0^2}{dt^2} f\left(\beta\,\theta_{t\varrho(e_j)}^+\right).$$

In particular, if ϱ is a rotation by an angle α in the plane $\{e_k, e_l\}$, we have

$$\sum_{j=2}^{d} \mathcal{L}_j^2 f\left(\beta\,\varrho\right) = \frac{d_0^2}{dt^2}\left[\sum_{j \neq k, l} f\left(\beta\,\theta_{te_j}^+\right) + f\left(\beta\,\theta_{t(\cos\alpha)e_k - t(\sin\alpha)e_l}^+\right)\right.$$

$$\left. + f\left(\beta\,\theta_{t(\sin\alpha)e_k + t(\cos\alpha)e_l}^+\right)\right]$$

$$= \left[\sum_{j \neq k, l} \mathcal{L}_j^2 f + \left[(\cos\alpha)\mathcal{L}_k - (\sin\alpha)\mathcal{L}_l\right]^2 f + \left[(\sin\alpha)\mathcal{L}_k + (\cos\alpha)\mathcal{L}_l\right]^2 f\right](\beta)$$

$$= \sum_{j=2}^{d} \mathcal{L}_j^2 f\,(\beta).$$

As any $\varrho \in SO(d-1)$ is a finite product of planar rotations, this proves the wanted invariance.

It remains to show that Kf is Hölderian. By eqn (8.10), this amounts to verifying that the $\mathcal{U}^2 \mathcal{L}_j^2 f$ are Hölderian. This verification is provided by Lemma 8.6.1 (whose proof can be applied directly to the Hölderian function $\mathcal{L}_j^2 f$). $\diamond$

Proposition 8.6.4 *We have $\int Kf\, d\tilde{\lambda}^\Gamma = 0$ (where K is defined in eqn (8.11)).*

Proof By using eqn (8.8) to define $\mathcal{U}^2$, for $2 \leq j \leq d$, $\beta \in \mathbb{F}^d$ and $S > 0$ we have

$$\mathcal{U}^2 \mathcal{L}_j^2 f(\beta) = \mathcal{O}\left(e^{-2S}\right) + \int_0^S e^{-2s}\,\mathcal{L}_j^2 f\left(\beta\,\theta_s\right)\,ds \quad \text{and}$$

$$e^{-2s}\,\mathcal{L}_j^2 f\left(\beta\,\theta_s\right) = e^{-2s}\frac{d_0}{dt}\,\mathcal{L}_j f\left(\beta\,\theta_s\,\theta_{te_j}^+\right) = e^{-2s}\frac{d_0}{dt}\,\mathcal{L}_j f\left(\beta\,\theta_{e^s te_j}^+\,\theta_s\right)$$

$$= e^{-s}\frac{d_0}{dt}\,\mathcal{L}_j f\left(\beta\,\theta_{te_j}^+\,\theta_s\right) = \mathcal{L}_j\left(e^{-s}\,\mathcal{L}_j f(\cdot\,\theta_s)\right)(\beta) = \mathcal{L}_j^2\left(f(\cdot\,\theta_s)\right)(\beta).$$

Hence,

$$\sum_{j=2}^{d} \mathcal{U}^2 \mathcal{L}_j^2 f = \mathcal{O}\left(e^{-2S}\right) + \int_0^S \left(\mathcal{D} - \mathcal{L}_0^2 + (d-1)\mathcal{L}_0\right)f\left(\cdot\,\theta_s\right)\,ds.$$

Now,

$$\int_0^S \mathcal{L}_0 f\left(\cdot\,\theta_s\right)\,ds = \int_0^S \frac{d}{ds}\,f\left(\cdot\,\theta_s\right)\,ds = f\left(\cdot\,\theta_S\right) - f$$

and

$$\int_0^S \mathcal{L}_0^2 f\left(\cdot\,\theta_s\right)\,ds = \mathcal{L}_0 f\left(\cdot\,\theta_S\right) - \mathcal{L}_0 f.$$

Therefore, by the definition of Kf in eqn (8.11) we obtain

$$Kf = -\mathcal{D}\left(\int_0^S f(\cdot\theta_s)\,ds\right) + \mathcal{L}_0 f(\cdot\theta_S) + (1-d)f(\cdot\theta_S) + \mathcal{O}(e^{-2S}).$$

Hence, using the right invariance of $\tilde{\lambda}^\Gamma$ and the hypothesis made about f, we find

$$\int Kf\,d\tilde{\lambda}^\Gamma = -\int \mathcal{D}\left(\int_0^S f(\cdot\theta_s)\,ds\right)d\tilde{\lambda}^\Gamma$$

$$+ \int \mathcal{L}_0 f\,d\tilde{\lambda}^\Gamma + (1-d)\int f\,d\tilde{\lambda}^\Gamma + \mathcal{O}(e^{-2S})$$

$$= -\int\left(\int_0^S f(\cdot\theta_s)\,ds\right)\mathcal{D}^*1\,d\tilde{\lambda}^\Gamma + \mathcal{O}(e^{-2S}) = \mathcal{O}(e^{-2S}),$$

which gives the result by letting S go to infinity. $\diamondsuit$

8.7   Sinai's central limit theorem

The aim of this section is to complete the proof of the following theorem, essentially due to Sinai ([Si1], [Si2]), who treated the cocompact case. (See also [Ran] for the compact case, but with non-constant curvature). The cofinite case appeared in [LJ2], and the infinite case in [EFLJ3].

Theorem 8.7.1 *Let f be a Γ-invariant, $SO(d-1)$-invariant bounded real function on $\mathbb{F}^d$, of class C^2 with bounded and Hölderian derivatives, such that $\int f\,d\tilde{\lambda}^\Gamma = 0$. Then, for all $a \in \mathbb{R}$, we have*

$$\lim_{t\to\infty}\int \exp\left(\frac{a\sqrt{-1}}{\sqrt{t}}\int_0^t f(\beta\theta_s)\,ds\right)\tilde{\lambda}^\Gamma(d\beta) = \text{covol}(\Gamma) \times \exp\left(-\frac{a^2}{2}\,\mathcal{V}(f)\right),$$

$$(8.13)$$

where $\big(Kf$ being given by eqn $(8.11)\big)$

$$\mathcal{V}(f) := \left[(d-1)\,\text{covol}(\Gamma)\right]^{-1}\int\left[(f + \tfrac{1}{2}\mathcal{L}_0 VKf)^2 + \sum_{j=2}^d (f_j + \tfrac{1}{2}\mathcal{L}_j VKf)^2\right]d\tilde{\lambda}^\Gamma$$

$$(8.14)$$

vanishes if and only if f equals $\mathcal{L}_0 h$, for some $h \in L^2(\mathbb{F}^d, \tilde{\lambda}^\Gamma)$.

Note The first claim $\big(\text{eqn } (8.13)\big)$ reads equivalently as follows: *for all real a, we have*

$$\lim_{t\to\infty}\tilde{\lambda}^\Gamma\left[\beta \in \mathbb{F}^d \,\Big|\, \int_0^t f(\beta\theta_s)\,ds \le a\sqrt{t\,\mathcal{V}(f)}\right] = \frac{\text{covol}(\Gamma)}{\sqrt{2\pi}}\int_{-\infty}^a e^{-s^2/2}\,ds.$$

The kernel V which appears in Theorem 8.7.1, in the definition of the variance $\mathcal{V}$, is the potential kernel of the semi-group (P_t):

$$V := \int_0^\infty P_t \, dt, \tag{8.15}$$

which, as we saw in Theorem 8.3.4, makes sense on centred Γ-invariant functions which are regular enough, such as the Kf of Section 8.6 (by Lemma 8.6.3 and Proposition 8.6.4). By Remark 8.3.3, V preserves $SO(d-1)$-invariance too.

To give a meaning to eqn (8.14) in Theorem 8.7.1, we need on the one hand to justify the existence of L^2-derivatives $\mathcal{L}_j V K f$. On the other hand, to prove Theorem 8.7.1 we want to use Proposition 8.6.2 and its consequence eqn (8.10), but with $\left[\omega^f + \frac{1}{2} d(VKf)\right]$ instead of ω^f, in order to get rid of the problematic part $\int_0^t Kf(\xi_s) \, ds$ in eqn (8.10). Hence we must express this term in the form of

$$\int_0^t Kf(\xi_s)ds = VKf(\xi_t) - VKf(\xi_0) + \text{a martingale term.}$$

To do this, the idea is to show the following, which will apply to $F = Kf$, thanks to Lemma 8.6.3 and Proposition 8.6.4.

Proposition 8.7.2 *Any centred and bounded rotationally r-Hölderian function F on $\mathcal{M}$ admits a bounded potential VF with L^2-derivatives $\mathcal{L}_j VF$ such that, setting $V_b := \int_0^b P_t \, dt$ (for $j \in \{0, 2, \dots, d\}$),*

$$(i) \; \left\| VF - V_bF \, dt \right\|_{L^2(\tilde{\lambda}^\Gamma)} \leq \left(\|F\|_\infty + \|F\|_r \right) \times \mathcal{O}(e^{-\delta b});$$

$$(ii) \; \left\| \mathcal{L}_j VF - \mathcal{L}_j V_bF \, dt \right\|_{L^2(\tilde{\lambda}^\Gamma)} \leq \left(\|F\|_\infty + \|F\|_r \right) \times \mathcal{O}(e^{-\delta b/2}).$$

Proof

(1) Assume first that F has continuous bounded derivatives $\mathcal{L}_j F$, $\mathcal{L}_j^2 F$, and set

$$\mathcal{E}(F) := \int \sum_{j=0,\, 2 \leq j \leq d} |\mathcal{L}_j F|^2 \, d\tilde{\lambda}^\Gamma = -\int F \mathcal{D}F \, d\tilde{\lambda}^\Gamma, \tag{8.16}$$

the last equality holding since

$$\int \sum_{j=0,\, 2 \leq j \leq d} |\mathcal{L}_j F|^2 \, d\tilde{\lambda}^\Gamma = -\int \left(\mathcal{L}_0^2 F + \sum_{j=2}^d \mathcal{L}_j^2 F \right) F \, d\tilde{\lambda}^\Gamma$$

$$= -\frac{1}{2} \int (\mathcal{D}F + \mathcal{D}^* F) \, F \, d\tilde{\lambda}^\Gamma.$$

Observe now that by Lemma 8.1.2, for any $b > 0$, we have

$$P_b F - F = \int_0^b \frac{d}{dt} P_t F \, dt = \frac{1}{2} \int_0^b DP_t F \, dt = \frac{1}{2} DV_b F. \qquad (8.17)$$

Then, for any $b \geq c \geq 0$, using eqns (8.16) and (8.17) and Theorem 8.3.4, we obtain

$$\left\| (V_b - V_c) F \right\|_2 \leq C \left(\|F\|_\infty + \|F\|_r \right) \left(e^{-\delta c} - e^{-\delta b} \right) / \delta$$

and

$$\mathcal{E} \left((V_b - V_c) F \right) = \int (V_c - V_b) F \times D(V_b - V_c) F \, d\tilde{\lambda}^\Gamma$$

$$= 2 \int (V_c - V_b) F \times (P_b - P_c) F \, d\tilde{\lambda}^\Gamma \leq 2 \left\| (V_b - V_c) F \right\|_2 \left(\|P_b F\|_2 + \|P_c F\|_2 \right)$$

$$\leq 2 C^2 \delta^{-1} \left(\|F\|_\infty + \|F\|_r \right)^2 \left(e^{-\delta b} + e^{-\delta c} \right).$$

This shows that $VF := \lim\limits_{b \to \infty} V_b F$ and $\mathcal{L}_j VF := \lim\limits_{b \to \infty} \mathcal{L}_j V_b F$ are well defined in $L^2(\tilde{\lambda}^\Gamma)$ and that (i) and (ii) hold. Moreover, we have (with $C' := (C + 2 C^2) \delta(r, \Gamma)^{-1}$)

$$\|V_b F\|_2 \leq C' \left(\|F\|_\infty + \|F\|_r \right) \qquad \text{and}$$

$$\|VF\|_2 = \lim\limits_{b \to \infty} \|V_b F\|_2 \leq C' \left(\|F\|_\infty + \|F\|_r \right), \qquad (8.18)$$

and

$$\mathcal{E}(VF) = \lim\limits_{b \to \infty} \mathcal{E}(V_b F) \leq C' \left(\|F\|_\infty + \|F\|_r \right)^2. \qquad (8.19)$$

By the definition of $\mathcal{L}_j$, showing that $\mathcal{L}_j VF$ defined as above is indeed an L^2-derivative amounts to proving that $s^{-1} \int_0^s \theta_t^j \mathcal{L}_j VF \, dt$ goes to $\mathcal{L}_j VF$ in L^2 as $s \searrow 0$, where (to avoid handling complicated expressions), for $s \geq 0, \beta \in \mathbb{F}^d$, we set

$$\theta_s^0 F(\beta) := F(\beta \theta_s) \qquad \text{and} \qquad \theta_s^j F(\beta) := F(\beta \theta_{s e_j}^+) \qquad \text{for } 1 \leq j \leq d.$$

Note that by Remark 8.1.3, $V_b F$ admits bounded $\mathcal{L}_j$- and $\mathcal{L}_j^2$-derivatives, since for example

$$\|\mathcal{L}_j V_b F\|_\infty = \left\| \mathcal{L}_j \int_0^b P_s F \, ds \right\|_\infty = \left\| \int_0^b \mathcal{L}_j P_s F \, ds \right\|_\infty \leq \int_0^b \|\mathcal{L}_j P_s F\|_\infty ds < \infty.$$

Now, for $j \in \{0, 2, \ldots, d\}$ and $b, s \geq 0$, the regularity of $V_b F$ ensures that we have

$$\int_0^s \theta_t^j \, \mathcal{L}_j V_b F \, dt = s^{-1} \int_0^s \frac{d}{dt} (\theta_t^j \, V_b F) \, dt = \theta_s^j \, V_b F - V_b F,$$

and then we can let $b \nearrow \infty$, provided the convergences in $L^2(\tilde{\lambda}^\Gamma)$ of $\theta_s^j V_b F$ to $\theta_s^j V F$ and of $\theta_s^j \mathcal{L}_j V_b F$ to $\theta_s^j \mathcal{L}_j V F$ are uniform in $s \in [0, 1]$. This is ensured by (i) and (ii) proved above for F and by the invariance of $\tilde{\lambda}^\Gamma$ with respect to θ_s^j (recall Proposition 5.1.1(iv)).

(2) For a generic F as in the statement, as is easily seen using a convolution and a partition of unity, we can find a sequence (F_n) of centred and bounded functions on $\mathcal{M}$ that have continuous bounded derivatives $\mathcal{L}_j F$, $\mathcal{L}_j^2 F$, such that

$$\|F - F_n\|_{L^2(\tilde{\lambda}^\Gamma)} \longrightarrow 0 \quad \text{and} \quad \|F_n\|_\infty \leq 2 \|F\|_\infty, \|F_n\|_r \leq 2 \|F\|_r.$$

Then, on the one hand, using Remark 8.3.3 and Theorem 8.3.4, we have:

$$\|VF - VF_n\|_{L^2(\tilde{\lambda}^\Gamma)} \leq \int_0^\infty \|P_t(F - F_n)\|_2 \, dt$$

$$\leq \int_0^\infty \min\left\{ \|F - F_n\|_2, 3C(\|F\|_\infty + \|F\|_r) \, e^{-\delta t} \right\} dt$$

$$= 2C(\|F\|_\infty + \|F\|_r)\delta^{-1}\|F - F_n\|_2 \times \log\left[3e \, C(\|F\|_\infty + \|F\|_r)\|F - F_n\|_2^{-1} \right],$$

which goes to 0, and similarly for $\|V_b F - V_b F_n\|_2$ (uniformly with respect to b), so that (i) holds.

On the other hand, in a way similar to 1) above, for $b, n, p \in \mathbb{N}^*$ we have

$$\mathcal{E}(V_b(F_n - F_p)) = \int V_b(F_p - F_n) \mathcal{D}V_b(F_n - F_p) \, d\tilde{\lambda}^\Gamma$$

$$= 2 \int V_b(F_p - F_n) \left[P_b(F_n - F_p) - (F_n - F_p) \right] d\tilde{\lambda}^\Gamma$$

$$\leq 2(\|V_b F_p\|_2 + \|V_b F_n\|_2)(\|P_b(F_n - F_p)\|_2 + \|F_n - F_p\|_2)$$

$$\leq 4(\|V_b F_p\|_2 + \|V_b F_n\|_2)\, \|F_n - F_p\|_2,$$

and then, using eqn (8.18), we obtain via $b \nearrow \infty$

$$\mathcal{E}(V(F_n - F_p)) \leq 8\, C'(\|F\|_\infty + \|F\|_r) \times \|F_n - F_p\|_2.$$

This shows that (for $j \in \{0, 2, \ldots, d\}$) $(\mathcal{L}_j V F_n)$ is a Cauchy sequence in $L^2(\tilde{\lambda}^\Gamma)$, so that it converges to a limit, which we denote by $\mathcal{L}_j V F$. Moreover, for $n, b \in \mathbb{N}^*$ we have

$$\mathcal{E}(V(F_n - F)) \le 8\,C'(\|F\|_\infty + \|F\|_r) \times \|F_n - F\|_2 \qquad (8.20)$$

and

$$\mathcal{E}(V_b(F_n - F)) \le 8\,C'(\|F\|_\infty + \|F\|_r) \times \|F_n - F\|_2,$$

which entail readily that (ii) holds.

We are now left with the verification that $\mathcal{L}_j V F$ is indeed the L^2-derivative at 0 of $\theta_s^j V F$. As in 1) above, since we know that this holds if F_n is used instead of F, it is enough to make sure that the convergences in $L^2(\tilde\lambda^\Gamma)$ of $\theta_s^j V F_n$ to $\theta_s^j V F$ and of $\theta_s^j \mathcal{L}_j V F_n$ to $\theta_s^j \mathcal{L}_j V F$ are uniform with respect to $s \in [0,1]$. This is ensured by the above $L^2(\tilde\lambda^\Gamma)$-estimates and, again, by the invariance of $\tilde\lambda^\Gamma$ with respect to θ_s^j. $\diamond$

As already stated Lemma 8.6.3 and Proposition 8.6.4 allow us to apply Proposition 8.7.2 to the function $Kf = \operatorname{div} \omega^f$ in eqn (8.10), which gives full meaning to eqn (8.14) in Theorem 8.7.1. This allows us also to apply Itô's formula to VKf or, alternatively, to apply eqn (8.10) with $[\omega^f + \frac{1}{2} d(VKf)]$ instead of ω^f. This yields the following result.

Proposition 8.7.3 *For any $t \ge 0$, we have*

$$\int_{\xi[0,t]} \omega^f = \frac{1}{2} VKf(\xi_0) - \frac{1}{2} VKf(\xi_t) + \overline{M}_t^f,$$

where $\left(\overline{M}_t^f\right)$ is a continuous martingale having quadratic variation

$$\left\langle \overline{M}^f \right\rangle_t = \int_0^t \left[\left(f + \frac{1}{2}\mathcal{L}_0 VKf\right)^2 + \sum_{j=2}^d \left(f_j + \frac{1}{2}\mathcal{L}_j VKf\right)^2 \right](\xi_s)\ ds.$$

Proof As the existence of $\operatorname{div}\left[d(VKf)\right]$ is not clear, we again use an approaching sequence (F_n) for $F \equiv Kf$ as in the proof of Proposition 8.7.2, and apply Proposition 8.6.2 and eqn (8.10) to $[\omega^f + \frac{1}{2} d(V_b F_n)]$ instead of ω^f or $[\omega^f + \frac{1}{2} d(VKf)]$.
By Proposition 8.6.2 and eqn (8.17), we have

$$\operatorname{div}\left[d(V_b F_n)\right] = \mathcal{D}(V_b F_n) = 2\,P_b F_n - 2\,F_n,$$

so that we obtain in this way

$$\int_{\xi[0,t]} \omega^f + \frac{1}{2} V_b F_n(\xi_t) - \frac{1}{2} V_b F_n(\xi_0) = M_t^{f,b,n} + \frac{1}{2} \int_0^t [Kf + P_b F_n - F_n](\xi_s)\ ds,$$

$$(8.21)$$

where, by eqn (8.12), the continuous martingale $(M_t^{f,b,n})$ has quadratic variation

$$\langle M^{f,b,n} \rangle_t = \int_0^t \left[\left(f + \frac{1}{2} \mathcal{L}_0 V_b F_n \right)^2 + \sum_{j=2}^d \left(f_j + \frac{1}{2} \mathcal{L}_j V_b F_n \right)^2 \right] (\xi_s) \, ds. \quad (8.22)$$

The result follows from Proposition 8.7.2 (including eqn (8.20)) by letting b, n go to infinity. $\diamond$

It remains to prove a central limit theorem for the martingale $(\overline{M}_t^f)$. Recall from Definition 8.2.1 and Proposition 8.2.2 that the diffusion (ξ_s) is stationary, with semi-group (P_t) and law μ (proportional to $\tilde{\lambda}^\Gamma$) on $\Gamma \backslash \mathbb{F}^d$.

Lemma 8.7.4 *Let* (M_t) *be a continuous real martingale of the form*

$$M_t = \sum_{j=0,2,\dots,d} \int_0^t \psi_j(\xi_s) \, dW_s^j,$$

for some $\psi_j \in L^2(\mathcal{M}, \mu)$. *Then the law of* $M_t/\sqrt{t}$ *converges, as* $t \to \infty$, *towards a centred Gaussian law with variance* $\int_{\Gamma \backslash \mathbb{F}^d} H^2 \, d\mu$, *where* $H^2 := \sum_{j=0,2,\dots,d} \psi_j^2$.

Proof (1) By Proposition 5.3.2, Theorem 5.2.3 (the ergodic theorem) applies to (ξ_t), which goes to infinity in $\mathrm{PSO}(1,d)$ by eqn (8.3) and Proposition 6.3.3. This shows that $\langle M \rangle_t / t$ goes to $\int_{\Gamma \backslash \mathbb{F}^d} H^2 \, d\mu$ almost surely as $t \to \infty$.

(2) Suppose here that $H^2 \in L^\infty(\mathcal{M}, \mu)$, and note that for any real a,

$$N_t := \exp \left(\frac{a\sqrt{-1}}{\sqrt{t}} M_t + \frac{a^2}{2t} \langle M \rangle_t \right)$$

defines a local martingale, which is bounded since its modulus is plainly dominated by $e^{a^2 \, \|H^2\|_\infty / 2}$. Hence it is a bounded martingale. Then, taking advantage of the non-negativity of H^2 and setting $F := H^2 - \int H^2 \, d\mu$, we have

$$\left| \mathbb{E} \left[N_t \, \exp \left(-\frac{a^2}{2t} \int_0^t F(\xi_s) \, ds \right) \right] - 1 \right| = \left| \mathbb{E} \left[N_t \left[\exp \left(-\frac{a^2}{2t} \int_0^t F(\xi_s) \, ds \right) - 1 \right] \right] \right|$$

$$\leq \|N_t\|_2 \times \left\| \exp \left(-\frac{a^2}{2t} \int_0^t F(\xi_s) \, ds \right) - 1 \right\|_2$$

$$\leq e^{(a^2/2)\|H^2\|_\infty} \times e^{(a^2/2)\|H^2\|_1} a^2/2 \left\| t^{-1} \int_0^t F(\xi_s) \, ds \right\|_2,$$

which goes to zero by (1) above. Therefore we finally obtain the convergence of

$$\mathbb{E} \left[\exp \left(a\sqrt{-1} \, \frac{M_t}{\sqrt{t}} \right) \right] = \mathbb{E} \left[N_t \, \exp \left(-\frac{a^2}{2t} \int_0^t F(\xi_s) \, ds \right) \right] \times \exp \left(-\frac{a^2}{2} \int H^2 \, d\mu \right)$$

towards $\exp(-a^2/2 \int H^2 \, d\mu)$, which concludes the proof in the case of a bounded integrand H^2.

(3) To complete the proof, we reduce the general case of $H^2 \in L^1(\mathcal{M}, \mu)$, i.e., of $\psi_j \in L^2(\mathcal{M}, \mu)$, to the above case. For any $n \in \mathbb{N}^*$, we set $\psi_j^n := \min\{n, \max\{-n, \psi_j\}\}$ and $M_t^n := \int_0^t \sum_{j=0,2,\dots,d} \psi_j^n(\xi_s) \, dW_s^j$. Then

$$\left\| \frac{M_t - M_t^n}{\sqrt{t}} \right\|_2^2 = \sum_{j=0,2,\dots,d} \|\psi_j - \psi_j^n\|_2^2$$

goes to 0 as $n \to \infty$, and therefore

$$\lim_{n \to \infty} \sup \left| \mathbb{E}\left[\exp\left(a \sqrt{-1} \frac{M_t}{\sqrt{t}} \right) - \exp\left(a \sqrt{-1} \frac{M_t^n}{\sqrt{t}} \right) \right] \right|^2$$

$$\leq \lim_{n \to \infty} \sup \mathbb{E}\left[\left| \exp\left(a \sqrt{-1} \frac{M_t - M_t^n}{\sqrt{t}} \right) - 1 \right|^2 \right]$$

$$= 4 \lim_{n \to \infty} \sup \mathbb{E}\left[\sin^2\left(a \frac{M_t - M_t^n}{2\sqrt{t}} \right) \right] = 0,$$

uniformly with respect to t. Hence we can apply (2) above to each (M_t^n) and go to the limit

$$\lim_{t \to \infty} \mathbb{E}\left[\exp\left(a \sqrt{-1} \frac{M_t}{\sqrt{t}} \right) \right] = \lim_{n \to \infty} \lim_{t \to \infty} \mathbb{E}\left[\exp\left(a \sqrt{-1} \frac{M_t^n}{\sqrt{t}} \right) \right]$$

$$= \lim_{n \to \infty} \exp\left(-\frac{a^2}{2} \int \sum_{j=0,2,\dots,d} (\psi_j^n)^2 \, d\mu \right) = \exp\left(-\frac{a^2}{2} \int H^2 \, d\mu \right). \qquad \diamond$$

Lemma 8.7.5 *As $t \to \infty$, the law of $t^{-1/2} \int_{\xi[0,t]} \omega^f$ converges towards a centred Gaussian law with the following variance (recall from Definition 8.2.1 that $\mu := \mathrm{covol}(\Gamma)^{-1} \tilde{\lambda}^\Gamma$):*

$$\mathcal{V}_0(f) := \int \left[(f + \tfrac{1}{2} \mathcal{L}_0 V K f)^2 + \sum_{j=2}^{d} (f_j + \tfrac{1}{2} \mathcal{L}_j V K f)^2 \right] d\mu.$$

Proof

(1) We can apply Lemma 8.7.4 to the martingale $(\overline{M}_t^f)$ appearing in Proposition 8.7.3. Indeed, on the one hand, as already stated Lemmas 8.4.2.2 and 8.6.3 and Propositions 8.6.4 and 8.7.2 show that $H^2 \in L^1(\mathcal{M}, \mu)$. And, on the other hand, the continuous martingale $(\overline{M}_t^f)$ of Proposition 8.7.3 is of the Brownian form considered in Lemma 8.7.4, since (by the proof of Proposition 8.7.3) it actually reads

$$\overline{M}_t^f = M_t^f + \frac{1}{2}\sum_j \int_0^t \mathcal{L}_j VKf(\xi_s)\, dW_s^j,$$

with $\left(M_t^f\right)$ of the wanted form, by eqns (8.10) and (8.9) and the proof of Proposition 8.6.2.

(2) Then, $t^{-1/2}\left(VKf(\xi_0) - VKf(\xi_t)\right)$ goes to zero in probability as $t \to \infty$, since $VKf(\xi_t)$ is stationary (recall Definition 8.2.1 and Proposition 8.2.2). Hence, applying Proposition 8.7.3, we see that we have to deal with $\lim_{t\to\infty} \mathbb{E}\left[\exp\left(a\sqrt{-1}\,\overline{M}_t^f/\sqrt{t}\right)\right]$. Now, by (1) above we have

$$\lim_{t\to\infty} \mathbb{E}\left[\exp\left(a\sqrt{-1}\,\overline{M}_t^f/\sqrt{t}\right)\right] = \exp\left(-\frac{a^2}{2}\,\mathcal{V}_0(f)\right).$$

$\diamondsuit$

Finally, Theorem 8.5.1 and Lemma 8.7.5 imply at once that

$$\lim_{t\to\infty}\int \exp\left(\frac{a\sqrt{-1}}{\sqrt{t}}\int_0^{(d-1)t/2} f(\beta\,\theta_s)\, ds\right) d\mu(\beta) = \exp\left(-\frac{a^2\,\mathcal{V}_0(f)}{2}\right).$$

Recalling that (in Definition 8.2.1) $\mu = \mathrm{covol}(\Gamma)^{-1}\tilde{\lambda}^{\Gamma}$, and changing a to $a/\sqrt{d-1}$, which changes $\mathcal{V}_0(f)$ to the $\mathcal{V}(f)$ of eqn (8.14), we deduce immediately the first claim of Sinai's central limit theorem (Theorem 8.7.1).

The last assertion of Theorem 8.7.1 is easy: it is clear from eqn (8.14) that if $\mathcal{V}(f) = 0$, then $f = \mathcal{L}_0(-\frac{1}{2}VKf)$ $\tilde{\lambda}^{\Gamma}$-almost everywhere, and VKf indeed belongs to $L^2(\tilde{\lambda}^{\Gamma})$ by Proposition 8.7.2. Conversely, if $f = \mathcal{L}_0 h$, then by eqn (8.1)

$$t^{-1/2}\int_0^t f(\beta\theta_s)\, ds = t^{-1/2}\left[h(\beta\theta_t) - h(\beta)\right]$$

goes to zero in μ-probability, by the invariance of Proposition 5.1.1 (iv). This clearly forces $\mathcal{V}(f) = 0$.

The proof of Theorem 8.7.1 is now complete.

Remark 8.7.6 To establish Theorem 8.7.1, we actually used only the following slightly weaker assumption: f and $\mathcal{L}_0 f$ are bounded, rotationally Hölderian on $\mathbb{H}^d \times \partial\mathbb{H}^d$ and continuous along the stable leaves, and, for $2 \le j \le d$, $\mathcal{L}_j f$ and $\mathcal{L}_j^2 f$ are bounded and Hölderian on $\mathbb{F}^d$.

Remark 8.7.7 The same argument applied to a function f that satisfies the same assumptions except that its average is not necessarily zero yields the convergence in probability of the ergodic mean towards the average. This is enough to establish the above central limit theorem (recall that the ergodicity was used in the contour deformation argument or, more precisely, in the proof of Theorem 8.5.1). But this result is clearly weaker than the mixing theorem established in Section 5.3.

8.8 Notes and comments

The central limit theorem for the geodesic flow appeared originally in the articles of Sinai ([Si1], [Si2]) and Ratner [Ran] for the cocompact case. The stochastic analysis method that we have used in this chapter originates from the article [LJ2], and differs from the standard method, which is based on Markovian partitions and coding. Other noteworthy articles on this subject are [L2], [L3] and [CL].

This type of study was then extended to some related questions:

- the central limit theorem for the geodesic flow in hyperbolic manifolds of infinite volume [EFLJ3];
- singular windings of geodesics (in particular, near cusps in manifolds of constant negative curvature) in the case of finite volume ([LJ1], [GLJ], [LJ3], [F1], [ELJ], [F2], [F3], [EFLJ4]) and in the case of infinite volume ([EFLJ1], [EF]);
- exit laws [EFLJ2] and counting closed geodesics ([BP1], [BP2]).

Appendix A
Geometry

A.1   Structure of pseudo-symmetric matrices

We describe here the reduction of symmetric endomorphisms in Minkowski space $\mathbb{R}^{1,d}$. This is the Lorentzian counterpart of a well-known result in the Euclidean framework. We considered in Section 1.1.5 the diagonal matrix $J = \mathrm{diag}(1, -1, \ldots, -1) \in \mathcal{M}(d+1)$, which has diagonal entries $(1, -1, \ldots, -1)$, the matrix of the Minkowski pseudo-metric $\langle \cdot, \cdot \rangle$ in $\mathbb{R}^{1,d}$. In other words, $J_{ij} = J^{ij} = 1_{\{i=j=0\}} - 1_{\{i=j\neq 0\}}$.

Definition A.1.1 *An endomorphism T of $\mathbb{R}^{1,d}$ is (pseudo-)**symmetric** if for any $u, v \in \mathbb{R}^{1,d}$, $\langle Tu, v \rangle = \langle u, Tv \rangle$; i.e., if the associated bilinear form $\mathbb{R}^{1,d} \times \mathbb{R}^{1,d} \ni (u, v) \mapsto \langle Tu, v \rangle$ is symmetric. Given any pseudo-orthonormal basis $b = (b_0, b_1, \ldots, b_d)$ of $\mathbb{R}^{1,d}$, we set $T_{ij} := \langle Tb_i, b_j \rangle$, and call $((T_{ij}))$ the matrix of T in the basis b.*

Thus the symmetry of T reads, as usual, $T_{ij} = T_{ji}$ in terms of its matrix (in any pseudo-orthonormal basis). Note that the j-th coordinate of any $u \in \mathbb{R}^{1,d}$ is $u^j := J^{jj} \langle b_j, u \rangle$, so that $\langle Tu, v \rangle = \sum_{0 \leq i, j \leq d} T_{ij} u^i v^j$; while $(Tu)^j = J^{jj} \langle b_j, Tu \rangle = \sum_{i=0}^{d} J^{jj} T_{ji} u^i$.

Our aim here is to establish the following reduction theorem for symmetric endomorphisms of $\mathbb{R}^{1,d}$ (stated without proof in [HE], pp. 89–90). This is actually a result of simultaneous reduction of such an endomorphism and the Minkowski pseudo-metric. It is more complicated than its classical Euclidean analogue, as was already the case with Theorem 1.5.1 for the reduction of isometries.

Theorem A.1.2 *([HE], pp. 89–90) Let T be a real symmetric endomorphism of $\mathbb{R}^{1,d}$. Then there exists a pseudo-orthonormal basis $(b_0, \ldots, b_d)$ in which the matrix $((T_{ij}))$ of T reduces to one of the following types (with real λ, λ_j, μ):*

$$(i1) = \begin{pmatrix} \lambda & 0 & \cdots & 0 \\ 0 & \lambda_1 & \cdots & 0 \\ \vdots & \vdots & \ddots & \vdots \\ 0 & 0 & \cdots & \lambda_d \end{pmatrix}, \quad (i2) = \begin{pmatrix} \varepsilon + \mu & \varepsilon & 0 & \cdots & 0 \\ \varepsilon & \varepsilon - \mu & 0 & \cdots & 0 \\ 0 & 0 & \lambda_2 & \cdots & 0 \\ \vdots & \vdots & \vdots & \ddots & \vdots \\ 0 & 0 & 0 & \cdots & \lambda_d \end{pmatrix} \text{ (with } \varepsilon = \pm 1\text{)},$$

$$(i3) = \begin{pmatrix} -\lambda & \mu & 0 & \cdots & 0 \\ \mu & \lambda & 0 & \cdots & 0 \\ 0 & 0 & \lambda_2 & \cdots & 0 \\ \vdots & \vdots & \vdots & \ddots & \vdots \\ 0 & 0 & 0 & \cdots & \lambda_d \end{pmatrix} \text{ (with } \mu \neq 0\text{)}, \quad (ii) = \begin{pmatrix} -\lambda & 0 & 1 & 0 & \cdots & 0 \\ 0 & \lambda & 1 & 0 & \cdots & 0 \\ 1 & 1 & \lambda & 0 & \cdots & 0 \\ 0 & 0 & 0 & \lambda_3 & \cdots & 0 \\ \vdots & \vdots & \vdots & \vdots & \ddots & \vdots \\ 0 & 0 & 0 & 0 & \cdots & \lambda_d \end{pmatrix}.$$

We begin by quoting two easy facts about $\mathbb{R}^{1,d}$.

Remark A.1.3 (i) If $\langle u, u \rangle = 1$ and $w \in u^\perp \smallsetminus \{0\}$, then $\langle w, w \rangle < 0$.

(ii) If two lightlike vectors u, v are linearly independent, then $\langle u, v \rangle \neq 0$, and $u', v' := u \pm (2\langle u, v \rangle)^{-1} v$ generate the same plane as $\{u, v\}$ and are such that $\langle u', u' \rangle = -\langle v', v' \rangle = 1$ and $\langle u', v' \rangle = 0$.

The main step of the reduction is the following lemma.

Lemma A.1.4 *Let T be a (real) symmetric endomorphism in Minkowski space $\mathbb{R}^{1,d}$. Then there exists a pseudo-orthonormal basis $(b_0, \ldots, b_d)$ of $\mathbb{R}^{1,d}$ such that one of the following two mutually exclusive cases occurs;*

- *Either (i) $\{b_2, \ldots, b_d\}$ are eigenvectors of T, and there exist real α, β, γ such that*

$$Tb_0 = \alpha b_0 - \beta b_1 \quad and \quad Tb_1 = \gamma b_1 + \beta b_0$$

- *or (ii) $\{b_3, \ldots, b_d\}$ are eigenvectors of T, and there exist real $q \neq 0$ and a, λ such that*

$$Tb_0 = ab_0 - (\lambda + a)b_1 - q\, b_2, Tb_1 = (a + \lambda)b_0 - (2\lambda + a)b_1 - q\, b_2, \; Tb_2$$
$$= q(b_0 - b_1) - \lambda b_2.$$

Remark A.1.5 The two types of symmetric endomorphisms in Lemma A.1.4 correspond to the following two types of symmetric matrices $((T_{ij}))$ (with $q \neq 0$):

$$\begin{pmatrix} \alpha & \beta & 0 & \cdots & 0 \\ \beta & \gamma & 0 & \cdots & 0 \\ 0 & 0 & \lambda_2 & \cdots & 0 \\ \vdots & \vdots & \vdots & \ddots & \vdots \\ 0 & 0 & 0 & \cdots & \lambda_d \end{pmatrix}, \quad \begin{pmatrix} a & \lambda + a & q & 0 & \cdots & 0 \\ \lambda + a & 2\lambda + a & q & 0 & \cdots & 0 \\ q & q & \lambda & 0 & \cdots & 0 \\ 0 & 0 & 0 & \lambda_3 & \cdots & 0 \\ \vdots & \vdots & \vdots & \vdots & \ddots & \vdots \\ 0 & 0 & 0 & 0 & \cdots & \lambda_d \end{pmatrix}.$$

Proof T necessarily admits one complex-valued eigenvalue at least: there exist real numbers x, y and real vectors u, v (both non-vanishing), such that $T(u + \sqrt{-1}\, v) = (x + \sqrt{-1}\, y)(u + \sqrt{-1}\, v)$ or, equivalently, $Tu = xu - yv$ and $Tv = yu + xv$. Then $x\langle u, v \rangle - y\langle v, v \rangle = \langle Tu, v \rangle = \langle u, Tv \rangle = y\langle u, u \rangle + x\langle u, v \rangle$, whence either $y = 0$ and then u or v is an eigenvector associated with the eigenvalue x, or $\langle u, u \rangle + \langle v, v \rangle = 0$ and u, v span a stable plane. (In fact, if $y \neq 0$ and $\alpha u + \beta v = 0$, then $0 = \alpha Tu + \beta Tv = (\alpha x + \beta y)u + (\beta x - \alpha y)v$, whence $0 = (\alpha x + \beta y)\beta - (\beta x - \alpha y)\alpha = (\alpha^2 + \beta^2)y$ and then $\alpha = \beta = 0$.)

In the latter case, if the linearly independent vectors u, v are neither lightlike nor orthogonal, up to a multiplication and an exchange, we can suppose that $\langle u, u \rangle =$

$1 = -\langle v, v \rangle$ and then, replacing v by $(1 + \langle u, v \rangle^{-2})^{-1/2}(u - \langle u, v \rangle^{-1}v)$, that they are orthogonal. Also in the case $\langle u, u \rangle + \langle v, v \rangle = 0$, but if the linearly independent vectors u, v are lightlike, then they cannot be orthogonal, by Remark A.1.3 (ii), so that by replacing $\{u, v\}$ by $\{u + v, u - v\}$, we are brought back to the preceding case.

So far, we have found that either T admits a real eigenvector $b \ne 0$ (with pseudo-norm 1, 0 or -1) or it admits a real stable plane $\{b_0, b_1\}$ such that $\langle b_0, b_0 \rangle - 1 = \langle b_0, b_1 \rangle = \langle b_1, b_1 \rangle + 1 = 0$, and T necessarily operates in $\{b_0, b_1\}$ as in the statement.

Up to replacing T by some $T + \alpha \mathbf{1}$, we can restrict ourselves to the case of an invertible T. Note now that the orthogonal space $V^\perp$ of any subspace V preserved by T (i.e., $TV = V$) has to be preserved by T, and that eigensubspaces associated with distinct eigenvalues are orthogonal. So that we can decompose $\mathbb{R}^{1,d} = V_0 \oplus \ldots \oplus V_k$ into a direct and orthogonal sum of stable subspaces V_j, such that: any V_j contains a real non-zero eigenvector b_j, or is a plane $\{b_0, b_1\}$ as described in the statement; if b_j is a non-lightlike eigenvector, then $V_j = \mathbb{R}b_j$ is a line; at most one V_j can contain a non-spacelike vector (by Remark A.1.3); some V_j must contain a timelike vector.

In particular, if V_0 is a plane $\{b_0, b_1\}$ as described in the statement, or if each V_j is a line, then the decomposition (i) of the statement follows directly. Let us therefore focus on the remaining case, where any V_j may contain a real non-zero eigenvector b_j, V_j is a line for $0 \le j < k < d$, and the b_k are lightlike and do not span V_k, and are such that $Tb_k = \lambda b_k$. We can assume, moreover, by taking k maximal, that V_k does not contain any non-lightlike eigenvector.

By Remark A.1.3, $\mathbb{R}^{1,d} \setminus \mathbb{R}b_k$ cannot contain any non-spacelike vector orthogonal to b_k, so that we can suppose that $\langle b_j, b_j \rangle = -1$ for $0 \le j < k$. Moreover, V_k must contain a vector v_k such that $\langle v_k, v_k \rangle = 1$. By Remark A.1.3$(i)$, we must have $\langle b_k, v_k \rangle \ne 0$, and then $v_{k+1} := v_k - (\langle b_k, v_k \rangle)^{-1}b_k$ is such that $\langle v_{k+1}, v_{k+1} \rangle + 1 = \langle v_k, v_{k+1} \rangle = 0$, and (up to a multiplication) $b_k = v_k - v_{k+1}$. We can then complete $\{v_k, v_{k+1}\}$ into a pseudo-orthonormal basis $\{v_k, v_{k+1}, \ldots, v_d\}$ of V_k, with $\langle v_i, v_i \rangle = -1$ for $k < i \le d$. We set, for $k \le j \le d$, $Tv_j = \sum_{i=0}^d q_{ij}v_i$.

Since $T(v_k - v_{k+1}) = v_k - v_{k+1}$, and by the symmetry of T, we must furthermore have $Tv_k = av_k + (\lambda - a)v_{k+1} + \sum_{i=k+2}^d q_i v_i$ and $Tv_{k+1} = (a - \lambda)v_k + (2\lambda - a)v_{k+1} + \sum_{i=k+2}^d q_i v_i$, and then, for $k + 2 \le j \le d$, we have $Tv_j = q_j(v_{k+1} - v_k) + \sum_{i=k+2}^d q_{ij}v_i$, with $q_{i\ell} = q_{\ell i}$ for $k + 2 \le i < \ell \le d$. By the usual reduction of symmetric real matrices in a Euclidean space, we can assume that $q_{i\ell} = \delta_{i\ell}\lambda_i$ for $k + 2 \le i \le \ell \le d$.

Note now that for any $k + 2 \le j \le d$, the vector $v_j' := a_j(v_k - v_{k+1}) + (\lambda - \lambda_j)v_j$ satisfies $Tv_j' := \lambda_j v_j'$ and $\langle v_j', v_j' \rangle = -(\lambda - \lambda_j)^2$, and hence is a spacelike eigenvector if $\lambda_j \ne \lambda$. By the maximality of k, this forces $\lambda_j = \lambda$ for $k + 2 \le j \le d$. Again by the maximality of k, we can restrict ourselves to the case $q_j \ne 0$ for $k + 2 \le j \le d$. Now λ is the only eigenvalue of the restriction of T to V_k (it can be immediately seen that $\det\left[(T - \mu\mathbf{1})_{|V_k}\right] = -(\lambda - \mu)^{d-k+1}$, by subtracting the first column from the second column and adding the first row to the second row).

Furthermore, for $k < d - 2$, the vector $(v_k - v_{k+1} + q_{k+3}v_{k+2} - q_{k+2}v_{k+3})$ would be a spacelike eigenvector. Hence we must have $k \ge d - 2$. Since $k = d - 1$ would yield only a subcase of case (i), we can suppose that $k = d - 2$ (i.e., V_k has three dimensions). Then $(v_k - v_{k+1})$ is the only eigenvector of the restriction of T to V_k. Note that this case is not reducible to case (i), since it would then have at least two non-collinear (complex-valued) eigenvectors. Finally, by reordering indices and changing (λ, q) to $(-\lambda, -q)$, we obtain the form (ii) of the statement.   ◇

End of the proof of Theorem A.1.2   A change of the pseudo-orthonormal basis is given by a matrix $\theta \in \mathrm{PSO}(1, d)$, and corresponds to changing the symmetric matrix $((T_{ij}))$ to $\theta((T_{ij}))\theta^{-1}$. In fact, if $\theta b'_j = b_j$, then $T'_{ij} = \langle Tb'_i, b'_j \rangle = \langle T\theta^{-1}b_i, \theta^{-1}b_j \rangle = \langle \theta T\theta^{-1}b_i, b_j \rangle$. By Remark A.1.5, it is clearly enough to consider the cases

$$((T_{ij})) = \begin{pmatrix} \alpha & \beta \\ \beta & \gamma \end{pmatrix}$$

and

$$((T_{ij})) = \begin{pmatrix} a & \lambda + a & q \\ \lambda + a & 2\lambda + a & q \\ q & q & \lambda \end{pmatrix}.$$

In the first case, using

$$\theta = \begin{pmatrix} \mathrm{ch}\, t & \mathrm{sh}\, t \\ \mathrm{sh}\, t & \mathrm{ch}\, t \end{pmatrix} \in \mathrm{PSO}(1, 1),$$

with $\mathrm{th}\,(2t) = -2\beta/\alpha + \gamma$ if $|\alpha + \gamma| > 2\,|\beta|$, we obtain the diagonal form $(i1)$. If $|\alpha + \gamma| = 2\,|\beta| \neq 0$, with $e^{-2t} = |\beta|$ (and $-\mathrm{ch}\, t$ instead of $\mathrm{ch}\, t$ for $(\alpha + \gamma)\beta < 0$), we obtain the form $(i2)$. Finally, if $|\alpha + \gamma| < 2\,|\beta|$, with $e^{4t} = 2\beta - \alpha - \gamma/2\beta + \alpha + \gamma$ and, moreover,

$$\mu = \beta\sqrt{1 - \left(\frac{\alpha + \gamma}{2\beta}\right)^2}$$

and $\lambda = \gamma - \alpha/2$, we obtain the form $(i3)$.
    In the second case, using first

$$\theta = \begin{pmatrix} \varepsilon\,\mathrm{ch}\, t & \varepsilon\,\mathrm{sh}\, t & 0 \\ \varepsilon\,\mathrm{sh}\, t & \varepsilon\,\mathrm{ch}\, t & 0 \\ 0 & 0 & 1 \end{pmatrix} \in \mathrm{PSO}(1, 2)$$

with $t := -\log|q|$ and ε being the sign of q, we obtain a reduction of $((T_{ij}))$ to

$$\begin{pmatrix} u - \lambda & u & 1 \\ u & u + \lambda & 1 \\ 1 & 1 & \lambda \end{pmatrix},$$

with

$$u := (a + \lambda)q^{-2}.$$

Then, using

$$\theta' = \begin{pmatrix} 1 + u^2/8 & -u^2/8 & -u/2 \\ u^2/8 & 1 - u^2/8 & -u/2 \\ -u/2 & u/2 & 1 \end{pmatrix} \in \mathrm{PSO}(1, 2),$$

we obtain the last form (ii) of the statement.   ◇

As in ([HE], Section 4.3), the following is easily deduced from Theorem A.1.2.

Remark A.1.6 ([HE], Section 4.3) *Let T be a real symmetric endomorphism of $\mathbb{R}^{1,d}$ which satisfies the 'Weak Energy Condition' $\langle T\zeta, \zeta \rangle \geq 0$ for any timelike vector ζ. Then, in some pseudo-orthonormal basis, its matrix $((T_{ij}))$ is of one of the following types described in Theorem A.1.2 (with λ, λ_j, μ real):*

- *either $((T_{ij}))$ is of type $(i1)$, with $\lambda \geq 0$ and $\lambda + \lambda_i \geq 0$ for $1 \leq i \leq d$,*
- *or $((T_{ij}))$ is of type $(i2)$, with $\varepsilon = 1$, $\mu \geq 0$ and $\lambda_i \geq 0$ for $2 \leq i \leq d$.*

The 'Dominant Energy Condition' requires that, moreover $T\zeta$ is non-spacelike. This happens for type $(i1)$ if and only if $|\lambda_i| \leq \lambda$ for $1 \leq i \leq d$, and for type $(i2)$ if and only if $\varepsilon = 1$ and $0 \leq \lambda_i \leq \mu$ for $2 \leq i \leq d$.

The following variant of the reduction of symmetric endomorphisms can be also deduced from Lemma A.1.4 and Remark A.1.5.

Remark A.1.7 *Any symmetric matrix $((T_{ij}))$ can be decomposed into a linear combination of (at most $d+1$) rank-1 matrices, $T_{ij} = \sum_{n=0}^{r} q_n\, U_i(n) U_j(n)$, with $0 \leq r \leq d$. Moreover, it is always possible that at most $U(0)$ is non-spacelike in this decomposition.*

A.2   The full commutation relation in PSO$(1, d)$

We have already seen the fundamental commutation relations (1.16) between the matrices θ_t and θ_x^+ of PSO$(1, d)$ (recall Proposition 1.4.3), and between these and a rotation matrix $\varrho \in$ SO(d) (recall Theorem 2.6.1). We did not need the full expression for the relation in the latter case, so the description of Theorem 2.6.1 was not complete. Here we specify this commutation relation completely. Since $r \in$ SO$(d-1)$ commutes with the matrices θ_t and satisfies $r\theta_x^+ r^{-1} = \theta_{rx}^+$ for any $x \in \mathbb{R}^{d-1}$ (by Proposition 1.4.3), it is sufficient to consider the planar rotation matrices (with $\alpha \neq 0$) $\mathcal{R}_{\alpha,\sigma}$ introduced in Section 2.6 (after Theorem 2.6.1), and also considered in Proposition 3.3.2.

As the general commutation relation between $\mathcal{R}_{\alpha,\sigma}$ and any $\theta_x^+ \theta_{\log y}$ is somewhat complicated, it is more tractable to split this relation into three cases, namely when $x = 0$, when $y = 1$ and x is collinear with σ, and when $y = 1$ and x is orthogonal to σ. The corresponding three commutation relations are as follows.

Proposition A.2.1 *Take any $\sigma \in \mathbb{S}^{d-2}$ and any non-zero $\alpha \in \mathbb{R}/2\pi\mathbb{Z}$, determining a generic planar rotation $\mathcal{R}_{\alpha,\sigma}$ (using the notation of Section 2.6). We then have the following three commutation relations, determining how $\mathcal{R}_{\alpha,\sigma}$ commutes with the elements of $\mathbb{A}^d$.*

(i) For any $y > 0$, we have

$$\mathcal{R}_{\alpha,\sigma}\, \theta_{\log y} = \theta_{x'}^+\, \theta_{\log y'}\, \mathcal{R}_{\alpha',\sigma}, \tag{A.1}$$

where

$$\frac{y}{y'} = y^2 \sin^2\left(\tfrac{\alpha}{2}\right) + \cos^2\left(\tfrac{\alpha}{2}\right), \quad x' = \left(1 - \left(\tfrac{y'}{y}\right)\right)\cotg\left(\tfrac{\alpha}{2}\right)\sigma$$

$$\text{and} \quad \cotg\left(\tfrac{\alpha'}{2}\right) = y^{-1}\cotg\left(\tfrac{\alpha}{2}\right).$$

(ii) *For any real t, setting* $x := \left(t - \cotg\left(\alpha/2\right)\right)\sigma$, *we have*

$$\mathcal{R}_{\alpha,\sigma}\theta_x^+ = \theta_{x'}^+\theta_{\log y'}\mathcal{R}_{\alpha',\sigma}, \tag{A.2}$$

where

$$y' = (1+t^2)^{-1}\sin^{-2}\left(\tfrac{\alpha}{2}\right), \quad x' = \left(\cotg\left(\tfrac{\alpha}{2}\right) - ty'\right)\sigma \text{ and } \cotg\left(\tfrac{\alpha'}{2}\right) = t.$$

(iii) *For any vector* $x \in \mathbb{R}^{d-1}$ *which is orthogonal to* σ, *we have*

$$\mathcal{R}_{\alpha,\sigma}\theta_x^+ r_{\sigma,x}^{\alpha} = \theta_{x'}^+\theta_{\log y'}\mathcal{R}_{\alpha',\sigma}, \tag{A.3}$$

where

$$\frac{1}{y'} = |x|^2\,\sin^2\left(\tfrac{\alpha}{2}\right) + 1, \quad x' = (1-y')\cotg\left(\tfrac{\alpha}{2}\right)\sigma + y'x,$$

$\mathcal{R}_{\alpha',\sigma'}$ *is determined by the relation*

$$\cotg\left(\tfrac{\alpha'}{2}\right)\sigma' = \cotg\left(\tfrac{\alpha}{2}\right)\sigma - x,$$

and $r_{\sigma,x}^{\alpha} \in \mathrm{SO}(d-1)$ *is the rotation in the plane* (σ, x) *defined by*

$$r_{\sigma,x}^{\alpha}\sigma = \frac{2\,\cotg\left(\alpha/2\right)x + \left(\cotg^2(\alpha/2) - |x|^2\right)\sigma}{|x|^2 + \cotg^2(\alpha/2)}.$$

Remark A.2.2 The three commutation relations of Proposition A.2.1 can be gathered together into a unique commutation relation, as follows.

For any $x \in \mathbb{R}^{d-1}$, $y > 0$, $\sigma \in \mathbb{S}^{d-2}$ *and non-zero* $\alpha \in \mathbb{R}/2\pi\mathbb{Z}$, *we have*

$$\mathcal{R}_{\alpha,\sigma}\theta_x^+\theta_{\log y} r_{\sigma,x}^{\alpha} = \theta_{x'}^+\theta_{\log y'}\mathcal{R}_{\alpha',\sigma'}, \tag{A.4}$$

where

$$\frac{y}{y'} = (y^2 + |x|^2)\,\sin^2\left(\tfrac{\alpha}{2}\right) + \cos^2\left(\tfrac{\alpha}{2}\right) - \langle x, \sigma\rangle\,\sin\alpha,$$

$$x' = \left(1 - \frac{y'}{y}\right)\cotg\left(\tfrac{\alpha}{2}\right)\sigma + \left(\frac{y'}{y}\right)(x + 2\langle x, \sigma\rangle\,\sigma),$$

$\mathcal{R}_{\alpha',\sigma'}$ *is determined by the relation*

$$\cotg\left(\tfrac{\alpha'}{2}\right)\sigma' = \left[\cotg\left(\tfrac{\alpha}{2}\right)\sigma - x - 2\langle x, \sigma\rangle\,\sigma\right]/y,$$

and $r_{\sigma,x}^{\alpha} \in \mathrm{SO}(d-1)$ *is the rotation in the plane* (σ, x) *defined by*

$$r_{\sigma,x}^{\alpha}\sigma = \frac{2\left(\cotg\left(\alpha/2\right) - \langle x, \sigma\rangle\right)x + \left(\cotg^2(\alpha/2) - |x|^2\right)\sigma}{|x|^2 - 2\langle x, \sigma\rangle\,\cotg\left(\alpha/2\right) + \cotg^2(\alpha/2)}.$$

In the degenerate case $x = -\cotg(\alpha/2)\sigma$, *we have* $x' = -x, \alpha' = \pi,$ $\sigma' = \sigma$ *and* $r^\alpha_{\sigma,x} = 1$. *In all cases we have* $\cotg(\alpha'/2)\sigma' = (\cotg(\alpha/2)\sigma - x')/y'$, $\sin^2(\alpha'/2) = y'y\sin^2(\alpha/2)$ *and*

$$\frac{(y')^2 + |x'|^2 + 1}{y'} = \frac{y^2 + |x|^2 + 1}{y}.$$

Proof (i) We take a direct pseudo-orthonormal basis of $\mathcal{M}$, $\mathcal{B} := (e_0, e_1, \sigma, v_3, \dots, v_d)$. We have to verify that eqn (A.1) holds true on each vector of $\mathcal{B}$. As we are dealing with elements of $PSO(1, d)$, it is in fact sufficient to consider only d vectors of $\mathcal{B}$.

For $3 \le j \le d$, we first have $\theta^+_{x'} v_j = v_j - \langle x', v_j\rangle(e_0 + e_1) = v_j$, whence

$$\mathcal{R}_{\alpha,\sigma}\theta_{\log\, y} v_j = v_j = \theta^+_{x'}\theta_{\log\, y'}\ \mathcal{R}_{\alpha',\sigma}\ v_j.$$

Then

$$\theta^+_{x'}\theta_{\log\, y'} R_{\alpha',\sigma'} e_0 = \left[\frac{(y')^2 + |x'|^2 + 1}{2y'}\right]e_0 + \left[\frac{(y')^2 + |x'|^2 - 1}{2y'}\right]e_1 + \frac{x'}{y'}$$

$$= \left[\frac{y^4\ \sin^2(\alpha/2) + \cos^2(\alpha/2) + y^2}{2y[y^2\sin^2(\alpha/2) + \cos^2(\alpha/2)]}\right]e_0 + \left[\frac{y^4\sin^2(\alpha/2) - \cos^2(\alpha/2) + y^2\cos\alpha}{2y[y^2\sin^2(\alpha/2) + \cos^2(\alpha/2)]}\right]$$

$$[\cos\alpha]e_1 + \left[\frac{1}{y'} - \frac{1}{y}\right]\cotg(\alpha/2)\sigma$$

$$= \frac{y^2 + 1}{2y}e_0 + \frac{y^2 - 1}{2y}[\cos\alpha]e_1 + \frac{y^2 - 1}{2y}[\sin\alpha]\sigma = \mathcal{R}_{\alpha,\sigma}\theta_{\log\, y}e_0,$$

and

$$\theta^+_{x'}\theta_{\log\, y'} R_{\alpha',\sigma}\sigma = \theta^+_{x'}\theta_{\log\, y'}[\cos\alpha'\sigma - \sin\alpha'\, e_1]$$

$$= \cos\alpha'[\sigma - \langle x',\sigma\rangle(e_0 + e_1)] - \sin\alpha'\left(\left[\frac{(y')^2 - |x'|^2 - 1}{2y'}\right]e_0\right.$$

$$\left. + \left[\frac{(y')^2 - |x'|^2 + 1}{2y'}\right]e_1 - \frac{x'}{y'}\right)$$

$$= \left[\cos\alpha' + \frac{y^2 - 1}{2y}\sin\alpha\sin\alpha'\right]\sigma - \left[\langle x',\sigma\rangle\cos\alpha' + \sin\alpha'\left[\frac{(y')^2 - |x'|^2 - 1}{2y'}\right]\right]$$

$$(e_0 + e_1) - \frac{\sin\alpha'}{y'}e_1$$

$$= \left[\frac{\cotg^2(\alpha/2) - y^2 + 2(y^2 - 1)\cos^2(\alpha/2)}{y^2 + \cotg^2(\alpha/2)}\right]\sigma - [\sin\alpha]e_1$$

$$= [\cos\alpha]\sigma - [\sin\alpha]e_1 = \mathcal{R}_{\alpha,\sigma}\theta_{\log\, y}\sigma,$$

since the coefficient of $-(e_0 + e_1)$, equal to

$$\left[\frac{y'}{y}\right]^2 \left[\frac{\sin\alpha}{2}\right] \left[y^4 \sin^2(\tfrac{\alpha}{2}) - y^2 + \cos^2(\tfrac{\alpha}{2}) + y^2 - (1 - y^2)^2 \cos^2(\tfrac{\alpha}{2}) \sin^2(\tfrac{\alpha}{2}) \right.$$
$$\left. - [y^2 \sin^2(\tfrac{\alpha}{2}) + \cos^2(\tfrac{\alpha}{2})]^2 \right],$$

vanishes.

(ii) We proceed in the same way, to verify that eqn (A.2) holds true. Firstly, $\mathcal{R}_{\alpha,\sigma}\theta_x^+$ and $\theta_{x'}^+\theta_{\log y'}\mathcal{R}_{\alpha',\sigma}$ act identically on $v_3, \ldots, v_d$. Then

$$(y')^2 + |x'|^2 = y'\left[(1 + t^2)\cos^2(\tfrac{\alpha}{2}) - 2t \cot g(\tfrac{\alpha}{2}) + 1 + \cot g^2(\tfrac{\alpha}{2})\right]$$

implies

$$(y')^2 + |x'|^2 + 1 = y'\left[2 + (\cot g(\tfrac{\alpha}{2}) - t)^2\right] = 2y'\left[1 + \tfrac{1}{2}|x'|^2\right]$$

and

$$\frac{(y')^2 + |x'|^2 - 1}{y'} = (\cot g(\tfrac{\alpha}{2}) - t)\left[\cot g(\tfrac{\alpha}{2}) + \sin\alpha - t\cos\alpha\right]$$
$$= \left[|x|^2 \cos\alpha + 2(\cot g(\tfrac{\alpha}{2}) - t)\sin\alpha\right],$$

while

$$\frac{x'}{y'} = \left[(1 + t^2)\cos(\tfrac{\alpha}{2})\sin(\tfrac{\alpha}{2}) - t\right]\sigma = \left[\cos(\tfrac{\alpha}{2}) - t\sin(\tfrac{\alpha}{2})\right]\left[\sin(\tfrac{\alpha}{2}) - t\cos(\tfrac{\alpha}{2})\right]\sigma$$
$$= \left[\tfrac{1}{2}|x|^2 \sin\alpha - (\cot g(\tfrac{\alpha}{2}) - t)\cos\alpha\right]\sigma,$$

so that

$$\theta_{x'}^+\theta_{\log y'}\mathcal{R}_{\alpha',\sigma}e_0 = \left[\frac{(y')^2 + |x'|^2 + 1}{2y'}\right]e_0 + \left[\frac{(y')^2 + |x'|^2 - 1}{2y'}\right]e_1 + \frac{x'}{y'}$$
$$= \left[1 + \tfrac{1}{2}|x|^2\right]e_0 + \left[\tfrac{1}{2}|x|^2 \cos\alpha + (\cot g(\tfrac{\alpha}{2}) - t)\sin\alpha\right]e_1 + \left[\tfrac{1}{2}|x|^2 \sin\alpha - (\cot g(\tfrac{\alpha}{2}) - t)\cos\alpha\right]\sigma$$
$$= \left[1 + \tfrac{1}{2}|x|^2\right]e_0 + \tfrac{1}{2}|x|^2[\cos\alpha\; e_1 + \sin\alpha\;\sigma] + \left[t - \cot g(\tfrac{\alpha}{2})\right]\left[\cos\alpha\;\sigma - \sin\alpha\;e_1\right]$$
$$= \mathcal{R}_{\alpha,\sigma}\theta_x^+ e_0.$$

Finally, we have

$$\theta_{x'}^{+}\theta_{\log\ y'}\mathcal{R}_{\alpha',\sigma}\sigma = \theta_{x'}^{+}\theta_{\log\ y'}[\cos\alpha'\sigma - \sin\alpha'e_1]$$

$$= \cos\alpha'\left[\sigma - \langle x',\sigma\rangle(e_0+e_1)\right] - \sin\alpha'\left(\left[\frac{(y')^2-|x'|^2-1}{2y'}\right]e_0\right.$$

$$\left.+\left[\frac{(y')^2-|x'|^2+1}{2y'}\right]e_1 - \frac{x'}{y'}\right)$$

$$= \left(\frac{t^2-1}{1+t^2}+t\left[\sin\ \alpha - \frac{2t}{1+t^2}\right]\right)\sigma - 2t\sin^2(\tfrac{\alpha}{2})e_1$$

$$-\left([ty'-\cotg(\tfrac{\alpha}{2})]\frac{t^2-1}{1+t^2}+\frac{2t}{1+t^2}\left[\frac{(y')^2-|x'|^2-1}{2y'}\right]\right)(e_0+e_1)$$

$$= [t\sin\alpha-1]\,\sigma - 2t\sin^2(\tfrac{\alpha}{2})\,e_1 - \left(t^3y'-\cotg(\tfrac{\alpha}{2})(t^2-1)-t\left[\frac{|x'|^2+1}{y'}\right]\right)\frac{e_0+e_1}{1+t^2}$$

$$= [t\sin\alpha-1]\,\sigma - 2t\,\sin^2(\tfrac{\alpha}{2})\,e_1 - \left(\cotg(\tfrac{\alpha}{2})(1-t^2)-t[1+t^2-2t\,\cotg(\tfrac{\alpha}{2})]\right)\frac{e_0+e_1}{1+t^2}$$

$$= [t\sin\alpha-1]\,\sigma - 2t\sin^2(\tfrac{\alpha}{2})\,e_1 + [t-\cotg(\tfrac{\alpha}{2})](e_0+e_1)$$

$$= [t-\cotg(\tfrac{\alpha}{2})]\,e_0 - \left[\sin\alpha - [t-\cotg(\tfrac{\alpha}{2})]\cos\alpha\right]e_1 + \left[\cos\alpha + [t-\cotg(\tfrac{\alpha}{2})]\sin\alpha\right]\sigma$$

$$= \cos\alpha\ \sigma - \sin\alpha\ e_1 - \langle x,\sigma\rangle(e_0+\cos\alpha\ e_1+\sin\alpha\ \sigma) = \mathcal{R}_{\alpha,\sigma}\theta_x^{+}\sigma.$$

(iii)] We take a direct pseudo-orthonormal basis of $\mathcal{M}$, $\mathcal{B} := (e_0,e_1,\sigma,u,v_4,\ldots,v_d)$, with $u := x/|x|$. We again have to verify that eqn (A.3) holds true on d vectors of $\mathcal{B}$. Firstly, $\mathcal{R}_{\alpha,\sigma}\theta_x^{+}r_{\sigma,x}^{\alpha}$ and $\theta_{x'}^{+}\theta_{\log y'}\mathcal{R}_{\alpha',\sigma}$ act identically on $v_4,\ldots,v_d$. Then we have

$$\frac{(y')^2+|x'|^2+1}{y'} = y'\left[|x|^2+\sin^{-2}(\tfrac{\alpha}{2})\right] - 2\cotg^2(\tfrac{\alpha}{2}) + \frac{\sin^{-2}(\alpha/2)}{y'} = |x|^2+2,$$

so that

$$\theta_{x'}^{+}\theta_{\log\ y'}\mathcal{R}_{\alpha',\sigma}e_0 = \left[\tfrac{1}{2}|x|^2+1\right](e_0+e_1) - \left[|x|^2\sin^2(\tfrac{\alpha}{2})+1\right]e_1$$

$$+\left[|x|^2\sin(\tfrac{\alpha}{2})\cos(\tfrac{\alpha}{2})\sigma + x\right]$$

$$= \left[1+\tfrac{1}{2}|x|^2\right]e_0 + \tfrac{1}{2}|x|^2[\cos\alpha\ e_1+\sin\alpha\ \sigma]+x = \mathcal{R}_{\alpha,\sigma}\ \theta_x^{+}r_{\sigma,x}^{\alpha}e_0.$$

Then

$$\theta_{x'}^{+}\theta_{\log\ y'}\mathcal{R}_{\alpha',\sigma'}(e_0+e_1) = \theta_{x'}^{+}\theta_{\log y'}\left[(1-\cos\alpha')e_0+\cos\alpha'(e_0+e_1)+\sin\alpha'\sigma'\right]$$

$$= \frac{2\theta_{x'}^{+}\theta_{\log y'}e_0}{|x|^2+\sin^{-2}(\alpha/2)} + \left[\frac{|x|^2+\cotg^2(\alpha/2)-1}{|x|^2+\sin^{-2}(\alpha/2)}\right]y'(e_0+e_1) + 2\theta_{x'}^{+}\left[\frac{\cotg(\alpha/2)\sigma-x}{|x|^2+\sin^{-2}(\alpha/2)}\right]$$

$$= \left[\frac{2+|x|^2}{|x|^2+\sin^{-2}(\alpha/2)}\right]e_0 + \left[\frac{|x|^2\cos\alpha}{|x|^2+\sin^{-2}(\alpha/2)}\right]e_1 + 2\sin^2(\alpha/2)x'$$

$$+ \left[\frac{|x|^2+\cotg^2(\alpha/2)-1}{|x|^2+\sin^{-2}(\alpha/2)}\right]y'(e_0+e_1)$$

$$+ \left[\frac{2\cotg(\alpha/2)}{|x|^2+\sin^{-2}(\alpha/2)}\right]\left[\sigma+(1-y')\cotg(\alpha/2)(e_0+e_1)\right] - \frac{2x+2y'|x|^2(e_0+e_1)}{|x|^2+\sin^{-2}(\alpha/2)}$$

$$= \left[1+y'-\frac{2\cotg^2(\alpha/2)}{|x|^2+\sin^{-2}(\alpha/2)}\right]e_0 + \left[\frac{|x|^2\cos\alpha}{|x|^2+\sin^{-2}(\alpha/2)}\right]e_1$$

$$- \left[\frac{|x|^2+\cotg^2(\alpha/2)+1}{|x|^2+\sin^{-2}(\alpha/2)}\right]y'(e_0+e_1)+(1-y')[\sin\alpha]\sigma$$

$$+ \left[\frac{2\cotg(\alpha/2)}{|x|^2+\sin^{-2}(\alpha/2)}\right]\left[\sigma+\cotg\left(\tfrac{\alpha}{2}\right)(e_0+e_1)\right]$$

$$= [1+y']e_0 + \left[\frac{|x|^2\cos\alpha}{|x|^2+\sin^{-2}(\alpha/2)}\right]e_1 - y'(e_0+e_1)+[\sin\alpha]\sigma + \left[\frac{2\cotg^2(\alpha/2)}{|x|^2+\sin^{-2}(\alpha/2)}\right]e_1$$

$$= e_0 + [\cos\alpha]\,e_1 + [\sin\alpha]\,\sigma = \mathcal{R}_{\alpha,\sigma}\theta_x^+ r_{\sigma,x}^\alpha(e_0+e_1).$$

So far, we have proved the existence of $r_{\sigma,x}^\alpha \in \mathrm{SO}(d-1)$ such that eqn (A.3) holds, with the given expressions for x', y', σ'. It remains only to establish the exact expression for $r_{\sigma,x}^\alpha$. We leave this verification to the interested reader, as an exercise below, which completes the proof of Proposition A.2.1.   ◇

Exercise   Perform the necessary computations to verify that the expression given in the statement of Proposition A.2.1(*iii*) indeed defines the rotation $r_{\sigma,x}^\alpha \in \mathrm{SO}(d-1)$ appearing in eqn (A.3). Check also that Remark A.2.2 is correct.

A.3   The d'Alembertian □ on $\mathbb{R}^{1,d}$

The d'Alembert operator is naturally associated with the Lorentz–Möbius group $\mathrm{PSO}(1,d)$, and induces the hyperbolic Laplacian in a natural way. To specify this second point, we use polar coordinates in the interior $\overset{o}{\mathcal{C}}$ of the solid light cone $\overline{\mathcal{C}}$: any future-oriented vector $\xi \in \mathbb{R}^{1,d}$ that has a positive pseudo-norm can be written in a unique way, namely $\xi = r\,p$, with $(r,p) \in \mathbb{R}_+^* \times \mathbb{H}^d$. In these polar coordinates, the d'Alembertian □ splits in a simple way, very similar to the polar splitting of the Euclidean Laplacian. In the following statement, which is the purpose of this section, we introduce the d'Alembertian □ by its $\mathrm{PSO}(1,d)$-invariance, and we present its polar splitting, yielding the hyperbolic Laplacian.

Proposition A.3.1
(*i*) *There exists a unique (up to a multiplicative constant) second-order differential operator* □ *without a zero-order term, called the **d'Alembertian** of $\mathbb{R}^{1,d}$ (defined on*

C^2 functions on $\mathbb{R}^{1,d}$), which is $\mathrm{PSO}(1,d)$-invariant, in the sense that it satisfies $\Box(F \circ \gamma) = (\Box F) \circ \gamma$ for any $F \in C^2(\mathbb{R}^{1,d})$ and any $\gamma \in \mathrm{PSO}(1,d)$. In any Lorentz basis $\beta \equiv (\beta_0, \ldots, \beta_d) \in \mathbb{F}^d$ with corresponding coordinates $(\xi_0, \ldots, \xi_d)$, it reads

$$\Box = \frac{\partial^2}{(\partial \xi_0)^2} - \sum_{j=1}^{d} \frac{\partial^2}{(\partial \xi_j)^2}. \tag{A.5}$$

(ii) By Decomposing any $\xi \in \overset{o}{\mathcal{C}}$ according to $\xi = r\,p$ with $(r,p) \in \mathbb{R}_+^* \times \mathbb{H}^d$, we have the following decomposition of the d'Alembertian $\Box$:

$$\Box = \frac{\partial^2}{\partial r^2} + \frac{d}{r}\frac{\partial}{\partial r} - \frac{1}{r^2}\Delta, \tag{A.6}$$

where the hyperbolic Laplacian Δ operates on the p-coordinate.

Proof (i) We write, in some Lorentz basis,

$$\Box = \sum_{i,j=0}^{d} a_{ij}\frac{\partial^2}{\partial \xi_i\,\partial \xi_j} + \sum_{j=0}^{d} b_j \frac{\partial}{\partial \xi_j}.$$

Applying a matrix $\gamma \in \mathrm{PSO}(1,d)$ amounts to changing the Lorentz basis, thereby mapping the coordinate system ξ to another coordinate system $\xi' = \gamma\xi$, so that we have $\partial/\partial\xi_j = \gamma_j^k\,\partial/\partial\xi_k'$. This entails directly that the invariance of $\Box$ under γ is equivalent to the two conditions $\sum_{i,j=0}^{d} a_{ij}\gamma_k^i\gamma_\ell^j = a_{k\ell}$ and $\sum_{j=0}^{d} b_j\gamma_k^j = b_k$, for any $0 \le k,\ell \le d$. The only eigenvector b common to the whole of $\mathrm{PSO}(1,d)$ is plainly 0. Hence we are left with the symmetric matrix $a = ((a_{ij}))$, which has to satisfy $^t\gamma\,a\gamma = a$ for all $\gamma \in \mathrm{PSO}(1,d)$. Particularizing to $\varrho \in \mathrm{SO}(d)$, we see that $a' := ((a_{ij}))_{1 \le i,j \le d}$ must satisfy $^t\varrho\,a'\varrho = a'$ for all rotations, and must therefore be the negative of the unit matrix in $\mathcal{M}(d)$, up to a multiplicative constant. Hence we must have

$$a = \begin{pmatrix} a_0 & a_1 & a_2 & \cdots & a_d \\ a_1 & -1 & 0 & \cdots & 0 \\ a_2 & 0 & -1 & \cdots & 0 \\ \vdots & \vdots & \vdots & \ddots & \vdots \\ a_d & 0 & 0 & \cdots & -1 \end{pmatrix}.$$

Particularizing to the boosts θ_r then yields the conditions $a_2 = \ldots = a_d = 0$ and $(a_0 - 1)\,\mathrm{sh}\,r + 2a_1\,\mathrm{ch}\,r = 0 = (a_0 - 1)\,\mathrm{ch}\,r + 2a_1\,\mathrm{sh}\,r$, whence $a_0 - 1 = a_1 = 0$, which yields precisely the result that $a = J = \mathrm{diag}(1, -1, \ldots, -1)$ or, equivalently, that $\Box$ must read as written in eqn (A.5). Conversely, we have already noticed in Section 1.1.5 that $^t\gamma\,J\gamma = J$ holds for all $\gamma \in \mathrm{PSO}(1,d)$, so that $\Box$ satisfies the required invariance.

(ii) Consider, on $\overset{o}{\mathcal{C}}$, the canonical coordinates $(\xi_0, \ldots, \xi_d)$ of $\mathbb{R}^{1,d}$ and the alternative coordinate system

$$\left(r := \sqrt{\xi_0^2 - \xi_1^2 - \cdots - \xi_d^2},\; p_1 := \frac{\xi_1}{r}, \ldots, p_d := \frac{\xi_d}{r}\right),$$

in which we have $\mathbb{H}^d \equiv \{r = 1\}$, and $(p_1, \dots, p_d)$ are coordinates on $\mathbb{H}^d$. Performing this change of coordinates, we have

$$\frac{\partial}{\partial \xi_0} = \frac{\xi_0}{r}\left[\frac{\partial}{\partial r} - \sum_{j=1}^{d} \frac{p_j}{r}\frac{\partial}{\partial p_j}\right] \quad \text{and, for } 1 \le j \le d,$$

$$\frac{\partial}{\partial \xi_j} = \frac{1}{r}\frac{\partial}{\partial p_j} + \sum_{k=1}^{d} \frac{p_j\, p_k}{r}\frac{\partial}{\partial p_k} - p_j\frac{\partial}{\partial r}.$$

Hence

$$\frac{\partial^2}{\partial \xi_0^2} = \frac{\xi_0^2}{r^2}\frac{\partial^2}{\partial r^2} + \frac{r^2 - \xi_0^2}{r^3}\frac{\partial}{\partial r} - 2\frac{\xi_0^2}{r^3}\sum_{j=1}^{d} p_j\frac{\partial^2}{\partial p_j\,\partial r} + \frac{\xi_0^2}{r^4}\sum_{j,k=1}^{d} p_j p_k\frac{\partial^2}{\partial p_j\,\partial p_k}$$

$$- \frac{r^2 - 3\xi_0^2}{r^4}\sum_{j=1}^{d} p_j\frac{\partial}{\partial p_j},$$

and for $1 \le j \le d$,

$$\frac{\partial^2}{\partial \xi_j^2} =$$

$$= \frac{1}{r^2}\frac{\partial^2}{\partial p_j^2} + \sum_{k,\ell=1}^{d} \frac{p_j^2 p_k p_\ell}{r^2}\frac{\partial^2}{\partial p_k\,\partial p_\ell} + 2\sum_{k=1}^{d} \frac{p_j p_k}{r^2}\frac{\partial^2}{\partial p_j\,\partial p_k} + p_j^2\frac{\partial^2}{\partial r^2} - 2\frac{p_j}{r}\frac{\partial^2}{\partial p_j\,\partial r}$$

$$- 2\sum_{k=1}^{d} \frac{p_j^2 p_k}{r}\frac{\partial^2}{\partial p_k\,\partial r} + \sum_{k=1}^{d} \frac{p_k}{r^2}\frac{\partial}{\partial p_k} - \frac{1}{r}\frac{\partial}{\partial r} + 2\frac{p_j^2}{r^2}\sum_{k=1}^{d} p_k\frac{\partial}{\partial p_k} - \frac{p_j^2}{r}\frac{\partial}{\partial r} + 2\frac{p_j}{r^2}\frac{\partial}{\partial p_j}$$

$$+ \sum_{k=1}^{d} \frac{p_j^2 p_k}{r^2}\frac{\partial}{\partial p_k}.$$

Hence we obtain

$$\Box = \frac{\partial^2}{\partial r^2} + \frac{d}{r}\frac{\partial}{\partial r} - \frac{1}{r^2}\tilde{\Delta}, \quad \text{with} \quad \tilde{\Delta} := \sum_{j,k=1}^{d} (\delta_{jk} + p_j p_k)\frac{\partial^2}{\partial p_j\,\partial p_k} + d\sum_{j=1}^{d} p_j\frac{\partial}{\partial p_j}.$$

Consider now the polar coordinates (ϱ, ϕ) of $\mathbb{H}^d$, as for eqn (3.11) in Proposition 3.5.4. We have $p = (\operatorname{ch}\varrho)e_0 + (\operatorname{sh}\varrho)\phi$ and, for $1 \le j \le d$, $p_j = (\operatorname{sh}\varrho)\phi_j$. Proceeding similarly to the proof of Proposition 3.5.4, We perform a change of variables from $(p_1, \dots, p_d)$ to $(\varrho, \phi_2, \dots, \phi_d)$, in order to verify that $\tilde{\Delta} = \Delta$, as claimed. We obtain

$$\frac{\partial}{\partial p_j} = \frac{\phi_j}{\operatorname{ch}\varrho}\frac{\partial}{\partial \varrho} + \frac{1}{\operatorname{sh}\varrho}\sum_{\ell=2}^{d}(\delta_{j\ell} - \phi_j\phi_\ell)\frac{\partial}{\partial \phi_\ell},$$

whence

$$\sum_{j=1}^{d} p_j \frac{\partial}{\partial p_j} = \operatorname{th} \varrho \, \frac{\partial}{\partial \varrho}$$

and

$$\frac{\partial^2}{\partial p_j \, \partial p_k} = \frac{\phi_j \phi_k}{\operatorname{ch}^2 \varrho} \frac{\partial^2}{\partial \varrho^2} + \frac{1}{\operatorname{ch} \varrho \, \operatorname{sh} \varrho} \sum_{\ell=2}^{d} \left(\delta_{j\ell} \phi_k + \delta_{k\ell} \phi_j - 2\phi_j \phi_k \phi_\ell \right) \frac{\partial^2}{\partial \phi_\ell \, \partial r}$$

$$+ \frac{1}{\operatorname{sh}^2 \varrho} \sum_{\ell,m=2}^{d} (\delta_{j\ell} - \phi_j \phi_\ell)(\delta_{km} - \phi_k \phi_m) \frac{\partial^2}{\partial \phi_\ell \, \partial \phi_m} - \frac{\operatorname{sh} \varrho}{\operatorname{ch}^3 \varrho} \phi_j \phi_k \frac{\partial}{\partial \varrho}$$

$$- \frac{\phi_j}{\operatorname{sh}^2 \varrho} \sum_{\ell=2}^{d} (\delta_{k\ell} - \phi_k \phi_\ell) \frac{\partial}{\partial \phi_\ell} + \frac{1}{\operatorname{ch} \varrho \, \operatorname{sh} \varrho} \Big((1 - \delta_{1k})(\delta_{jk} - \phi_j \phi_k)$$

$$+ \delta_{1k}(\delta_{1j} - \phi_1^2) \frac{\phi_j}{\phi_1} \Big) \frac{\partial}{\partial \varrho} - \frac{1}{\operatorname{sh}^2 \varrho} \sum_{\ell=2}^{d} \Big((1 - \delta_{1k})(\delta_{jk} - \phi_j \phi_k) \phi_\ell$$

$$+ (\delta_{j\ell} - \phi_j \phi_\ell) \phi_k + \delta_{1k}(\delta_{1j} - \phi_1^2) \frac{\phi_j \phi_\ell}{\phi_1} \Big) \frac{\partial}{\partial \phi_\ell},$$

whence

$$\tilde{\Delta} = \frac{\partial^2}{\partial \varrho^2} + (d-1) \coth \varrho \, \frac{\partial}{\partial \varrho} + \frac{1}{\operatorname{sh}^2 \varrho} \left[\sum_{j,k=2}^{d} (\delta_{jk} - \phi_j \phi_k) \frac{\partial^2}{\partial \phi_j \, \partial \phi_k} - (d-1) \sum_{k=2}^{d} \phi_k \frac{\partial}{\partial \phi_k} \right],$$

which is indeed the expression for Δ in these coordinates, as we saw in the proof of Proposition 3.5.4. $\diamond$

A.4 Core–cusp decomposition

Here we prove the core–cusp decomposition theorem (Theorem A.4.5) in full generality, i.e., for a geometrically finite and cofinite Kleinian group Γ. Recall that it allowed us to establish the Poincaré inequality of Theorem 5.4.2.5. In fact, we shall reduce it to the following result on discrete subgroups of Euclidean isometries; see for example [Rac], Section 5.4.

Theorem A.4.1 *Any discrete subgroup Γ of Euclidean isometries in $\mathbb{R}^d$ has a free Abelian normal subgroup $\tilde{\Gamma}$ of finite index, containing all translations in Γ, and acting properly as a group of translations on some affine subspace of $\mathbb{R}^d$ whose dimension is the rank of $\tilde{\Gamma}$.*

We have already mentioned (in Corollary 4.4.1.6 and page 100) that a boundary point which is fixed by some parabolic element of a given Kleinian group Γ is called parabolic. The following is also classical.

Definition A.4.2 *A **parabolic (sub)group** is a non-trivial Kleinian (sub)group that fixes a parabolic point, and no other point in $\overline{\mathbb{H}^d} = \mathbb{H}^d \cup \partial \mathbb{H}^d$.*

The group $\mathcal{S}_\eta := \{h \in \mathrm{PSO}(1,d) | h(\eta) = \eta\}$ of hyperbolic isometries fixing a given light ray $\eta \in \partial \mathbb{H}^d$ is conjugate to the group of Euclidean similarities of $\mathbb{R}^{d-1}$. Any Kleinian group fixing η is conjugate to a group made up of Euclidean similarities, acting discontinuously (in the sense of Lemma 4.2.1) on $\mathbb{R}^{d-1}$. More precisely, we have the following.

Proposition A.4.3 *Consider $\eta \in \partial \mathbb{H}^d$ and any reference Lorentz frame $\beta \in \mathbb{F}^d(\eta)$, which fixes a Poincaré model for $\mathbb{H}^d$ by identification of the current point $q = \pi_0(\beta\, T_{x,y}) \in \mathbb{H}^d$ with its Poincaré coordinates $(x,y) \in \mathbb{R}^{d-1} \times \mathbb{R}_+^*$ (with respect to β). Then any hyperbolic isometry $g \in \mathrm{PSO}(1,d)$ fixes η if and only if it can be identified with a Euclidean similarity of the Poincaré upper half-space $\mathbb{R}^{d-1} \times \mathbb{R}_+^*$.*

Proof We denote by (x_j, y_j) $(1 \le j \le 2)$ the Poincaré coordinates (with respect to β) of $q_j \in \mathbb{H}^d$, and by (x_j', y_j') those of $q_j' := g(q_j)$. We use Proposition 2.1.3 and the beginning of its proof. We have $y_j = \langle q_j, \eta_{\beta_0} \rangle^{-1}$ and $y_j' = \langle q_j', \eta_{\beta_0} \rangle^{-1} = \langle q_j, g^{-1}(\eta_{\beta_0}) \rangle^{-1}$.

Suppose that $g(\eta) = \eta$. We then have $g^{-1}(\eta_{\beta_0}) = \eta_{g^{-1}(\beta_0)} = \langle g(\beta_0), \eta_{\beta_0} \rangle \eta_{\beta_0}$, whence $y_j' = \langle g(\beta_0), \eta_{\beta_0} \rangle^{-1} y_j$. By applying eqn (2.1) to dist (q_1, q_2) and dist (q_1', q_2'), which are equal, and replacing y_j' as computed above, we obtain at once $|x_1' - x_2'| = \langle g(\beta_0), \eta_{\beta_0} \rangle^{-1} |x_1 - x_2|$. This proves that g can indeed be identified with a similarity, having dilatation coefficient $\langle g(\beta_0), \eta_{\beta_0} \rangle^{-1}$.

Conversely, it is clear from eqn (2.1) that a similarity of $\mathbb{R}^{d-1} \times \mathbb{R}_+^*$ defines a hyperbolic isometry fixing η. $\diamond$

We have already met the notion of a cusp in Corollary 4.4.1.6.

Definition A.4.4 *A **cusp** of a Kleinian group Γ is a conjugation class (in Γ) of a maximal parabolic subgroup of Γ, so that the set of cusps of Γ is in one-to-one correspondence with the set of Γ-inequivalent parabolic points, and can be identified with it. The **rank** of a cusp is the rank of the corresponding parabolic subgroup (that is, the rank of any free Abelian subgroup having finite index, according to Theorem A.4.1).*

Given a fundamental polyhedron $\mathcal{P}$ of a Kleinian group Γ, each cusp η of Γ is represented by an ideal vertex of $\mathcal{P}$, say $v \in \partial \mathcal{P} \cap \partial \mathbb{H}^d$. Since we can suppose that $\mathcal{P}$ has only a finite number of vertices (recall that we are considering only geometrically finite Kleinian groups Γ), it is always possible to associate with each vertex v an open horoball B_v (based at v), intersecting only the sides of $\mathcal{P}$ incident to v, in such a way that these different horoballs are pairwise disjoint. Then, necessarily, if v represents a cusp η, $\mathcal{P} \cap B_v$ is a fundamental domain of the stabilizer Γ_η of η acting on B_v (this should be clear, using Proposition A.4.3). This implies that η is represented by a unique ideal vertex of $\mathcal{P}$, which we identify henceforth with η; accordingly, we shall write $B_v = B_\eta$.

For any given cusp η, we fix a reference frame $\beta \in \mathbb{F}^d(\eta)$ such that $\beta_0 \in \partial H_\eta$, and consider the corresponding Poincaré coordinates (x, y). This amounts to choosing a Poincaré model $\mathbb{R}^{d-1} \times \mathbb{R}_+^*$, such that $\eta \equiv \infty$ and $\partial B_\eta \equiv \{y = 1\}$.

We can describe (of course, up to a hyperbolic isometry) the shape of the solid cusp $\mathcal{P} \cap B_\eta$ (recall Section 5.4.2, page 117). In fact, the sides of $\mathcal{P}$ incident to η are now made up of vertical geodesics, so that we have in this model simply $\mathcal{P} \cap B_\eta = P \times]1, \infty[$, where $P := \mathcal{P} \cap \partial B_\eta$ is the bottom of the solid cusp $\mathcal{P} \cap B_\eta$. Moreover, P has also to be a fundamental domain of the stabilizer Γ_η of η in Γ, to which we apply Theorem A.4.1 now. Thus, denoting by $k \in \{1, \ldots, d-1\}$ the rank of η, we obtain a lattice group $\mathbb{Z}^k$ (conjugate to a finite-index subgroup of Γ_η), which is generated by k independent translations in the horosphere $\{y = 1\} \equiv \mathbb{R}^{d-1}$. Now, this means precisely that P is a finite quotient of a fundamental domain P' of $\mathbb{Z}^k$, which is a parallelepiped in $\{y = 1\}$. Finally, according to the expression for the hyperbolic metric in the Poincaré half-space model $\mathbb{R}^{d-1} \times \mathbb{R}^*_+$ (recall Proposition 2.1.3), the solid cusp $\mathcal{P} \cap B_\eta$ has a finite volume if and only if P has, and then if and only if P' has, and hence if and only if η has full rank $k = d - 1$.

Now, it turns out that all boundary points belonging to $\partial P \cap \partial \mathbb{H}^d$ either are cusps (recall that above, we identified a cusp with the unique ideal vertex of $\mathcal{P}$ representing it) or are not limit points of (any orbit of) Γ, as is guaranteed by [Rac], Theorem 12.3.4. Then, since the set $O(\Gamma)$ of ordinary (non-limit) points is open in $\partial \mathbb{H}^d$, it turns out that any point of $P \cap O(\Gamma)$ is Γ-equivalent to another point which belongs to the closure of an open subset of $\partial P \cap \partial \mathbb{H}^d$. This latter corresponds to a funnel (recall Remark 4.4.1.9), responsible for an infinite volume. Hence, clearly, if Γ is cofinite then $\Gamma \backslash \mathbb{H}^d$ cannot have a funnel, so that in this case all boundary points belonging to $\partial P \cap \partial \mathbb{H}^d$ are cusps.

Hence, once the finite set of horoballs B_η associated with the cusps as above have been removed from the fundamental polyhedron $\mathcal{P}$, $\mathcal{P}$ reduces to its so-called core, which is a relatively compact subset of $\mathbb{H}^d \cup O(\Gamma)$, and therefore a compact subset of $\mathbb{H}^d$ since Γ is cofinite. Therefore, we have the following theorem; see also [Rac], Theorem 12.6.6 and Corollary 4, and [B1], GF3 $\Rightarrow$ GF1.

Theorem A.4.5 *A convex fundamental polyhedron $\mathcal{P}$ of a cofinite and geometrically finite Kleinian group Γ is the disjoint union of a finite number of solid cusps, which are the intersections of $\mathcal{P}$ with open horoballs based at the cusps bounding $\mathcal{P}$, and of a* **core**, *which is a compact subset of $\mathbb{H}^d$. Furthermore, any solid cusp is the quotient by some finite subgroup of Γ of a solid cusp isometric to $P' \times]1, \infty[$, where $P' \subset \mathbb{R}^{d-1}$ is the compact fundamental parallelepiped of some lattice isomorphic to $\mathbb{Z}^{d-1}$.*

Appendix B
Stochastic calculus

B.1 A simple construction of a real Brownian motion

In order to construct a real Brownian motion simply and thereby justify Definition 6.2.1, let us consider the Haar basis of $L^2([0,1])$,

$$\varphi_{k,j} := 2^{k/2}\left(1_{\left[(j-1)2^{-k},(j-1/2)2^{-k}\right]} - 1_{\left[(j-1/2)2^{-k},j2^{-k}\right]}\right), \text{ for } k \in \mathbb{N}, 1 \le j \le 2^k.$$

We denote by $\phi_{k,j} := \int_0 \varphi_{k,j}$ the primitive of $\varphi_{k,j}$ which vanishes at 0. We thus have

$$\phi_{k,j}(t) = 2^{-k/2}\Phi(2^k t - j + 1), \text{ with}$$

$$(\forall u \in \mathbb{R}) \ \Phi(u) := 1_{[0,1/2]}(u)u + 1_{]1/2,1]}(u)(1-u).$$

Note that for a fixed $k \in \mathbb{N}$, the supports $\mathrm{Supp}(\phi_{k,j}) = \mathrm{Supp}(\varphi_{k,j}) = [(j-1)2^{-k}, j2^{-k}]$ have pairwise disjoint interiors, for $1 \le j \le 2^k$. Hence, the following series converges uniformly on $[0,1]$:

$$\sum_{k,j} \phi_{k,j}^2(t) = \sum_{k\in\mathbb{N}}\sum_{j=1}^{2^k} \phi_{k,j}^2(t) \le \sum_{k\in\mathbb{N}} 2^{-k} \times \frac{1}{4} = \frac{1}{2}.$$

Let $\{\xi_0\} \cup \{\xi_{k,j} \,|\, k \in \mathbb{N}, 1 \le j \le 2^k\}$ be a sequence of independent identically distributed $\mathcal{N}(0,1)$ (i.e., normalized centred Gaussian) random variables, and set

$$B_t := \xi_0 t + \sum_{k\in\mathbb{N}}\sum_{j=1}^{2^k} \xi_{k,j}\phi_{k,j}(t), \qquad \text{for any } 0 \le t \le 1. \tag{B.1}$$

We now verify that this series provides the wanted process.

Proposition B.1.1

    (i) *Almost surely, the series (B.1) converges uniformly on $[0,1]$, defining thereby an almost surely continuous process $(B_t)_{0\le t\le 1}$, vanishing at 0.*

    (ii) *For $0 \le s \le t \le 1$, the random variable $(B_t - B_s)$ is Gaussian and centred, with variance $(t-s)$.*

    (iii) *For $0 = t_0 \le t_1 \le \ldots \le t_n \le 1$, the increments $(B_{t_1} - B_{t_0}), \ldots, (B_{t_n} - B_{t_{n-1}})$ are independent (for any $n \in \mathbb{N}^*$).*

Proof (i) We set

$$\beta_k := \sup_{0\le t\le 1}\left|\sum_{j=1}^{2^k} \xi_{k,j}\phi_{k,j}(t)\right| \le \sup_{1\le j\le 2^k}|\xi_{k,j}| \times 2^{-k/2-1}.$$

We have

$$\mathbb{P}\left[\beta_k^2 > 2^{-k} \log 2^k\right] \leq \sum_{j=1}^{2^k} \mathbb{P}\left[|\xi_{k,j}| > \sqrt{4k \log 2}\right]$$

$$= 2^{k+1}(2\pi)^{-1/2} \int_{\sqrt{4k \log 2}}^{\infty} e^{-x^2/2} \, dx \leq 2^{-k},$$

whence $\sum_{k \in \mathbb{N}} \beta_k < \infty$ almost surely, by the Borel–Cantelli lemma, and then we have the wanted almost sure uniform convergence. The almost sure continuity follows at once.

(ii) By (i), the choice of the independent identically distributed sequence $\{\xi_{k,j}\}$ and the convergence of the series $\sum_{k,j} \phi_{k,j}^2$, for any real α we have

$$\mathbb{E}\left[e^{\sqrt{-1} \, \alpha \, (B_t - B_s)}\right] = \lim_{N \to \infty} \exp\left[-\frac{\alpha^2}{2}\left((t-s)^2 + \sum_{k=0}^{N}\sum_{j=1}^{2^k}\left(\phi_{k,j}(t) - \phi_{k,j}(s)\right)^2\right)\right].$$

Now, since $\{1\} \cup \{\varphi_{k,j} | k \in \mathbb{N}, 1 \leq j \leq 2^k\}$ is a complete orthonormal system in $L^2([0,1])$, by Parseval's formula we have

$$(t-s)^2 + \sum_{k \in \mathbb{N}}\sum_{j=1}^{2^k}\left(\phi_{k,j}(t) - \phi_{k,j}(s)\right)^2 = \left[\int_0^1 1_{[s,t]}\right]^2 + \sum_{k \in \mathbb{N}}\sum_{j=1}^{2^k}\left[\int_0^1 1_{[s,t]}\varphi_{k,j}\right]^2$$

$$= \int_0^1 1_{[s,t]} = t - s.$$

Hence we obtain $\mathbb{E}\left[e^{\sqrt{-1} \, \alpha(B_t - B_s)}\right] = e^{-\alpha^2(t-s)/2}$, as wanted.

(iii) As in (ii), setting $h := \sum_{\ell=1}^{n} \alpha_\ell 1_{[t_{\ell-1}, t_\ell]}$ for any real $\alpha_1, \ldots, \alpha_n$, we have

$$\sum_{\ell=1}^{n} \alpha_\ell(B_{t_\ell} - B_{t_{\ell-1}}) = \xi_0 \int_0^1 h + \sum_{k \in \mathbb{N}}\sum_{j=1}^{2^k} \xi_{k,j} \int_0^1 h\phi_{k,j},$$

and then

$$\mathbb{E}\left[e^{\sqrt{-1} \sum\limits_{\ell=1}^{n} \alpha_\ell(B_{t_\ell} - B_{t_{\ell-1}})}\right] = \exp\left[-\frac{1}{2}\left(\left[\int_0^1 h\right]^2 + \sum_{k \in \mathbb{N}}\sum_{j=1}^{2^k}\left[\int_0^1 h\varphi_{k,j}\right]^2\right)\right] = e^{-\int_0^1 h^2/2}$$

$$= \exp\left[-\sum_{\ell=1}^{n} \frac{\alpha_\ell^2(t_\ell - t_{\ell-1})}{2}\right] = \prod_{\ell=1}^{n} \mathbb{E}\left[e^{\sqrt{-1} \, \alpha_\ell(B_{t_\ell} - B_{t_{\ell-1}})}\right],$$

which concludes the proof. $\diamond$

Note   The process $(B_t)_{0 \leq t \leq 1}$ of eqn (B.1) and Proposition B.1.1 is a realization of a real Brownian motion on $[0,1]$. If the series in eqn (B.1) is considered without the additional term $\xi_0 t$, the resulting process is the real **Brownian bridge** on $[0,1]$.

To complete the construction of a real Brownian motion on the whole of $\mathbb{R}_+$, we consider now a sequence of independent processes $(B_t^n)_{0 \le t \le 1, n \in \mathbb{N}}$ as in Proposition B.1.1, which is obtained at once by using a sequence of independent reduced centred Gaussian variables

$$\{\xi_0^n | n \in \mathbb{N}\} \cup \{\xi_{k,j}^n | k, n \in \mathbb{N}, 1 \le j \le 2^k\}.$$

We define the process $(B_t)_{t \ge 0}$ by

$$B_t := \sum_{0 \le n \le t-1} B_1^n + B_{t-[t]}^{[t]}, \quad \text{where } [t] := \max\{n \in \mathbb{N} | n \le t\}. \tag{B.2}$$

It is immediately apparent from Proposition B.1.1 that this defines an almost surely continuous process $(B_t)_{t \ge 0}$, which has independent increments and is such that the random variable $(B_t - B_s)$ is Gaussian, and centred, with variance $(t - s)$, for any $0 \le s \le t$.

This completes the construction.

Remark B.1.2 The same construction as above, using, instead of the Haar basis, any uniformly bounded complete orthonormal system $\{\varphi_k | k \in \mathbb{N}^*\}$ of

$$L_0^2([0,1], \mathbb{R}) = \left\{ f \in L^2([0,1], \mathbb{R}) \middle| \int_0^1 f(t)\, dt = 0 \right\},$$

the primitive ϕ_k of φ_k which vanishes at 0 and 1, and a sequence $\{\xi_k | k \in \mathbb{N}\}$ of independent $\mathcal{N}(0,1)$ random variables, will yield a standard real Brownian motion on $[0,1]$ as well, by setting

$$B_t := \xi_0 t + \sum_{k \in \mathbb{N}^*} \xi_k \phi_k(t) \quad \text{for all } t \in [0,1].$$

In particular, taking the trigonometric sequence, we get the following Fourier expansion of the real Brownian motion on $[0,1]$: almost surely,

$$B_t = \xi_0 t + \frac{1}{\pi\sqrt{2}} \sum_{k \in \mathbb{N}^*} \frac{1}{k} \left(\xi_{2k} \left(1 - \cos[2\pi kt]\right) + \xi_{2k-1} \sin[2\pi kt] \right), \quad \text{for any } 0 \le t \le 1.$$

Another similar Fourier expansion is

$$B_t = \xi_0 t + \frac{\sqrt{2}}{\pi} \sum_{k \in \mathbb{N}^*} \xi_k \frac{\sin(\pi kt)}{k}, \quad \text{for any } 0 \le t \le 1.$$

Exercise   Justify this last expression (use Proposition 6.2.2 and compute

$$\sum_{k \ge 1} \frac{\sin(\pi ks) \sin(\pi kt)}{k^2},$$

by expanding $t \mapsto t(1-t)$ $\left(\text{or } t \mapsto \min\{s,t\} - st\right)$ in Fourier series on $[0,1]$).

Another classical way of obtaining a Brownian motion is as a limit in law of a random walk, as in the following theorem. See, for example, [RY], Theorem XIII.1.9, [RW1], Theorem I.8.2, and [Bi], Theorem 37.8.

Theorem B.1.3 *(Donsker) Let* $S_k = X_1 + \cdots + X_k$ *be a random walk on* $\mathbb{Z}$, *with elementary step* $X_j \in L^2$, *centred with variance* σ^2. *For* $n \in \mathbb{N}^*$ *and* $t \in \mathbb{R}_+$, *set*

$$S_t^n := \frac{1}{\sigma\sqrt{n}} \left(S_{[nt]} + \left(nt - [nt] \right) X_{[nt]+1} \right).$$

There exist a real Brownian motion (B_t) *and a sequence of processes* $(\tilde{S}^n)$ *such that (i) for each* n, $\tilde{S}^n$ *and* S^n *have the same law; (ii) for any* $T \in \mathbb{R}_+$, *we almost surely have* $\lim_{n\to\infty} \sup_{0\le t\le T} \left| \tilde{S}_t^n - B_t \right| = 0$.

B.2   Chaos expansion

We consider here **multiple Itô integrals**, which are iterated stochastic Itô integrals with respect to an $\mathbb{R}^k$-valued standard Brownian motion $W_t = (W_t^1, \ldots, W_t^k)$, of the form

$$\int_0^s dW_{s_1}^{j_1} \left[\int_0^{s_1} dW_{s_2}^{j_2} \left[\cdots \left[\int_0^{s_{n-1}} dW_{s_n}^{j_n} \, F_s(s_1, \ldots, s_n) \right] \cdots \right] \right]$$

(where $1 \le j_1, \ldots, j_n \le k$, $s > 0$, and F_s is any $\mathcal{M}(d)$-valued deterministic function, square integrable on $\{(s_1, \ldots, s_n) \in \mathbb{R}^n \, | \, 0 < s_n < \ldots < s_1 < s\}$), which we shall write in the more pleasant form

$$\int_{0<s_n<\ldots<s_1<s} F_s(s_1, \ldots, s_n) \, dW_{s_1}^{j_1} \ldots dW_{s_n}^{j_n}.$$

Lemma B.2.1 *The above multiple Itô integrals are pairwise orthogonal in* L^2. *More precisely, writing* $F_s \equiv F_s(s_1, \ldots, s_n), H_s \equiv H_s(s_1, \ldots, s_m)$,

$$FW_s^j := \int_{0<s_n<\ldots<s_1<s} F_s \, dW_{s_1}^{j_1} \ldots dW_{s_n}^{j_n},$$

$$HW_s^i := \int_{0<s_m<\ldots<s_1<s} H_s \, dW_{s_1}^{i_1} \ldots dW_{s_m}^{i_m}$$

and $\phi_s^{j,i} := FW_s^j \times HW_s^i$, *we have*

$$\mathbb{E}\left(\phi_s^{j,i}\right) = 1_{\{n=m\}} 1_{\{(j_1,\ldots,j_n)=(i_1,\ldots,i_n)\}} \times \int_{0<s_n<\ldots<s_1<s} F_s H_s \, ds_1 \ldots ds_n.$$

Proof   Using the Itô isometry identity (6.2) and setting $dW_{s_2}^{j_2} \ldots dW_{s_n}^{j_n} =: dW_{s_2\ldots s_n}^{j_2\ldots j_n}$, we have

$$\mathbb{E}\big(\phi_s^{j,i}\big) = \mathbb{E}\left[\int_0^s \left(\int_{0<s_n<\ldots<s_1} F_s \, dW_{s_2\ldots s_n}^{j_2\ldots j_n}\right) dW_{s_1}^{j_1}\right.$$

$$\times \left.\int_0^s \left(\int_{0<s_m<\ldots<s_1} H_s \, dW_{s_2\ldots s_m}^{i_2\ldots i_m}\right) dW_{s_1}^{i_1}\right]$$

$$= 1_{\{j_1=i_1\}} \int_0^s \mathbb{E}\left[\int_{0<s_n<\ldots<s_1} F_s \, dW_{s_2\ldots s_n}^{j_2\ldots j_n} \times \int_{0<s_n<\ldots<s_1} H_s \, dW_{s_2\ldots s_m}^{i_2\ldots i_m}\right] ds_1.$$

By induction, this leads for $n \leq m$ to

$$E\big(\phi_s^{j,i}\big) = 1_{\big\{(j_1,\ldots,j_n)=(i_1,\ldots,i_n)\big\}}$$

$$\times \int_{0<s_n<\ldots<s_1<s} \mathbb{E}\left[\int_{0<s_m<\ldots<s_{n+1}<s_n} F_s H_s \, dW_{s_{n+1}\ldots s_m}^{i_{n+1}\ldots i_m}\right] ds_1 \ldots ds_n,$$

which vanishes for $n < m$. $\diamond$

Theorem B.2.2 *For any $s \geq 0$, set $\Lambda_s := \exp\left[s\left(\frac{1}{2}\sum_{j=1}^k A_j^2 + A_0\right)\right]$. Then the solution (X_s) to the linear SDE (7.1) is given by the* **Wiener chaos expansion**

$$X_s = \Lambda_s + \sum_{n\geq 1} \sum_{1\leq j_1,\ldots,j_n \leq k} \int_{0<s_n<\ldots<s_1<s} \Lambda_{s_n} A_{j_n} \Lambda_{s_{n-1}-s_n} \cdots$$

$$A_{j_2}\Lambda_{s_1-s_2} A_{j_1} \Lambda_{s-s_1} \, dW_{s_1}^{j_1} \ldots dW_{s_n}^{j_n},$$

valid for any $s \geq 0$. This series converges in the L^2-norm, uniformly for bounded s, and all multiple Itô integrals which constitute it are pairwise orthogonal in L^2.

Proof Set $W_t^A := \sum_{j=1}^k A_j W_t^j$ and $B := \left(\frac{1}{2}\sum_{j=1}^k A_j^2 + A_0\right)$, and consider $M_t := X_t \Lambda_{s-t}$, for fixed $s > 0$. Then, by Itô's formula and eqn (7.1), we have

$$dM_t = -X_t B\Lambda_{s-t} \, dt + X_t(dW_t^A + B \, dt)\Lambda_{s-t} = X_t \, dW_t^A \, \Lambda_{s-t},$$

so that

$$(*) \qquad X_s - \Lambda_s = M_s - M_0 = \int_0^s dM_{s_1} = \int_{0<s_1<s} X_{s_1} \, dW_{s_1}^A \, \Lambda_{s-s_1}.$$

Iterating this, we obtain by induction

$$X_s - \Lambda_s = \sum_{n=1}^{N-1} \mathcal{J}_n^A(s) + \mathcal{J}_N^A(s)x \qquad \text{for any } N \geq 1,$$

where

$$\mathcal{J}_n^A(s) := \int_{0<s_n<\ldots<s_1<s} \Lambda_{s_n} \, dW_{s_n}^A \, \Lambda_{s_{n-1}-s_n} \cdots dW_{s_2}^A \, \Lambda_{s_1-s_2} \, dW_{s_1}^A \, \Lambda_{s-s_1}$$

$$= \sum_{1 \le j_1,\ldots,j_n \le k} \int_{0<s_n<\ldots<s_1<s} \Lambda_{s_n} A_{j_n} \Lambda_{s_{n-1}-s_n} \cdots A_{j_2} \Lambda_{s_1-s_2} A_{j_1} \Lambda_{s-s_1} \, dW_{s_1}^{j_1} \ldots dW_{s_n}^{j_n},$$

and where the remainder $\mathcal{J}_n^A(s)_X$ is obtained by replacing the term Λ_{s_n} by X_{s_n} in the above integral defining $\mathcal{J}_n^A(s)$.

Indeed, for $N = 1$, this is the above formula $(*)$, and if this holds for some $N \ge 1$, then, using formula $(*)$ again, we obtain

$$\mathcal{J}_N^A(s)_X = \int_{0<s_N<\ldots<s_1<s} \left[\Lambda_{s_N} + \int_{0<s_{N+1}<s_N} X_{s_{N+1}} \, dW_{s_{N+1}}^A \, \Lambda_{s_N-s_{N+1}} \right]$$

$$dW_{s_N}^A \, \Lambda_{s_{N-1}-s_N} \ldots \Lambda_{s-s_1} = \mathcal{J}_N^A(s) + \mathcal{J}_{N+1}^A(s)_X.$$

Then, using Lemma B.2.1, Lemma 7.2.12, and Remark 7.2.13 repeatedly, we have

$$\mathbb{E}\left[\left\| \mathcal{J}_n^A(s) \right\|_{HS}^2 \right]$$

$$= \sum_{1 \le j_1,\ldots,j_n \le k} \int_{0<s_n<\ldots<s_1<s} \left\| \Lambda_{s_n} A_{j_n} \Lambda_{s_{n-1}-s_n} \cdots A_{j_2} \Lambda_{s_1-s_2} A_{j_1} \Lambda_{s-s_1} \right\|_{HS}^2 \, ds_n \ldots ds_1$$

$$\le \sum_{1 \le j_1,\ldots,j_n \le k} \| A_{j_1} \|_{HS}^2 \cdots \| A_{j_n} \|_{HS}^2 \int_{0<s_n<\ldots<s_1<s} \| \Lambda_{s_n} \|_{HS}^2 \cdots \| \Lambda_{s-s_1} \|_{HS}^2 \, ds_n \ldots ds_1.$$

Observe that, using Lemma 7.2.12, for any $s \ge 0$ we have

$$\| \Lambda_s \|_{HS} \le \sum_{n \ge 0} \frac{s^n}{n!} \| B^n \|_{HS} \le e^{s \| B \|_{HS}} \le \exp\left[\left(\| A_0 \|_{HS} + \frac{1}{2} \sum_{j=1}^k \| A_j \|_{HS}^2 \right) s \right],$$

whence

$$\mathbb{E}\left[\left\| \mathcal{J}_n^A(s) \right\|_{HS}^2 \right] \le \sum_{1 \le j_1,\ldots,j_n \le k} \| A_{j_1} \|_{HS}^2 \cdots \| A_{j_n} \|_{HS}^2 \frac{s^n}{n!} e^{2s \, \| B \|_{HS}}$$

$$= \frac{s^n}{n!} \left[\sum_{j=1}^k \| A_j \|_{HS}^2 \right]^n e^{2s \, \| B \|_{HS}}.$$

This proves the convergence of the series in the L^2-norm, uniformly for bounded s.

Finally, we deduce from the preceding and from Lemma 7.2.12 the following control of the remainder in the induction formula:

$$\mathbb{E}\left[\left\|\mathcal{J}_N^A(s)x\right\|_{HS}^2\right]$$

$$\leq \left[\sum_{j=1}^k \|A_j\|_{HS}^2\right]^N \int_{0<s_N<\ldots<s_1<s} e^{2\,\|B\|_{HS}(s-s_N)}\mathbb{E}\left[\|X_{s_N}\|_{HS}^2\right]\,ds_N\ldots ds_1$$

$$\leq \frac{d}{N!}\left[\sum_{j=1}^k \|A_j\|_{HS}^2 \times s\right]^N \times \exp\left[2\Big(\|A_0\|_{HS} + \sum_{j=1}^k \|A_j\|_{HS}^2\Big)s\right] \xrightarrow[N\to\infty]{} 0. \qquad \diamondsuit$$

B.3 Brownian path and limiting geodesic

We saw in Proposition 7.6.7.3 that the hyperbolic Brownian motion (z_s) (starting at $p \in \mathbb{H}^d$) converges almost surely (as $s \to \infty$) to a boundary point $z_\infty \in \partial\mathbb{H}^d$, the law of which is the harmonic measure μ_p. Here we specify the asymptotic proximity between the typical Brownian path and its almost sure limiting geodesic (through p, ending at z_∞).

Let us begin with the disintegration of the law of the hyperbolic Brownian motion with respect to its end point $z_\infty \in \partial\mathbb{H}^d$.

Proposition B.3.1 *The law of the hyperbolic Brownian motion (Z_s) starting at $Z_0 = p \in \mathbb{H}^d$ disintegrates with respect to its end point $z_\infty \in \partial\mathbb{H}^d$ as follows: for any positive s and bounded measurable functions φ on $\partial\mathbb{H}^d$ and F on the space of trajectories (starting from p) $\mathcal{C}_p([0,s],\mathbb{H}^d)$, we have*

$$\mathbb{E}_p\left[F(z_{[0,s]})\varphi(z_\infty)\right] = \int_{\partial\mathbb{H}^d} \mathbb{E}_{e_0}\left[F \circ g_\eta^{-1}\big(\hat{z}_{[0,s]}\big)\right]\varphi(\eta)\mu_{e_0}(d\eta),$$

where $g_\eta \in \mathrm{PSO}(1,d)$ denotes any isometry such that $g_\eta p = e_0$ and $g_\eta \eta = e_0 + e_1$, and where $(\hat{z}_s)_{s\geq 0}$, whose law is described in Poincaré coordinates $(\hat{x}_s, \hat{y}_s)$ (using a standard Euclidean Brownian motion $(W_t^1, W_t') \in \mathbb{R} \times \mathbb{R}^{d-1}$) by

$$\hat{y}_s = e^{W_s^1 + (d-1)s/2} \quad and \quad \hat{x}_s = \int_0^s \hat{y}_t\,dW_t',$$

denotes the hyperbolic Brownian motion conditioned to exit $\mathbb{H}^d$ at $\mathbb{R}_+(e_0 + e_1) \in \partial\mathbb{H}^d$. The latter Brownian motion is also the process already encountered as associated with the dual process $(Z_s^* = T_{\hat{z}_s})$ of Section 8.1.

Proof Using the fact that the law of z_∞ conditionally on z_s is the harmonic measure μ_{z_s}, Fubini's theorem and Proposition 3.6.1.5, we have the following generalization of [F3], Section 14.1, Lemma 8:

$$\mathbb{E}_p\big[F(z_{[0,s]})\varphi(z_\infty)\big] = \int_{\partial\mathbb{H}^d} \mathbb{E}_p\Big[F(z_{[0,s]})\varphi(\eta)\mu_{z_s}(d\eta)\Big]$$

$$= \int_{\partial\mathbb{H}^d} \mathbb{E}_p\Big[F(z_{[0,s]})\langle z_s, \eta_p\rangle^{1-d}\Big]\varphi(\eta)\mu_p(d\eta)$$

$$= \int_{\partial\mathbb{H}^d} \mathbb{E}_p\Big[F(z_{[0,s]})e^{(d-1)B_\eta(p,z_s)}\Big]\varphi(\eta)\mu_p(d\eta).$$

This means that the law of the hyperbolic Brownian motion starting from p, conditionally on the event that its end point is $z_\infty \in \partial\mathbb{H}^d$, admits the density $\langle z_s, (z_\infty)_p\rangle^{1-d} = e^{(d-1)B_{z_\infty}(p,z_s)}$ with respect to its unconditioned law.[1]

By the invariance of the Brownian law under the isometry $g_\eta \in \mathrm{PSO}(1,d)$ (chosen as in the statement), we then obtain

$$\mathbb{E}_p\Big[F(z_{[0,s]})e^{(d-1)B_\eta(p,z_s)}\Big] = \mathbb{E}_{e_0}\Big[F\big(g_\eta^{-1}z_{[0,s]}\big)e^{(d-1)B_\eta(p,g_\eta^{-1}z_s)}\Big]$$

$$= \mathbb{E}_{e_0}\Big[F\circ g_\eta^{-1}\big(z_{[0,s]}\big)e^{(d-1)B_{g_\eta\eta}(g_\eta p,z_s)}\Big] = \mathbb{E}_{e_0}\Big[F\circ g_\eta^{-1}\big(z_{[0,s]}\big)e^{(d-1)B_{e_0+e_1}(e_0,z_s)}\Big]$$

$$= \mathbb{E}_{e_0}\Big[F\circ g_\eta^{-1}\big(z_{[0,s]}\big)\langle z_s, e_0+e_1\rangle^{1-d}\Big].$$

Otherwise, denoting by (x_s, y_s) the Poincaré coordinates of (z_s) and using eqn (1.18), we have

$$z_s = \Big(\frac{y_s^2 + |x_s|^2 + 1}{2y_s}\Big)e_0 + \Big(\frac{y_s^2 + |x_s|^2 - 1}{2y_s}\Big)e_1 + \sum_{j=2}^d \frac{x_s^j}{y_s}e_j,$$

whence $\langle z_s, e_0 + e_1\rangle = y_s^{-1}$. So far, we have obtained

$$\mathbb{E}_p\big[F(z_{[0,s]})\varphi(z_\infty)\big] = \int_{\partial\mathbb{H}^d} \mathbb{E}_{e_0}\Big[F\circ g_\eta^{-1}\big(z_{[0,s]}\big)y_s^{d-1}\Big]\varphi(\eta)\mu_{e_0}(d\eta)$$

$$= \int_{\partial\mathbb{H}^d} \mathbb{E}_{e_0}\Big[F\circ g_\eta^{-1}\big(\hat{z}_{[0,s]}\big)\Big]\varphi(\eta)\mu_{e_0}(d\eta),$$

where $(\hat{z}_s)$ denotes the hyperbolic Brownian motion conditioned to exit $\mathbb{H}^d$ at $\mathbb{R}_+(e_0 + e_1)$.

Denoting by $(\hat{x}_s, \hat{y}_s)$ its Poincaré coordinates, we deduce from the above that

$$\mathbb{E}_{(x,y)}[f(\hat{x}_s, \hat{y}_s)] = \mathbb{E}_{(x,y)}\Big[f(x_s, y_s)\Big(\frac{y_s}{y}\Big)^{d-1}\Big] = y^{1-d}P_s\big(y^{d-1}f\big)(x,y) = P_s^* f(x,y),$$

according to Remark 8.1.1(ii). Hence the law of $(\hat{z}_s)$ is computed in Poincaré coordinates $(\hat{x}_s, \hat{y}_s)$ by eqn (8.4) for the Poincaré coordinates associated with the dual process

---

[1] In other words, the conditioned hyperbolic Brownian motion is an 'h-process' (in the sense of [Do]) of the unconditioned motion, the harmonic function h here being the Poisson kernel appearing in Proposition 3.6.1.4.

$\left(Z_s^* = T_{\tilde{z}_s}\right)$ of Section 8.1, which are precisely the expressions of the statement. This proves the proposition. ◇

Of course, Proposition B.3.1 is valid for any test functional $\tilde{F}(z_{[0,s]}, z_\infty)$. We specialize now to the functional dist (z_s, γ_∞), where γ_∞ denotes the random limiting geodesic determined by p and z_∞ (see Figure B.1). We can also consider the very simply related horocyclic distance functional dist $_\mathcal{H}(z_s, \gamma_\infty)$, as specified by the following (the horocyclic distance between two points has already been encountered on page 58).

Lemma B.3.2 *Given an oriented geodesic* $\gamma_\infty \equiv (\eta', \eta)$ *and* $z \in \mathbb{H}^d$, *the* **horocyclic distance** dist $_\mathcal{H}(z, \gamma_\infty)$, *i.e., the hyperbolic length of the minimal curve which links* z *to* $\gamma_\infty \cap \mathcal{H}_\eta(z)$ *within the horocycle* $\mathcal{H}_\eta(z)$, *is equal to the hyperbolic sine* $\mathrm{sh}\big[\mathrm{dist}\,(z, \gamma_\infty)\big]$ *of the hyperbolic distance (and thus does not depend on the orientation of the geodesic* γ_∞*).*

Proof Consider (as in Proposition B.3.1) an isometry $g_\eta \in \mathrm{PSO}(1, d)$ such that $g_\eta \eta' = e_0 - e_1$ and $g_\eta \eta = e_0 + e_1$, so that $\gamma'_\infty := g_\eta \gamma_\infty$ is the vertical geodesic $(0, \infty)$ in Poincaré coordinates. We denote by (x, y) the Poincaré coordinates of $z' := g_\eta z$. Then, on the one hand, we have

$$\mathrm{dist}\,_\mathcal{H}(z, \gamma_\infty) = \mathrm{dist}\,_\mathcal{H}(z', \gamma'_\infty) = |x|/y.$$

On the other hand, using the fact that the minimizing geodesic from z' to γ'_∞ is orthogonal to γ'_∞ (recall Proposition 2.2.2.4) and eqn (2.1), we find at once that

$$\mathrm{dist}\,(z, \gamma_\infty) = \mathrm{dist}\,(z', \gamma'_\infty) = \mathrm{dist}\,\Big((x, y); \big(0, \sqrt{|x|^2 + y^2}\big)\Big) = \mathrm{argch}\left[\frac{\sqrt{|x|^2 + y^2}}{y}\right]$$

$$= \mathrm{argsh}\left[\frac{|x|}{y}\right] = \mathrm{argsh}\big[\mathrm{dist}\,_\mathcal{H}(z, \gamma_\infty)\big]. \qquad ◇$$

The asymptotic behaviour of dist (z_s, γ_∞) is given by the following proposition.

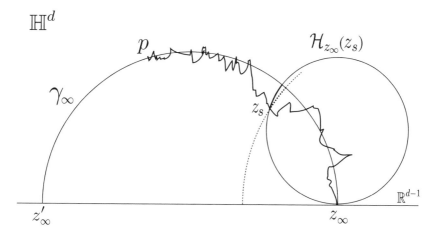

Fig. B.1 Brownian path and limiting geodesic

Proposition B.3.3 *Let γ_∞ denote the random limiting geodesic determined by p and z_∞. Then, as $s \to \infty$, the horocyclic distance* dist $_\mathcal{H}(z_s, \gamma_\infty)$ *(i.e., the hyperbolic sine* sh$\big[$dist $(z_s, \gamma_\infty)\big]$ *of the hyperbolic distance) from the hyperbolic Brownian motion z_s to γ_∞ converges in law to the Euclidean norm under the harmonic measure μ_{e_0}. Hence it admits the following limiting density (with respect to the Lebesgue measure of $\mathbb{R}_+$):*

$$r \longmapsto \frac{2^{d-1}\Gamma(d/2)}{\sqrt{\pi}\,\Gamma(d-1/2)}\,(1+r^2)^{1-d}r^{d-2}.$$

Proof We use Proposition B.3.1, and denote by γ'_∞ the vertical geodesic through e_0 (in the canonical frame). By the invariance of the hyperbolic metric under the isometry g_η, we obtain

$$\text{dist}\,_\mathcal{H}\big(g_\eta^{-1}\hat{z}_s, \gamma_\infty\big) = \text{dist}\,_\mathcal{H}\big(\hat{z}_s, g_\eta\gamma_\infty\big) = \text{dist}\,_\mathcal{H}\big(\hat{z}_s, \gamma'_\infty\big) = \frac{|\hat{x}_s|}{\hat{y}_s}.$$

Hence, by applying Proposition B.3.1, we obtain the following: for any bounded continuous function f,

$$\mathbb{E}_p\big[f\big(\text{dist}\,_\mathcal{H}(z_s, \gamma_\infty)\big)\big] = \int_{\partial\mathbb{H}^d}\mathbb{E}_{e_0}\Big[f\big(\text{dist}\,_\mathcal{H}(\hat{z}_s, \gamma'_\infty)\big)\Big]\mu_{e_0}(d\eta) = \mathbb{E}\left[f\left(\frac{|\hat{x}_s|}{\hat{y}_s}\right)\right].$$

Thus it is sufficient to specify the asymptotic law of $\hat{x}_s/\hat{y}_s$. Now, by changing the variable t to $(s-t)$, we obtain the following easy identity in law:

$$\frac{\hat{x}_s}{\hat{y}_s} = \int_0^s e^{W_t^1 - W_s^1 + ((d-1)/2)(t-s)}\,dW_t' \equiv \int_0^s e^{W_t^1 - ((d-1)/2)t}\,dW_t',$$

and the latter converges almost surely to $x_\infty := \int_0^\infty e^{W_t^1 - ((d-1)/2)t}\,dW_t'$, according to the proof of Proposition 7.6.7.3. Moreover, by this same proof, we almost surely also have $x_s \longrightarrow x_\infty$, so that x_∞ is equal to the non-trivial Poincaré coordinate of the end point z_∞ when the hyperbolic Brownian motion (z_s) is started from e_0. Therefore the law of x_∞ is given by the harmonic measure μ_{e_0}, whose density with respect to the Lebesgue measure of $\mathbb{R}^{d-1}$ is

$$\left[x \longmapsto \frac{2^{d-2}\Gamma(d/2)}{\pi^{d/2}}(1+|x|^2)^{1-d}\right],$$

according to Lemma 3.6.1.2. The limiting density of dist $_\mathcal{H}(\hat{z}_s, \gamma'_\infty)$ is easily deduced as that of $|x_\infty|$. ◇

References

[B1] Bowditch B.H. Geometrical finiteness for hyperbolic groups. *J. Funct. Anal.* 113, 245–317 (1993).

[B2] Bowditch B.H. Geometrical finiteness with variable negative curvature. *Duke Math. J.* 77(1), 229–274 (1995).

[Ba] Baxendale P.H. Asymptotic behaviour of stochastic flows of diffeomorphisms: two case studies. *Probab. Theory Related Fields* 73, 51–85 (1986).

[Be] Bernstein S. Équations différentielles stochastiques. Actualités Scientifiques et Industrielles 738, Conference de Mathématiques à Université de Genève, pp. 5–31. Hermann, Paris, 1938.

[Bi] Billingsley P. *Probability and measure*. Wiley Series in Probability Mathematics and Statistics. Wiley-Interscience, New York 1979, 1986, 1995.

[BF] Bailleul I., Franchi J. Non-explosion criteria for relativistic diffusions. ArXiv: http://arxiv.org/abs/1007.1893. To appear in Ann. Probab.

[Bk] Bourbaki N. *Éléments de mathématique. Fasc. XXXVII. Groupes et algèbres de Lie*. Actualités Scientifiques et Industrielles 1285 and 1349. Hermann, Paris, 1971–72.

[Bl] Bailleul I. Poisson boundary of a relativistic diffusion. *Probab. Theory Related Fields* 141(1–2), 283–329 (2008).

[Bn] Bourdon M. Structure conforme au bord et flot géodésique d'un CAT(-1)-espace. *Enseign. Math.* 41, 63–102 (1995).

[By] Bakry D. Functional inequalities for Markov semigroups. In *Probability measures on groups: recent directions and trends*, pp. 91–147, Tata Institute of Fundamental Research Mumbai, 2006.

[BP1] Babillot M., Peigné M. Homologie des géodésiques fermées sur des variétés hyperboliques avec bouts cuspidaux. *Ann. Sci. É.N.S.* 33(4), 81–120 (2000).

[BP2] Babillot M., Peigné M. Asymptotic laws for geodesic homology on hyperbolic manifolds with cusps. Bull. Soc. Math. France 134(1), 119–163 (2006).

[CE1] Carverhill A.P., Elworthy K.D. Flows of stochastic dynamical systems: the functional analytic approach. Z. Wahrsch. Verw. Geb. 65(2), 245–267 (1983).

[CE2] Carverhill A.P., Elworthy K.D. Lyapunov exponents for a stochastic analogue of the geodesic flow. *Trans. Am. Math. Soc.* 295(1), 85–105 (1986).

[CFS] Cornfeld I.P., Fomin S.V., Sinai Y.G. *Ergodic theory*. Grundlehren der mathematischen Wissenschaften 245, Springer, New York, 1982.

[CG] Chen S.S., Greenberg L. Hyperbolic spaces. In Contributions to analysis: a collection of papers dedicated to Lipman Bers, eds L.V. Ahlfors, I. Kra, B. Maskit and L. Nirenberg, pp. 49–87. Academic Press, New York, 1974.

[CK]   Carmona R., Klein A. Exponential moments for hitting times of uniformly ergodic markov processes. *Ann. Probab.* 11(3), 648–655 (1983).

[CL]   Conze J.P., Leborgne S. Méthode de martingales et flot géodésique sur une surface de courbure constante négative. Ergodic Theory Dyn. Syst. 21(2), 421–441 (2001).

[D1]   Dudley R.M. Lorentz-invariant Markov processes in relativistic phase space. *Ark. Mat.* 6(14), 241–268 (1965).

[Da]   Dal'bo F. *Trajectoires géodésiques et horocycliques.* Savoirs actuels, EDP Sciences. CNRS Éditions, Paris, 2007.

[Deb]   Debbasch F. A diffusion process in curved space–time. *J. Math. Phys.* 45(7), 2744–2760 (2004).

[Do]   Doob J.L. *Classical potential theory and its probabilistic counterpart.* Springer, New York, 1984: Berlin, 2001.

[D2]   Dudley R.M. A note on Lorentz-invariant Markov processes. *Ark. Mat.* 6(30), 575–581 (1967).

[D3]   Dudley R.M. Asymptotics of some relativistic Markov processes. *Proc. Natl. Acad. Sci. USA* 70, 3551–3555 (1973).

[Dt]   Durrett R. *Probability: theory and examples.* Wadsworth-Brooks/Cole, Pacific Grove, CA, 1991; Duxbury, Belmont, CA, 1996; Cambridge University Press, Cambridge, 2010.

[DFN]   Dubrovin B.A., Fomenko A.T., Novikov S.P. *Modern geometry; methods and applications.* Graduate Texts in Mathematics 93. Springer, New York, 1984, 1992.

[E]   Émery M. On some relativistic-covariant stochastic processes in Lorentzian space–times. *C. R. Math. Acad. Sci. Paris* 347(13–14), 817–820 (2009).

[EF]   Enriquez N., Franchi J. Masse des pointes, temps de retour et enroulements en courbure négative. *Bull. Soc. Math. France* 130(3), 349–386 (2002).

[EFLJ1] Enriquez N., Franchi J., Le Jan Y. Stable windings on hyperbolic surfaces. *Probab. Theory Related Fields* 119, 213–255 (2001).

[EFLJ2] Enriquez N., Franchi J., Le Jan Y. Canonical lift and exit law of the fundamental diffusion associated with a Kleinian group. In *Séminaire de Probabilités XXXV*, pp. 206–219. Springer, Berlin, 2001.

[EFLJ3] Enriquez N., Franchi J., Le Jan Y. Central limit theorem for the geodesic flow associated with a Kleinian group, case $\delta > d/2$. *J. Math. Pures Appl.* 80(2), 153–175 (2001).

[EFLJ4] Enriquez N., Franchi J., Le Jan Y. Enroulements browniens et subordination dans les groupes de Lie. In *Séminaire de Probabilités XXXIX*, pp. 357–380. Springer, Berlin, 2005.

[ELJ]   Enriquez N., Le Jan Y. Statistic of the winding of geodesics on a Riemann surface with finite volume and constant negative curvature. *Rev. Mat. Iberoam.* 13(2), 377–401 (1997).

[F1]   Franchi J. Asymptotic singular windings of ergodic diffusions. *Stoch. Process. Appl.* 62, 277–298 (1996).

[F2] Franchi J. Asymptotic singular homology of a complete hyperbolic 3-manifold of finite volume. *Proc. London Math. Soc.* 79(3), 451–480, (1999).

[F3] Franchi J. Asymptotic windings over the trefoil knot. *Rev. Mat. Iberoam.* 21(3), 729–770 (2005).

[F4] Franchi J. Relativistic diffusion in Gödel's universe. *Commun. Math. Phys.* 290(2), 523–555 (2009).

[FgK] Furstenberg H., Kesten H. Products of random matrices. *Ann. Math. Statist.* 31, 457–469 (1960).

[FK] Fricke R., Klein F. *Vorlesungen über die Theorie der automorphen Funktionen.* Band I: Die gruppentheoretischen Grundlagen. Band II: Die funktionentheoretischen Ausfhrungen und die Andwendungen. Bibliotheca Mathematica Teubneriana, Bände 3, 4. Johnson Reprint Corp., New York, 1965; Teubner, Stuttgart, 1965.

[FLJ1] Franchi J., Le Jan Y. Relativistic diffusions and Schwarzschild geometry. *Commun. Pure Appl. Math.* 60(2), 187–251 (2007).

[FLJ2] Franchi J., Le Jan Y. Curvature diffusions in general relativity. *Commun. Math. Phys.* 307(2), 351–382 (2011).

[Gh] Ghys É. Poincaré and his disk. In *The scientific legacy of Poincaré*, eds É. Charpentier, É. Ghys and A. Lesne, pp. 17–46. History Mathematics 36. American Mathematical Society, Providence, RI, 2010.

[Gi1] Gihman I.I. On some differential equations with random functions. [In Russian.] *Ukrain. Mat. Žurnal* 2(3), 45–69 (1950).

[Gi2] Gihman I.I. On the theory of differential equations of random processes. [In Russian.] *Ukrain. Mat. Žurnal* 2(4), 37–63 (1950).

[GLJ] Guivarc'h Y., Le Jan Y. Asymptotic windings of the geodesic flow on modular surfaces with continuous fractions. *Ann. Sci. É.N.S.* 26(4), 23–50 (1993).

[Gr] Gromov M. Structures métriques pour les variétés riemanniennes, ed. J. Lafontaine and P. Pansu. Cedic-F. Nathan, Paris, 1981.

[GS] Gihman I.I., Skorohod A.V. *The theory of stochastic processes, Vols. I, II, III.* Springer, Berlin, 2004, 2007.

[Gu] Guivarc'h Y. *Propriétés de mélange pour les groupes à un paramètre de Sl(d, R).* Fascicule de probababilités. Institut de Recherche Mathématique de Rennes, University of Rennes I, 1992.

[Ha] Has'minskiĭ R.Z. *Stochastic stability of differential equations.* [Transl. from Russian.] Monographs and Textbooks on the Mechanics of Solids and Fluids: Mechanics and Analysis 7. Sijthoff & Noordhoff, Alphen aan den Rijn, 1980.

[HE] Hawking S.W., Ellis G.F.R. *The large-scale structure of space–time.* Cambridge University Press, Cambridge, 1973.

[He] Helgason S. *Differential geometry and symmetric spaces.* Pure and Applied Mathematics Vol. 12. Academic Press, New York, 1962.

[HK] Hasselblatt B., Katok A. *Introduction to the modern theory of dynamical systems.* Encyclopedia of Mathematics and its Application 54. Cambridge University Press, Cambridge, 1995.

[Ho] Hochschild G. *The structure of Lie groups.* Holden-Day, San Francisco, 1965.

[Hop] Hopf E. Ergodicity theory and the geodesic flow on a surface of constant negative curvature. *Bull. Am. Math. Soc.* 77, 863–877 (1971).

[I] Itô K. Differential equations determining Markov processes. [In Japanese.] *Zenkoku Shijō Sūgaku Danwakai* 244(1077), 1352–1400 (1942).

[IMK] Itô K., McKean H.P. *Diffusion processes and their sample paths.* Second printing, corrected. Grandlehren der mathematischen Wissenschaften 125. Springer, Berlin, 1974.

[IW] Ikeda N., Watanabe S. *Stochastic differential equations and diffusion processes.* North-Holland Kodansha, Amsterdam, 1981.

[J] Jacod J. *Calcul stochastique et problèmes de martingales.* Lecture Notes in Mathematics 714. Springer, Berlin, 1979.

[Ka] Kallenberg O. *Foundations of modern probability.* Springer, Berlin, 1997, 2002.

[Kn] Knapp A.W. *Lie groups beyond an introduction.* Progress in Mathematics 140. Birkhäuser, Boston, 2002.

[KN] Kobayashi S., Nomizu K. *Foundations of differential geometry.* Interscience, New York, 1969.

[Kr] Kra I. *Automorphic forms and Kleinian groups.* Mathematics Lecture Note Series. W.A. Benjamin, Reading, MA, 1972.

[KS] Karatzas I., Shreve S. *Brownian motion and stochastic calculus.* Graduate Texts in Mathematics 113. Springer, New York, 1988, 1991.

[Ku] Kunita H. *Stochastic flows and stochastic differential equations.* Cambridge Studies in Advanced Mathematics 24. Cambridge University Press, Cambridge, 1990, 1997.

[KW] Kunita H., Watanabe S. On square integrable martingales. *Nagoya Math. J.* 30, 209–245 (1967).

[L1] Ledrappier F. *Quelques propriétés des exposants caractéristiques. École d'été de probabilités de St-Flour XII,* 1982, pp. 305–396. Lecture Notes in Mathematics 1097. Springer, Berlin, 1984.

[L2] Ledrappier F. Harmonic 1-forms on the stable foliation. *Bol. Soc. Bras. Math.* 25(2), 121–138 (1994).

[L3] Ledrappier F. Central limit theorem in negative curvature. *Ann. of Probab.* 23(3), 1219–1233 (1995).

[Le] Lehner J. *Discontinuous groups and automorphic functions.* Mathematical Survey 8. American Mathematical Society, Providence, RI 1964.

[Lév] Lévy P. *Processus stochastiques et mouvement brownien: suivi d'une note de M. Loève.* Gauthier-Villars, Paris, 1948.

[LJ1] Le Jan Y. Sur l'enroulement géodésique des surfaces de Riemann. *C. R. Acad. Sci. Paris* 314, Série I, 763–765 (1992).

[LJ2] Le Jan Y. The central limit theorem for the geodesic flow on non compact manifolds of constant negative curvature. *Duke Math. J.* 74(1), 159–175 (1994).

[LJ3]   Le Jan Y. Free energy for Brownian and geodesic homology. *Probab. Theory Related Fields* 102, 57–61 (1995).

[LJW]   Le Jan Y., Watanabe S. Stochastic flows of diffeomorphisms. In *Stochastic analysis: proceedings of the Taniqnchi International Symposium on Stochastic Analysis*, ed. K. Itô, Katata/Kyoto 1982, pp. 307–332. North-Holland Mathematical Library 32. North-Holland, Amsterdam, 1984.

[LL]   Landau L., Lifchitz E. *Physique théorique, tome II: Théorie des champs.* Éditions MIR Moscou, 1970.

[Mar]   Margulis G.A. *Discrete subgroups of semisimple Lie groups.* Ergebnisse der mathematik und ihrer Grenzgebiete, Series 3, Vol. 17. Springer, Berlin, 1991.

[Mas]   Maskit B. *Kleinian groups.* Grundlehren Mathematischen Wissenschaften 287. Springer, Berlin, 1988.

[Me]   Meyer P.A. Un cours sur les intégrales stochastiques. *Séminaire de probabilités X, Universite de Strasbourg*, ed. P.A. Meyer, pp. 245–400. Lecture notes in Mathematics 511. Springer, Berlin, 1976.

[Mi]   Miyake T. *Modular forms.* Springer, Berlin, 1989.

[MK]   McKean H.P. *Stochastic integrals.* Academic Press, New York, 1969.

[Mr]   Milnor J. Hyperbolic geometry: the first 150 years. *Bull. Am. Math. Soc. (N.S.)* 6(1), 9–24 (1982).

[MT]   Matsuzaki K., Taniguchi M. *Hyperbolic manifolds and Kleinian groups.* Oxford Mathematical Monographs, Oxford University Press, New York, 1998.

[N1]   Neveu J. *Martingales à temps discret.* Masson, Paris, 1972.

[N2]   Neveu J. Courte démonstration du théorème ergodique sur-additif. *Ann. Inst. H. Poincaré* 19(1), 87–90 (1983).

[Ni]   Nicholls P.J. *The ergodic theory of discrete groups.* London Mathematical Society Lecture Note Series 143. Cambridge University Press, Cambridge, 1989.

[O]   Oksendal B. *Stochastic differential equations: an introduction with applications.* Universitext. Springer, Berlin, 1985, 1989, 1992, 1995, 1998, 2003.

[Os]   Oseledets V.I. A multiplicative ergodic theorem. Characteristic Ljapunov exponents of dynamical systems. [In Russian.] *Trudy Moskov. Mat. Obšč.* 19, 179–210 (1968).

[Pa]   Pansu P. Métriques de Carnot-Carathéodory et quasiisométries des espaces symétriques de rang un. *Ann. of Math.* 2nd Series 129(1), 1–60 (1989).

[P1]   Patterson S.J. The Laplacian operator on a Riemann surface, III. Compositio Math. 33(3), 227–259 (1976).

[P2]   Patterson S.J. Lectures on measures on limit sets of Kleinian groups. In *Analytical and geometrical aspects of hyperbolic spaceed*. D. Epstein, pp. 281–323. London Mathematical Society Lecture Note Series 111. Cambridge University Press, Cambridge, 1987.

[Pr]   Protter J. *Stochastic integration and differential equations.* Stochastic Modelling and Applied Probability 21. Springer, Berlin, 1990, 2005.

[PS] Phillips R.S., Sarnak P. The Laplacean for domains in hyperbolic space and limit sets of Kleinian groups. *Acta Math.* 155, 173–241 (1985).

[PSt] Port S.C., Stone C.J. *Brownian motion and classical potential theory.* Academic Press, New York, 1978.

[Rac] Ratcliffe J.G. *Foundations of hyperbolic manifolds.* Springer, Berlin, 1994.

[Rag] Raghunathan M.S. *Discrete subgroups of Lie groups.* Ergebnisse der Mathematik und ihrer Grenzgebiete 68. Springer, New York, 1972.

[Ran] Ratner M. The central limit theorem for geodesic flows on n-dimensional manifolds of negative curvature. *Israel J. Math.* 16, 181–197 (1973).

[Ru] Ruelle D. Ergodic theory of differentiable dynamical systems. *Publ. Math. Inst. Hantes Études Sci.* 50, 27–58 (1979).

[RW1] Rogers L.C.G., Williams D. *Diffusions, Markov processes, and martingales. Vol. 1. Foundations.* Wiley, New York, 1994.

[RW2] Rogers L.C.G., Williams D. *Diffusions, Markov processes, and martingales. Vol. 2. Itô calculus.* Cambridge Mathematical Library, Cambridge University Press, Cambridge, 2000.

[RY] Revuz D., Yor M. *Continuous martingales and Brownian motion.* Springer, Berlin, 1991, 1994, 1999.

[Sc] Schoeneberg B. *Elliptic modular functions.* Springer, Berlin, 1974.

[Si1] de Saint-Gervais H.P. *Uniformisation des surfaces de Riemann. Retour sur un théorème centenaire.* É.N.S. Éditions, Lyon, 2010.

[Si1] Sinai Y.G. The central limit theorem for geodesic flows on manifolds of constant negative curvature. *Dokl. Akad. Nauk. SSSR* 133, 1303–1306 (1960).

[Si2] Sinai, Y.G. The central limit theorem for geodesic flows on manifolds of constant negative curvature. *Sov. Math. Dokl.* 1, 938–987 (1960).

[St] Steele J.M. Kingman's subadditive ergodic theorem. *Ann. I.H.P., Section B,* 25(1), 93–98 (1989).

[Sw] Sullivan D. Entropy, Hausdorff measures old and new, and limit sets of geometrically finite Kleinian groups. *Acta Math.* 153, 259–277 (1984).

[Sw] Swan R.G. Generators and relations for certain special linear groups. *Adv. Math.* 6, 1–77 (1971).

[W] Wald R.M. *General relativity.* University of Chicago Press, Chicago, 1984.

[Y] Yue C. The ergodic theory of discrete isometry groups on manifolds of variable negative curvature. *Trans. Am. Math. Soc.* 348(12), 4965–5005 (1996).

General notation

$\mathbb{N} = \{0, 1, 2, 3, \ldots\}$; $\mathbb{N}^* = \{1, 2, 3, \ldots\}$; $\mathbb{R}^* = \mathbb{R} \smallsetminus \{0\} = \{x \in \mathbb{R} \mid x \neq 0\}$;
$\mathbb{R}_+ = [0, \infty[= \{x \in \mathbb{R} \mid x \geq 0\}$; $\mathbb{R}_+^* =]0, \infty[= \{x \in \mathbb{R} \mid x > 0\}$;
$]a, b[= \{x \in \mathbb{R} \mid a < x < b\}$; $]a, b] = \{x \in \mathbb{R} \mid a < x \leq b\}$; $[a, b[= \{x \in \mathbb{R} \mid a \leq x < b\}$.

$[x] := \max\{n \in \mathbb{Z} \mid n \leq x\}$ denotes the integral part of $x \in \mathbb{R}$.

$\Re(z), \Im(z)$ denote the real and imaginary parts of a complex number $z \in \mathbb{C}$.

$|u|$, $u.v$ denote the usual Euclidean norm and inner product of $u, v \in \mathbb{R}^n$.

${}^t M$ denotes the transpose of a matrix M.

$\mathrm{Tr}(M) \equiv \mathrm{Tr}[M]$ denotes the trace of a square matrix M.

$\mathrm{Tr}_{\mathcal{G}}(\phi)$ denotes the trace of an endomorphism ϕ of $\mathcal{G}$.

$\dfrac{d_0}{dt} f(t) \equiv f'(0)$ denotes the derivative of f at 0 with respect to t.

1_A denotes the indicator function of the set A: $1_A(x) = 1$ if $x \in A$, $= 0$ if not.

$\mathrm{tg} := \dfrac{\sin}{\cos}$; $\mathrm{cotg} := \dfrac{\cos}{\sin}$; $\mathrm{arctg} := \mathrm{tg}^{-1}$; $\mathrm{arccos} := \cos^{-1}$.

$\mathrm{ch}\, t := \frac{1}{2}(e^t + e^{-t})$; $\mathrm{sh}\, t := \frac{1}{2}(e^t - e^{-t})$; $\mathrm{th} := \dfrac{\mathrm{sh}}{\mathrm{ch}}$; $\mathrm{coth} := \dfrac{\mathrm{ch}}{\mathrm{sh}}$.
$\mathrm{argch} := \mathrm{ch}^{-1}$; $\mathrm{argsh} := \mathrm{sh}^{-1}$; $\mathrm{argth} := \mathrm{th}^{-1}$.

$\mathbf{1}$ denotes the unit matrix, the unit element of $\mathrm{GL}(d)$ and, generally, the unit of a group.

$C(E)$ denotes the space of continuous real functions on a metric space E.

$C_b(E)$ denotes the space of bounded continuous real functions on a metric space E.

$C_b^2(E)$ denotes the set of $f \in C^2(E)$ which have bounded derivatives of order ≤ 2.

$\mathrm{dist}\,(p, q)$ denotes the hyperbolic distance between $p, q \in \mathbb{H}^d$
Proposition 2.1.1, page 23

$\langle X, Y \rangle_t$ denotes the quadratic covariation bracket   Section 6.5, page 141

$\int H\, dB, \int X\, dY$ denote stochastic Itô integrals   Section 6.4, page 132

$\int X \circ dY$ denotes a stochastic Stratonovich integral   Section 6.6, page 144

$\langle \xi, \xi' \rangle$ denotes the Lorentz quadratic form on $\mathbb{R}^{1,d}$ Section 1.2, page 10

Other notation

λ	denotes the Haar measure of $\mathrm{PSO}(1, d)$	Section 3.3, page 55
$\tilde{\lambda}$	denotes the Liouville measure of $\mathbb{F}^d$	Definition 3.6.2.1, page 72
$\tilde{\lambda}^\Gamma$	denotes the Liouville measure of $\Gamma\backslash\mathbb{F}^d$	Proposition 5.1.1, page 105
$\mathcal{M}$	denotes the two-sided quotient $\Gamma\backslash\mathbb{F}^d/\mathrm{SO}(d-1)$	Chapter 8, page 195
$\mathcal{M}_c, \mathcal{M}_c^\infty$	denote spaces of continuous L^2 martingales	Prop. 6.3.7, page 131
$\mathcal{M}(d)$	denotes the algebra of real $d \times d$ square matrices	Section 1.1.1, page 1
μ	denotes an invariant probability law on $\Gamma\backslash\mathbb{F}^d$	Definition 8.2.1, page 199
μ_p	denotes the harmonic measure at $p \in \mathbb{H}^d$	Definition 3.6.1.1, page 69
$\mathcal{N}(0, V)$	denotes the centred normal law with variance V	Definition 6.2.1, page 127
ω^f	denotes a 1-form on $\mathbb{F}^d$, associated with f	Equation (8.9), page 207
$\mathcal{P}^{d+1}$	denotes the Poincaré group	Section 7.7.1, page 187
p_η	denotes a spacelike projection of $p \in \mathbb{H}^d$ in the direction $\eta \in \partial\mathbb{H}^d$	Proposition 2.2.2.1, page 26
π_0	denotes the canonical projection from $\mathbb{F}^d$ onto $\mathbb{H}^d$	Definition 1.2.2, page 11
π_1	denotes the projection from $\mathbb{F}^d$ onto $\mathbb{H}^d \times \partial\mathbb{H}^d$	Section 2.2.2, page 26
P_s, P_s^*	denote Brownian semi-groups on $\mathbb{A}^d$, $\mathbb{F}^d$, $\Gamma\backslash\mathbb{F}^d$	Section 8.1, pages 195, 198
$\mathrm{PSL}(2)$	denotes the quotient $\mathrm{SL}(2)/\pm\mathbf{1}$	Proposition 1.1.5.1, page 9
$\mathrm{PSO}(1, d)$	denotes the Lorentz-Möbius group	Section 1.3, page 12
Q_s	denotes the hyperbolic semi-group (on $\mathbb{H}^d$ or $\Gamma\backslash\mathbb{H}^d$)	Equation (7.14), page 185
$\mathbb{R}^{1,d}$	denotes Minkowski space	Section 1.2, page 10
$\mathcal{R}_{\alpha,\sigma} \equiv \mathcal{R}_{\tilde{\sigma}}$	denotes a planar rotation	Pages 45 and 56
$\mathcal{S}_b, \mathcal{S}$	denote spaces of semimartingales	Section 6.5, pages 137 and 139
$\mathrm{SL}(d)$	denotes the special linear subgroup of $\mathrm{GL}(d)$	Section 1.1.5, page 8

Index of terms